REAL ANALYSIS

Un Cig Ji

KM KYUNG MOON SA

Preface

This book is prepared with two purpose: one is for a course of university level on abstract measure, integration theory and functional analysis, and another one is for a self-contained reference for whom interested in real analysis. This book is made up with clear explanations and elementary proofs to study the measure, integration theory and functional alone easily. However, the required prerequisite (pre-study) to read this book is the first course in analysis and a foundation of topology.

The measure and integration theory becomes more important as one of most fundamental theories in the present century for wide areas of applications including mathematical physics, finance, biology and etc. There are several text books for the measure and integration theory with two different approaches. One is starting with the Lebesgue measure and integration theory to understand several motivations and examples, and then it is extended to abstract measure and integration theory. Another is starting abstract measure and integration theory and then the Lebesgue measure and integration theory is dealt and understood as an example. Both approaches have their own advantages and disadvantages. For this book, I have chosen the latter and so this book is starting with abstract measure theory and then dealing the Lebesgue measure and integration theory as an example, and so I hope that the readers can reach quickly to the measure and integration theory and then can apply it depending on their own interests.

Chapter 1 offers fundamental notions in the first course in analysis and topology, which are necessary to read the main parts of this book. Chapters 2 and 3 deals with the general measure theory including constructions of abstract measures and the Lebesgue measure as an example of abstract measure. The readers interesting to the integration theory can study only the definition of abstract measure with fundamental properties and then can skip the constructions of abstract measures and the Lebesgue measure. Chapter 4 deals with measurable functions which is necessary for the abstract integration theory Chapter 5 provides the general theory of abstract integration and considers the Lebesgue integral as an example of an abstract integral The Lebeasgue integral is considered as an extension of the Riemann integral. Chapter 6 offers the Radon-Nikodym theorem, the fundamental theorem in calculus and Riemann-Stieltjes integral. Chapters 7 and 8 provides the fundamental notions in functional analysis. Chapter 9 considers the convergence of measurable functions including pointwise convergence, convergence in measure, almost uniform convergence and convergence in mean, among them the convergence in measure is weakest. Chapter 10 provides the fundamental theorems in functional analysis. Exercises are given at the end of each chapter and solutions to some problems are given.

February 28, 2018
Un Cig Ji

Contents

Chapter 1

Preliminaries

1.1 Partially Ordered Set

Definition 1.1.1 Let X be a nonempty set and $\mathcal{R}$ is a relation on X. Then the relation $\mathcal{R}$ is called a **partial ordering** if it satisfies that

- **(reflexive)** $x\mathcal{R}x$ for all $x \in X$,
- **(transitive)** if $x\mathcal{R}y$ and $y\mathcal{R}z$, then $x\mathcal{R}z$,
- if $x\mathcal{R}y$ and $y\mathcal{R}x$, then $x = y$.

In this case, E is called a **partially ordered set**.

Let E be a nonempty subset of a partially ordered set X with a partial ordering $\mathcal{R}$. Then the partial ordering $\mathcal{R}$ is called a **linear** (or **total**) **ordering** on E if for any $x, y \in E$,

$$x\mathcal{R}y \quad \text{or} \quad y\mathcal{R}x.$$

In this case, E is called a **linear ordered set** (or **totally ordered set**).

Definition 1.1.2 Let X be a totally ordered set with a order relation $\leq$ and E a subset of X. Then an element x of X is called

(i) a **maximal element** of X if $x = z$ whenever there exists $z \in X$ such that $y \leq z$ for all $y \in X$,

(ii) an **upper bound** of E if $y \leq x$ for all $y \in E$,

(iii) a **lower bound** of E if $x \leq y$ for all $y \in E$.

Zorn's Lemma. Let X be a partially ordered set. If every linearly (totally) ordered subset of X has an upper bound, then X has a maximal element.

1.2 Completeness Axiom and Archimedian Property

Definition 1.2.1 Let E be a nonempty subset of $\mathbb{R}$. Then we say that

(i) E is **bounded above** if there exists $M \in \mathbb{R}$ such that $x \leq M$ for all $x \in E$. In this case, M is called an **upper bounded** for E;

(ii) E is **bounded below** if there exists $m \in \mathbb{R}$ such that $x \geq m$ for all $x \in E$ In this case, m is called a **lower bound** for E;

(iii) E is **bounded** if it is bounded above and bounded below, i.e., there exists a $M \geq 0$ such that $|x| \leq M$ for all $x \in E$.

[Axiom] [**Completeness Axiom**] Let E be a nonempty subset of $\mathbb{R}$ which is bounded above. Then there exists the smallest element in the set of all upper bounds for E. The smallest element of all upper bounds for a non-empty subset E of $\mathbb{R}$ which is bounded above is called the **supremum** of E and denoted by $\sup(E)$.

Remark 1.2.2 We can prove that there exists an ordered field containing the field $\mathbb{Q}$ of all rational numbers as a subfield which satisfies the completeness axiom (for the proof, we refer to [8]). Also, we note that any two ordered fields containing the field $\mathbb{Q}$ and satisfying the completeness axiom are isomorphic. Therefore, an ordered field containing $\mathbb{Q}$ and satisfying the completeness axiom is called the **real field** and denoted by $\mathbb{R}$.

Lemma 1.2.3 Let E be a nonempty subset of $\mathbb{R}$ which is bounded below. Then there exists a largest element in the set of all lower bounds for E. The largest element of all lower bounds for a non-empty subset E of $\mathbb{R}$ which is bounded below is called the ***infimum*** of E and denoted by $\inf(E)$.

[Proof] For the given nonempty subset E of $\mathbb{R}$ which is bounded below, put $-E = \{-x \mid x \in E\}$. Then $-E$ is nonempty and bounded above and so by the completeness axiom there exists the supremum $\sup(-E)$ of $-E$, and then we can easily see that $-\sup(-E)$ is the largest element of the set of all lower bounds. □

Remark 1.2.4 Let E be a nonempty subset of $\mathbb{R}$ which is bounded below. Then by the proof of Lemma 1.2.3, we see that

$$\inf(E) = -\sup(-E).$$

Definition 1.2.5 A subset E of $\mathbb{R}$ is said to be **inductive** if

(i) $1 \in E$;

(ii) for any $x \in E$, $x + 1 \in E$.

The set of all **natural numbers** is denoted by $\mathbb{N}$, i.e.,

$$\mathbb{N} = \{1, 2, 3, \cdots\}.$$

Theorem 1.2.6 [**Principle of Mathematical Induction**] For each $n \in \mathbb{N}$, we assign a mathematical statement $S(n)$. Suppose that the statement $S(1)$ is true. Also, suppose that if the statement $S(k)$ is true for a natural number $k \in \mathbb{N}$, then the statement $S(k+1)$ is true. Then the statement $S(n)$ is true for all $n \in \mathbb{N}$.

[Proof] Put $T = \{n \in \mathbb{N} \,|\, S(n) \text{ is true}\}$. Then by assumptions, we see that $1 \in T$ and if $k \in T$, then $k + 1 \in T$, which means that T is an inductive set. Since $\mathbb{N}$ is the smallest inductive set (see Exercise **1.1**), $\mathbb{N} \subset T$ which implies that the statement $S(n)$ is true for all $n \in \mathbb{N}$. □

Theorem 1.2.7 [**Archimedian Property**] The set $\mathbb{N}$ of all natural numbers is not bounded above.

[Proof] Suppose that $\mathbb{N}$ is bounded above. Then by the completeness axiom there exists a $m \in \mathbb{R}$ such that $n \leq m$ for all $n \in \mathbb{N}$ and m is the smallest element of the set of all upper bounds of $\mathbb{N}$ and so $m - 1$ is not upper bound of $\mathbb{N}$, i.e., there exists a $n \in \mathbb{N}$ such that $m - 1 < n$, equivalently, $m < n + 1$ and so $n + 1 \notin \mathbb{N}$ which is a contradiction with that N is inductive. □

Remark 1.2.8 Let $\epsilon > 0$ be given. Then by the Archimedian Property (Theorem 1.2.7) there exists a $n \in \mathbb{N}$ such that $n > 1/\epsilon$, equivalently, $1/n < \epsilon$, which is another version of the Archimedian Property and is used frequently in the study of Analysis.

1.3 Countable and Uncountable Sets

For given two sets A and B, we say that A and B are equivalent if there exists a one-to-one and onto function $f : A \to B$, in this case we write $A \sim B$. Then $\sim$ is an equivalent relation among sets, i.e.,

(i) $\sim$ is reflexive, i.e., for any set A (in our interest), there is a one-to-one and onto function f from A onto B,

(ii) $\sim$ is symmetric, i.e., if $A \sim B$, then $B \sim A$,

(iii) $\sim$ is transitive, i.e., if $A \sim B$ and $B \sim C$, then $A \sim C$.

Definition 1.3.1 Let A be a set. then A is said to be

(i) **finite** if A is the empty set or A is equivalent to $\{1, 2, \cdots, n\}$ for some $n \in \mathbb{N}$;

(ii) **countably infinite** if A is equivalent to $\mathbb{N}$,

(ii) **countable** if A is finite or countably infinite,

(ii) **uncountable** if A is not countable.

Theorem 1.3.2 Every subset of a countable set is countable.

[Proof] Let A be a countable set and B a subset of A.

(i) Let A be the empty set. Then B is only empty set and B is finite, and so it is countable.

(ii) Let A be a nonempty finite set and B a nonempty subset of A. By definition, there exists a one-to-one and onto function $f : \{1, 2, \cdots, n\} \to A$. Put

$$
\begin{aligned}
m_1 &= \min\{1 \le i \le n \,|\, f(i) \in B\}, \\
m_2 &= \min(\{1 \le i \le n \,|\, f(i) \in B\} \setminus \{m_1\}) \quad \text{if} \quad \{1 \le i \le n \,|\, f(i) \in B\} \setminus \{m_1\} \neq \emptyset, \\
&\vdots \\
m_k &= \min(\{1 \le i \le n \,|\, f(i) \in B\} \setminus \{m_1, \cdots, m_{k-1}\}) \\
&\qquad\qquad \text{if} \quad \{1 \le i \le n \,|\, f(i) \in B\} \setminus \{m_1, \cdots, m_{k-1}\} \neq \emptyset \\
&\qquad\qquad \text{and} \quad \{1 \le i \le n \,|\, f(i) \in B\} \setminus \{m_1, \cdots, m_k\} = \emptyset.
\end{aligned}
$$

Define a function $g : \{1, 2, \cdots, k\} \to B$ by $g(j) = f(m_j)$ and then g is one-to-one and onto, and so B is finite.

(iii) Let A be a countably infinite set and B a nonempty subset of A. Then by definition, there exists a one-to-one and onto function $f : \mathbb{N} \to A$, and then by similar arguments as used in (i), we define the sequence $\{m_i\}$ of natural numbers. If $\{m_i\}$ is a finite sequence, then as proved in (i), B is finite. If $\{m_i\}$ is an infinite sequence, then define a function $g : \mathbb{N} \to B$ by

$$g(j) = f(m_j), \qquad j \in \mathbb{N}.$$

Then g is one-to-one, in fact, for any $i \neq j$, $m_i \neq m_j$ and so $g(i) = f(m_i) \neq f(m_j) = g(j)$. Also, for any $x \in B$, there exists $l \in \mathbb{N}$ such that $f(l) = x \in B$ and so $l \in \{i \in \mathbb{N} \,|\, f(i) \in B\}$, which implies that there exists $k \in \mathbb{N}$ such that $1 \leq k \leq l$ and $m_k = l$. Therefore, $g(k) = f(m_k) = f(l) = x \in B$ and g is onto. Hence B is countably infinite. □

Theorem 1.3.3 The set $\mathbb{N} \times \mathbb{N}$ is countable.

[Proof] Define a function $g : \mathbb{N} \times \mathbb{N} \to \mathbb{N}$ by

$$g(m, n) = \frac{(m+n-1)(m+n-2)}{2} + n, \qquad (m, n) \in \mathbb{N} \times \mathbb{N}.$$

Suppose that $g(m, n) = g(m', n')$ and put $l = m + n$ and $l' = m' + n'$. Then we have

$$\frac{(m+n-1)(m+n-2)}{2} + n = \frac{(m'+n'-1)(m'+n'-2)}{2} + n',$$

equivalently, we have

$$\sum_{i=1}^{l-1} i + l - m = \sum_{i=1}^{l'-1} i + l' - m'.$$

Therefore, if $l < l'$, then we have

$$\begin{aligned}\left(\sum_{i=1}^{l'-1} i + l' - m'\right) - \left(\sum_{i=1}^{l-1} i + l - m\right) &= \sum_{i=l}^{l'-1} i + (l' - l) - m' + m \\ &\geq (l' - l) + (l' - 1 - m') + m \\ &> 0,\end{aligned}$$

which is impossible and similarly we see that if $l > l'$, then $g(m, n) > g(m', n')$ which is also impossible. Therefore, $m + n = l = l' = m' + n'$ and so $n = n'$ and $m = m'$. Hence g is one-to-one. For any given $k \in \mathbb{N}$ there exists $l \in \mathbb{N}$ such that

$$\sum_{i=1}^{l-1} i < k \leq \sum_{i=1}^{l} i,$$

where $\sum_{i=1}^{0} i = 0$, and then for the $n = k - \sum_{i=1}^{l-1} i$ and $m = l - n$, we have that $g(m, n) = k$ and so g is onto. Therefore, $g^{-1} : \mathbb{N} \to \mathbb{N} \times \mathbb{N}$ is one-to-one and onto, and so the set $\mathbb{N} \times \mathbb{N}$ is countable. □

Corollary 1.3.4 For any countable sets A and B, $A \times B$ is countable.

[Proof] There exist countably infinite sets $\widetilde{A}$ and $\widetilde{B}$ such that $A \subset \widetilde{A}$ and $B \subset \widetilde{B}$. By Theorem 1.3.3, $\widetilde{A} \times \widetilde{B}$ is countable. In fact, there exist two one-to-one and onto functions $f : \mathbb{N} \to \widetilde{A}$ and $h : \mathbb{N} \to \widetilde{B}$ and so the function $f \times h : \mathbb{N} \times \mathbb{N} \to \widetilde{A} \times \widetilde{B}$ is one-to-one and onto, where $f \times h(m, n) = (f(m), h(n))$ for any $(m, n) \in \mathbb{N} \times \mathbb{N}$. Therefore, with a one-to-one and onto function $g : \mathbb{N} \to \mathbb{N} \times \mathbb{N}$, the function $g \circ (f \times h) : \mathbb{N} \to \widetilde{A} \times \widetilde{B}$ is one-to-one and onto. □

Corollary 1.3.5 For any $n \in \mathbb{N}$, $\times_{i=1}^{n} \mathbb{N} = \left(\times_{i=1}^{n-1} \mathbb{N}\right) \times \mathbb{N}$ is countable.

[Proof] The proof is straightforward from Corollary 1.3.4 and the mathematical induction. □

Corollary 1.3.6 The set $\mathbb{Q}$ of rational numbers is countable.

[Proof] Exercise **1.2**. □

Corollary 1.3.7 The union of a countable collection of countable sets is countable.

[Proof] Exercise **1.3**. □

A countable family $\{E_n\}_{n=1}^{\infty}$ is said to be

(1) **descending** if for any $n \in \mathbb{N}$, $E_n \supset E_{n+1}$;

(2) **ascending** if for any $n \in \mathbb{N}$, $E_n \subset E_{n+1}$.

Theorem 1.3.8 For any $a < b$, (a, b) is uncountable.

[Proof] It is clear that (a, b) is not finite. Suppose that (a, b) is countably infinite. Then there is a one-to-one and onto function $f : \mathbb{N} \to (a, b)$. Let $\{[x_n, y_n]\}_{n=1}^{\infty}$ be a descending countable family of (nonempty) closed intervals such that for any $n \in \mathbb{N}$,

$$[x_n, y_n] \subset (a, b) \quad \text{and} \quad f(n) \notin [x_n, y_n]. \tag{1.1}$$

Then $\{x_n \,|\, n \in \mathbb{N}\}$ is bounded above by $y_1 < b$. Put $x = \sup\{x_n \,|\, n \in \mathbb{N}\}$ and then $a < x < b$. If there exists $\ell \in \mathbb{N}$ such that $f(\ell) = x$, then by (1.1),

$$x = f(\ell) \notin [x_\ell, y_\ell]. \tag{1.2}$$

But for any $n \in \mathbb{N}$, $x_n \leq x$ (since $x = \sup\{x_n \,|\, n \in \mathbb{N}\}$) and for any $n, m \in \mathbb{N}$, $x_m < y_n$, which implies that $x \leq y_n$ for any $n \in \mathbb{N}$ and so $x \in [x_n, y_n]$ for any $n \in \mathbb{N}$, which is a contradiction with (1.2). Therefore, (a, b) is uncountable. □

1.4 Open and Closed Sets

Definition 1.4.1 Let E be a subset of $\mathbb{R}$ and $x \in E$. Then x is called an **interior point** of E if there exists an $\epsilon > 0$ such that $(x - \epsilon, x + \epsilon) \subset E$.

Definition 1.4.2 A subset $\mathcal{O}$ of $\mathbb{R}$ is said to be **open** if every point $x \in \mathcal{O}$ is an interior point of $\mathcal{O}$.

Remark 1.4.3 Let $\mathcal{O}$ be a subset of $\mathbb{R}$. Then $\mathcal{O}$ is open if and only if for any $x \in \mathcal{O}$, there exists an open set U such that $x \in U \subset \mathcal{O}$.

Example 1.4.4

(1) For any $a, b \in \mathbb{R}$ with $a < b$, (a, b) is an open set. In fact, for any $x \in (a, b)$, by taking

$$\epsilon = \min\{b - x, x - a\},$$

it holds that $(x - \epsilon, x + \epsilon) \subset (a, b)$.

(2) $\mathbb{R}$ and $\emptyset$ are open.

(3) The intersection of any finite family of open sets is open.

(4) The union of any family of open sets is open.

Remark 1.4.5 The properties (2),(3) and (4) in Example 1.4.4 are characteristic properties of the Euclidean topology. In general, let X be a set and τ be a family of subsets of X. Then (X, τ) is called a **topological space** if τ satisfies

(1) $X, \emptyset \in \tau$;

(2) the intersection of any finite family of sets in τ is in τ;

(3) the union of any family of sets in τ is in τ.

An element of τ is called an **open set**.

Theorem 1.4.6 Every nonempty open subset of $\mathbb{R}$ is a countable union of disjoint open intervals.

[Proof] Let $\mathcal{O}$ be a nonempty open subset of $\mathbb{R}$ and let $x \in \mathcal{O}$. There exists an open interval (a_x, b_x) containing x such that $(a_x, b_x) \subset \mathcal{O}$, put

$$l_x = \inf\{a_x \mid (a_x, b_x) \text{ is an open interval such that } x \in (a_x, b_x) \subset \mathcal{O}\},$$
$$r_x = \sup\{b_x \mid (a_x, b_x) \text{ is an open interval such that } x \in (a_x, b_x) \subset \mathcal{O}\},$$

where l_x can be taken as $-\infty$ and r_x can be taken as ∞, and put $I_x = (l_x, r_x)$. Then we obtain that

(1) $l_x, r_x \notin \mathcal{O}$, in fact, if we assume that $l_x \in \mathcal{O}$, since $\mathcal{O}$ is open, there exists a sufficiently small $\epsilon > 0$ such that $(l_x - \epsilon, l_x - \epsilon) \subset \mathcal{O}$ and $x \in (l_x - \epsilon, b) \subset \mathcal{O}$ for a number $b > x$ and so $l_x \leq l_x - \epsilon$, equivalently, $\epsilon \leq 0$, which is a contradiction. Similarly, we see that $r_x \notin \mathcal{O}$;

(2) $I_x = (l_x, r_x) \subset \mathcal{O}$, in fact, if $y \in (l_x, r_x)$, then by definition there exist $l_x < z < y$ and $y < w < r_x$ such that $z < y < w$ and $(z, w) \subset \mathcal{O}$, which implies that $y \in \mathcal{O}$.

If $x, y \in \mathcal{O}$ with $x \neq y$, then $I_x = I_y$ or $I_x \cap I_y = \emptyset$. Note that $y \in I_x$ or $y \notin I_x$. Therefore, $I_x = I_y$ in case of $y \in I_x$, and if $y \notin I_x$, then $I_x \cap I_y = \emptyset$. Finally, we have

$$\mathcal{O} = \bigcup_{x \in \mathcal{O}} I_x,$$

which is a countable union since each I_x contains a rational number and the set of all rational numbers is countable.

□

Definition 1.4.7 Let E be a subset of $\mathbb{R}$ and $x \in \mathbb{R}$. Then x is called a **point of closure** (or, simply, **closure point**) of E if for any open set $\mathcal{O}$ containing x, $E \cap \mathcal{O} \neq \emptyset$. The set of all points of closure of E is called the **closure** of E and denoted by $\overline{E}$. If $E = \overline{E}$, then E is said to be **closed**.

Remark 1.4.8 Since $E \subset \overline{E}$, E is closed if and only if $\overline{E} \subset E$.

Proposition 1.4.9 Let E and F be subsets of $\mathbb{R}$. Then we have that

(i) $\overline{E}$ is closed;

(ii) if $E \subset F$, then $\overline{E} \subset \overline{F}$;

(iii) $\overline{E}$ is the smallest closed subset of $\mathbb{R}$ containing E.

[Proof] (i) Take any $x \in \overline{\overline{E}}$. For any open set $\mathcal{O}_x$ containing x, $\overline{E} \cap \mathcal{O}_x \neq \emptyset$ and so there exists $y \in \overline{E} \cap \mathcal{O}_x$, which means that y is a point of closure of E and O_x is an open set containing y. Therefore, $E \cap \mathcal{O}_x \neq \emptyset$, which implies that $x \in \overline{E}$. Finally, we proved that $\overline{\overline{E}} \subset \overline{E}$ and so $\overline{E}$ is closed.

(ii) The proof is obvious. In fact, for any $x \in \overline{E}$ and any open subset $\mathcal{O}_x$ containing x, it holds that

$$F \cap \mathcal{O}_x \supset E \cap \mathcal{O}_x \neq \emptyset,$$

which implies that $x \in \overline{F}$.

(iii) Suppose that F is a closed set containing E. Then we have $\overline{E} \subset \overline{F} = F$. Therefore, we prove that every closed set involving E also involves $\overline{E}$ and $E \subset \overline{E}$, and so $\overline{E}$ is the smallest closed set containing E. □

Proposition 1.4.10 $\mathbb{R}$ and $\emptyset$ are closed.

[Proof] By definition, for any subset E of $\mathbb{R}$, $\overline{E} \subset \mathbb{R}$, and so $\overline{\mathbb{R}} \subset \mathbb{R}$. Therefore, $\mathbb{R}$ is closed. Note that for any open subset $\mathcal{O}_x$ of $\mathbb{R}$ containing $x \in \mathbb{R}$,

$$\mathcal{O}_x \cap \emptyset = \emptyset.$$

Therefore, there is no element x in $\mathbb{R}$ such that for any open subset $\mathcal{O}_x$ of $\mathbb{R}$ containing x, $\mathcal{O}_x \cap \emptyset \neq \emptyset$. Hence $\overline{\emptyset} = \emptyset$ and so $\emptyset$ is closed. □

Proposition 1.4.11 A subset of $\mathbb{R}$ is open if and only if its complement in $\mathbb{R}$ is closed.

[Proof] $\Rightarrow$) Let $\mathcal{O}$ is open in $\mathbb{R}$ and x in $\overline{\mathcal{O}^c}$. Suppose that $x \in \mathcal{O}$. Since $\mathcal{O}$ is open, x is an interior point of $\mathcal{O}$. Then there exists $\epsilon > 0$ such that $(x-\epsilon, x+\epsilon) \subset \mathcal{O}$ and $(x-\epsilon, x+\epsilon) \cap \mathcal{O}^c = \emptyset$ and so x is not a closure point of $\mathcal{O}^c$, which is a contradiction.

$\Leftarrow$) Suppose that $\mathcal{O}^c$ is a closed set and $\mathcal{O}$ is not open. Then $\mathcal{O} \neq \emptyset$. In fact, if $\mathcal{O} = \emptyset$, then by Proposition 1.4.10, $\mathcal{O}^c = \mathbb{R}$ is closed, and by Example 1.4.4, $\mathcal{O}$ is open, which contradicts. Since $\mathcal{O}$ is not open, there exists $x \in \mathcal{O}$ such that x is not an interior point of $\mathcal{O}$. Therefore, for any $\epsilon > 0$, $(x-\epsilon, x+\epsilon) \cap \mathcal{O}^c \neq \emptyset$. Therefore, for any open set $\mathcal{O}_x$ containing x, $\mathcal{O}_x \cap \mathcal{O}^c \neq \emptyset$, which implies that $x \in \overline{\mathcal{O}^c} = \mathcal{O}^c$ since $\mathcal{O}^c$ is closed. Hence $x \in \mathcal{O} \cap \mathcal{O}^c$, which is a contradiction. □

Remark 1.4.12 Let (X, τ) be a topological space and $F \subset X$. Then, motivated by Proposition 1.4.11, we say that F is **closed** if F^c is open, i.e., $F^c \in \tau$.

Corollary 1.4.13 We have

(i) for any $a, b \in \mathbb{R}$ with $a < b$, $[a, b]$ is a closed set;

(ii) the union of any finite family of closed sets is closed;

(iii) the intersection of any family of closed sets is closed.

[Proof] The proof is straightforward by Example 1.4.4 and Proposition 1.4.11. □

Definition 1.4.14 Let X be a set and E a subset of X. Let $\{E_\lambda\}_{\lambda\in\Lambda}$ be a family of subsets of X.

(i) The family $\{E_\lambda\}_{\lambda\in\Lambda}$ is called a **cover** of E if $E \subset \cup_{\lambda\in\Lambda} E_\lambda$. If the index set Λ is finite, then $\{E_\lambda\}_{\lambda\in\Lambda}$ is called a **finite cover** of E.

(ii) If $\{E_\lambda\}_{\lambda\in\Lambda}$ is a **cover** of E and E_λ is open for any $\lambda \in \Lambda$, then $\{E_\lambda\}_{\lambda\in\Lambda}$ is called an **open cover** of E.

(iii) If $E \subset \cup_{\lambda\in I} E_\lambda$ for a subset $I \subset \Lambda$, then $\{E_\lambda\}_{\lambda\in I}$ is called a **subcover** of E.

Proposition 1.4.15 Let $a, b \in \mathbb{R}$ with $a < b$. Then every open cover of $[a, b]$ has a finite subcover.

[Proof] Let $\mathcal{U} = \{U_\lambda\}_{\lambda\in\Lambda}$ be an open cover of $[a,b]$. Put

$$E = \{x \in [a,b] \mid \{U_\lambda\}_{\lambda\in\Lambda} \text{ has a finite subcover of } [a,x]\}.$$

Then we obtain that

(i) there exists $\epsilon > 0$ such that $[a, a+\epsilon) \subset E$. In fact, since $a \in [a,b] \subset \cup_{\lambda\in\Lambda} U_\lambda$, there exists $\lambda_a \in \Lambda$ such that $a \in U_{\lambda_a}$ and, since U_{λ_a} is open, there exists a (sufficiently small) $\epsilon > 0$ such that $(a-\epsilon, a+\epsilon) \subset U_{\lambda_a}$. If we choose the $\epsilon > 0$ as $a + \epsilon < b$, then for any $a \le x < a+\epsilon$, $\{U_{\lambda_a}\}$ is a cover of $[a,x]$ and so $[a, a+\epsilon) \subset E$;

(ii) E is bounded since $E \subset [a,b]$.

Put $c = \sup(E)$. Then we obtain that

- $c \in E$ and $c = b$.

In fact, by (i) and (ii) $a < c \le b$ and so $c \in [a,b]$. Therefore, there exists $\lambda_c \in \Lambda$ such that $c \in U_{\lambda_c}$ and so, since U_{λ_c} is open, there exists a $\delta > 0$ such that $(c-\delta, c+\delta) \subset U_{\lambda_c}$ and for any x with $c - \delta < x < c$, $x \in E$ since $c = \sup(E)$, there exists a finite family $\{U_{\lambda_1}, U_{\lambda_2}, \cdots, U_{\lambda_k}\}$ covering $[a,x]$ and so

$$\text{the finite family } \{U_{\lambda_1}, U_{\lambda_2}, \cdots, U_{\lambda_k}, U_{\lambda_c}\} \text{ covers } [a, c+\delta) \tag{1.3}$$

In particular, $\{U_{\lambda_1}, U_{\lambda_2}, \cdots, U_{\lambda_k}, U_{\lambda_c}\}$ covers $[a,c]$. Therefore, $c \in E$. Suppose that $c < b$. Then by (1.3), $\{U_{\lambda_1}, U_{\lambda_2}, \cdots, U_{\lambda_k}, U_{\lambda_c}\}$ covers $[a,x']$ for any $x' \in (c, c+\delta) \cap [a,b]$, and so $c \neq \sup(E)$, which is a contradiction; □

Theorem 1.4.16 [**Heine-Borel Theorem**] Let F be a closed and bounded subset of $\mathbb{R}$. Then every open cover of F has a finite subcover.

[Proof] Let $\mathcal{U} = \{U_\lambda\}_{\lambda\in\Lambda}$ be an open cover of F. Since F is bounded, there exist $a, b \in \mathbb{R}$ such that $F \subset [a,b]$. For the given open cover $\mathcal{U} = \{U_\lambda\}_{\lambda\in\Lambda}$ of F, since F^c is open, $\mathcal{U} \cup \{F^c\}$ is an open cover of $[a,b]$. Therefore, by Proposition 1.4.15, there exists a finite subfamily $\{V_1, V_2, \cdots, V_k\} \subset \mathcal{U} \cup \{F^c\}$ covering $[a,b]$ and then $\{V_1, V_2, \cdots, V_k\} \backslash \{F^c\}$ covers F. □

Definition 1.4.17 Let K be a subset of $\mathbb{R}$. Then K is said to be **compact** if every open cover of K has a finite subcover.

Theorem 1.4.18 Let K be a subset of $\mathbb{R}$. Then K is compact if and only if K is closed and bounded.

[Proof] $\Rightarrow$) Let K be a compact set.
(i) Boundedness. For each $x \in \mathbb{R}$ and $\epsilon > 0$, put

$$B_\epsilon(x) = \{y \in \mathbb{R} \mid |x-y| < \epsilon\} = (x-\epsilon, x+\epsilon).$$

Since $K \subset \cup_{x\in K} B_\epsilon(x)$, there exists a finite set $\{x_1, \cdots, x_n\} \subset K$ such that

$$K \subset \bigcup_{i=1}^{n} B_\epsilon(x_i),$$

and so K is bounded.

(ii) Closedness. If $K = \emptyset$, then K is closed. Suppose that $K \neq \emptyset$. Also, by (i), K is bounded and so $K \subsetneq \mathbb{R}$, i.e., $K^c \neq \emptyset$. We prove that K^c is open. Take any point $x \in K^c$. Then for any $y \in K$ with $|y-x| > 0$, i.e., $x \neq y$ and so there exist open sets U_y and V_y such that

$$y \in U_y, \quad x \in V_y, \quad U_y \cap V_y = \emptyset.$$

Then $\{U_y | y \in K\}$ is an open cover of K. Since K is compact, there exists a finite set $\{y_1, \cdots, y_n\} \subset K$ such that $\{U_{y_1}, \cdots, U_{y_n}\}$ is a finite cover of K, i.e.,

$$K \subset \bigcup_{i=1}^{n} U_{y_i} \iff \left(\bigcup_{i=1}^{n} U_{y_i}\right)^c \subset K^c. \tag{1.4}$$

Put

$$V = \bigcap_{i=1}^{n} V_{y_i}$$

and then V is an open set and $x \in V$. Moreover, $V \cap U_{y_i} = \emptyset$ for any $i = 1, 2, \cdots, n$. Therefore, we obtain that

$$V \cap \left(\bigcup_{i=1}^{n} U_{y_i}\right) = \bigcup_{i=1}^{n} (V \cap U_{y_i}) = \emptyset \iff V \subset \left(\bigcup_{i=1}^{n} U_{y_i}\right)^c \tag{1.5}$$

and so, by (1.4) and (1.5), we have

$$x \in V \subset \left(\bigcup_{i=1}^{n} U_{y_i}\right)^c \subset K^c$$

and so K^c is open.

$\Leftarrow$) The proof is immediate from the Heine-Borel Theorem (Theorem 1.4.16). □

Theorem 1.4.19 [**Nested Set Theorem**] Let $\{F_n\}_{n=1}^{\infty}$ be a descending (countable) family of nonempty closed subsets of $\mathbb{R}$ such that F_1 is bounded. Then we have

$$\bigcap_{n=1}^{\infty} F_n \neq \emptyset.$$

[Proof] Suppose that the intersection is empty, i.e.,

$$\bigcap_{n=1}^{\infty} F_n = \emptyset \quad \Longrightarrow \quad F_1 \subset \mathbb{R} = \left(\bigcap_{n=1}^{\infty} F_n\right)^{\mathrm{c}} = \bigcup_{n=1}^{\infty} F_n^{\mathrm{c}},$$

i.e., $\{F_n^{\mathrm{c}}\}_{n=1}^{\infty}$ is an open cover of the closed and bounded set F_1 and so Heine-Borel theorem (Theorem 1.4.16), there exists a finite family $\{F_{\lambda_1}^{\mathrm{c}}, \cdots, F_{\lambda_k}^{\mathrm{c}}\}$ covering F_1. Suppose that $\{F_{\lambda_1}^{\mathrm{c}}, \cdots, F_{\lambda_k}^{\mathrm{c}}\}$ is ascending and so

$$F_1 \subset \bigcup_{\ell=1}^{k} F_{\lambda_\ell}^{\mathrm{c}} = F_{\lambda_k}^{\mathrm{c}}$$

and so for any $n \geq \lambda_k$, since $F_{\lambda_k} \supset F_n$, $F_1 \subset F_{\lambda_k}^{\mathrm{c}} \subset F_n^{\mathrm{c}}$, which implies that

$$F_n = F_n \cap F_1 \subset F_n \cap F_n^{\mathrm{c}} = \emptyset,$$

which is a contradiction. □

Example 1.4.20 For each $n \in \mathbb{N}$, put $F_n = [n, \infty)$. Then $\{F_n\}$ is a descending countable family of nonempty closed sets, but

$$\bigcap_{n=1}^{\infty} F_n = \emptyset.$$

1.5 Sequences

A sequence $\{a_n\}_{n=1}^{\infty}$ of real numbers is said to be

(i) **bounded** if there exists a constant $M \geq 0$ such that $|a_n| \leq M$ for all n;

(ii) **increasing** if for any n, $a_n \leq a_{n+1}$;

(iii) **decreasing** if for any n, $a_n \geq a_{n+1}$;

(iv) **monotone** if it is increasing or decreasing.

Definition 1.5.1 A sequence $\{a_n\}_{n=1}^{\infty}$ of real numbers is said to **converge** to a real number a if for any $\epsilon > 0$, there exists a $N \in \mathbb{N}$ such that for any $n \geq N$,

$$|a_n - a| < \epsilon.$$

In this case, a is called the **limit** of the sequence and we write

$$\lim_{n \to \infty} a_n = a.$$

If $\{a_n\}_{n=1}^{\infty}$ does not converge, then $\{a_n\}_{n=1}^{\infty}$ is said to **diverge**.

Example 1.5.2 Consider a sequence $\{a_n\}_{n=1}^{\infty}$ given by $a_n = (n-1)/(n+1)$ for any $n \in \mathbb{N}$. Let $\epsilon > 0$ be given. Note that

$$|a_n - 1| = \left|\frac{n-1}{n+1} - 1\right| = \frac{2}{n+1} < \epsilon \iff n > \frac{2}{\epsilon} - 1. \tag{1.6}$$

Therefore, for any given $\epsilon > 0$, if we take $N \in \mathbb{N}$ with $N > \frac{2}{\epsilon} - 1$ (such N exists by Archimedian Property, see Theorem 1.2.7), then for any $n \geq N$, $n > \dfrac{2}{\epsilon} - 1$ and, by (1.6), $|a_n - 1| < \epsilon$. Hence $\displaystyle\lim_{n\to\infty} \frac{n-1}{n+1} = 1$.

Remark 1.5.3 Let (X, τ) be a topological space and $\{a_n\}_{n=1}^{\infty} \subset X$ a sequence. Then we say that $\{a_n\}_{n=1}^{\infty} \subset X$ **converges** to $a \in X$ (with respect to τ) if for any $U_a \in \tau$ containing a, there exists $N \in \mathbb{N}$ such that for any $n \geq N$, $a_n \in U_a$.

Proposition 1.5.4 Every convergent sequence of real numbers is bounded.

[Proof] Exercise **1.9**. □

Proposition 1.5.5 Let $\{a_n\}_{n=1}^{\infty}$ and $\{b_n\}_{n=1}^{\infty}$ be sequences of real numbers. Suppose that $\lim_{n\to\infty} a_n = a$ and $\lim_{n\to\infty} b_n = b$. Then we have

(i) for any $\alpha \in \mathbb{R}$, $\{\alpha a_n\}_{n=1}^{\infty}$ converges to αa;

(ii) $\{a_n + b_n\}_{n=1}^{\infty}$ converges to $a + b$;

(iii) if $\lim_{n\to\infty} a_n = c$, then $a = c$;

(iv) if there exists $N \in \mathbb{N}$ such that for $n \geq N$, $a_n \leq b_n$, then $a \leq b$.

[Proof] (i) If $\alpha = 0$, then it is obvious. Suppose that $\alpha \neq 0$. Let $\epsilon > 0$ be given. Then since $\lim_{n\to\infty} a_n = a$, there exists a $N \in \mathbb{N}$ such that for any $n \geq N$,

$$|a_n - a| < \frac{\epsilon}{|\alpha|}$$

and so for any $n \geq N$, we have

$$|\alpha a_n - \alpha a| = |\alpha||a_n - a| < |\alpha|\frac{\epsilon}{|\alpha|} = \epsilon,$$

which implies that $\{\alpha a_n\}_{n=1}^{\infty}$ converges to αa.

(ii) Let $\epsilon > 0$ be given. There exist $N_1, N_2 \in \mathbb{N}$ such that for any $n \geq N_1$,

$$|a_n - a| < \frac{\epsilon}{2},$$

and for any $n \geq N_2$,

$$|b_n - b| < \frac{\epsilon}{2}.$$

Put $N = \max\{N_1, N_2\}$. Then for any $n \geq N$, since $n \geq N_1$ and $n \geq N_2$,

$$\begin{aligned} |(a_n + b_n) - (a + b)| &\leq |a_n - a| + |b_n - b| \\ &< \frac{\epsilon}{2} + \frac{\epsilon}{2} = \epsilon, \end{aligned}$$

which implies that $\{a_n + b_n\}_{n=1}^{\infty}$ converges to $a + b$.

(iii) By (i) and (ii), we obtain that

$$\begin{aligned} a - c &= \lim_{n\to\infty} a_n - \lim_{n\to\infty} a_n = \lim_{n\to\infty} (a_n - a_n) \\ &= 0, \end{aligned}$$

which gives the proof.

(iv) (1) Suppose that for $n \in \mathbb{N}$, $a_n \leq 0$. We prove that $a \leq 0$. In fact, if $a > 0$, then by the Archimedian property (Theorem 1.2.7, see Remark 1.2.8), there exists $0 < \epsilon < a$, and for the given $\epsilon > 0$ there exists $N \in \mathbb{N}$ such that for any $n \geq N$, $|a_n - a| < \epsilon$ and so $0 < a - \epsilon < a_n < a + \epsilon$, which is a contradiction.

(2) Without lose of generality, we may assume that for $n \in \mathbb{N}$, $a_n \leq b_n$. Then we obtain that

$$\begin{aligned} a - b &= \lim_{n\to\infty} a_n - \lim_{n\to\infty} b_n = \lim_{n\to\infty} (a_n - b_n) \\ &\leq 0, \end{aligned}$$

which gives the proof. □

Theorem 1.5.6 Let $\{a_n\}_{n=1}^{\infty}$ be a monotone sequence. Then $\{a_n\}_{n=1}^{\infty}$ converges if and only if $\{a_n\}_{n=1}^{\infty}$ is bounded.

[Proof] $\Rightarrow$) The proof is obvious by Proposition 1.5.4.

$\Leftarrow$) Suppose that $\{a_n\}_{n=1}^{\infty}$ is an increasing sequence. Put $a = \sup\{a_n \,|\, n \in \mathbb{N}\}$ which exists by the Completeness Axiom. We prove that $\{a_n\}_{n=1}^{\infty}$ converges to a. Let $\epsilon > 0$ be given. Then since $a - \epsilon$ is not an upper bound of $\{a_n \,|\, n \in \mathbb{N}\}$, there exists a $N \in \mathbb{N}$ such that $a - \epsilon < a_N < a$ and so for any $n \geq N$, $a - \epsilon < a_N \leq a_n \leq a < a + \epsilon$, which implies that $|a_n - a| < \epsilon$. Thus $\{a_n\}_{n=1}^{\infty}$ converges to a. □

Let $\{a_n\}_{n=1}^{\infty}$ be a sequence and $\{n_k\}_{k=1}^{\infty}$ a strictly increasing sequence of natural numbers, i.e., for $l < m$, $n_l < n_m$. Then $\{a_{n_k}\}_{k=1}^{\infty}$ is called a **subsequence** of $\{a_n\}_{n=1}^{\infty}$.

Theorem 1.5.7 **[Bolzano-Weierstrass Theorem]** Every bounded sequence of real numbers has a convergent subsequence.

[Proof] Let $\{a_n\}_{n=1}^{\infty}$ is a bounded sequence of real numbers such that $|a_n| \le M$ for all n and some $M \ge 0$. For each $n \in \mathbb{N}$, put

$$F_n = \overline{\{a_\ell \,|\, \ell \ge n\}} : \quad \text{the closure of } \{a_\ell \,|\, \ell \ge n\}.$$

Then $\{F_n\}_{n=1}^{\infty}$ is a descending countable family of nonempty closed sets such that F_1 is bounded. Therefore, by the Nested Set Theorem (Theorem 1.4.19), $\cap_{n=1}^{\infty} F_n \ne \emptyset$ and so there exists $a \in \cap_{n=1}^{\infty} F_n$. For each $k \in \mathbb{N}$, since $a \in F_k = \overline{\{a_\ell \,|\, \ell \ge k\}}$,

$$\left(a - \frac{1}{k}, a + \frac{1}{k}\right) \cap \{a_\ell \,|\, \ell \ge k\} \ne \emptyset$$

and so there exists

$$a_{n_k} \in \left(a - \frac{1}{k}, a + \frac{1}{k}\right) \cap \{a_\ell \,|\, \ell \ge k\}.$$

Then we have a subsequence $\{a_{n_k}\}_{k=1}^{\infty}$ and given $\epsilon > 0$ there exists a $N \in \mathbb{N}$ such that $1/N < \epsilon$ and for any $k \ge N$,

$$|a_{n_k} - a| < \frac{1}{k} \le \frac{1}{N} < \epsilon,$$

which proves that $\lim_{k\to\infty} a_{n_k} = a$. □

Definition 1.5.8 A sequence $\{a_n\}_{n=1}^{\infty}$ is called a **Cauchy sequence** if for any $\epsilon > 0$, there exists a $N \in \mathbb{N}$ such that for any $m, n \ge N$,

$$|a_m - a_n| < \epsilon.$$

Proposition 1.5.9 Every Cauchy sequence is bounded.

[Proof] Let $\{a_n\}_{n=1}^{\infty}$ be a Cauchy sequence. For a given $\epsilon = 1$, there exists a $N \in \mathbb{N}$ such that for any $m, n \ge N$, $|a_n - a_m| < 1$ and so for any $n \ge N$, $|a_n - a_N| < 1$, which implies that for any $n \ge N$, $|a_n| < |a_N| + 1$. Therefore, for $M = \max\{|a_1|, \cdots, |a_{N-1}|, |a_N| + 1\}$, $|a_n| \le M$ for all n, and so $\{a_n\}_{n=1}^{\infty}$ is bounded. □

Theorem 1.5.10 A sequence of real numbers converges if and only if it is a Cauchy sequence.

[Proof] $\Rightarrow$) Let $\{a_n\}_{n=1}^{\infty}$ be a sequence which converges to a. Then for any given $\epsilon > 0$, there exists a $N \in \mathbb{N}$ such that for any $n \geq N$, $|a_n - a| < \epsilon/2$ and so for any $m, n \geq N$,

$$\begin{aligned}|a_m - a_n| &\leq |a_n - a| + |a_m - a| \\ &< \frac{\epsilon}{2} + \frac{\epsilon}{2} = \epsilon,\end{aligned}$$

which implies that $\{a_n\}_{n=1}^{\infty}$ is a Cauchy sequence.

$\Leftarrow$) Let $\{a_n\}_{n=1}^{\infty}$ be a Cauchy sequence. Then by Proposition 1.5.9 $\{a_n\}_{n=1}^{\infty}$ is bounded. Then by the Bolzano-Weierstrass Theorem (Theorem 1.5.7), there exists a subsequence $\{a_{n_k}\}_{k=1}^{\infty}$ of $\{a_n\}_{n=1}^{\infty}$ such that $\{a_{n_k}\}_{k=1}^{\infty}$ converges to a real number $a \in \mathbb{R}$. Then for any given $\epsilon > 0$, there exist $N, K \in \mathbb{N}$ such that for any $m, n \geq N$, $|a_n - a_m| < \epsilon/2$ and for any $k \geq K$, $|a_{n_k} - a| < \epsilon/2$. Therefore, for any $n \geq N$, we have

$$\begin{aligned}|a_n - a| &\leq |a_n - a_{n_k}| + |a_{n_k} - a| \\ &< \frac{\epsilon}{2} + \frac{\epsilon}{2} = \epsilon,\end{aligned}$$

where $k \geq \max\{N, K\}$ and so $n_k \geq N$. Therefore, $\{a_n\}_{n=1}^{\infty}$ converges to the real number a. □

Let $\{a_n\}_{n=1}^{\infty}$ be a sequence of real numbers. If for each $M \geq 0$ there exists a $N \in \mathbb{N}$ such that for any $n \geq N$, $a_n \geq M$, then we say that $\{a_n\}_{n=1}^{\infty}$ converges to infinity and we write $\lim_{n\to\infty} a_n = \infty$. Similarly, we define $\lim_{n\to\infty} a_n = -\infty$.

Definition 1.5.11 Let $\{a_n\}_{n=1}^{\infty}$ be a sequence of real numbers.

(1) The limit superior, denoted by $\limsup_{n\to\infty} a_n$, of $\{a_n\}_{n=1}^{\infty}$ is defined by

$$\limsup_{n\to\infty} a_n = \lim_{n\to\infty} \left(\sup\{a_\ell | \ell \geq n\}\right).$$

(2) The limit inferior, denoted by $\liminf_{n\to\infty} a_n$, of $\{a_n\}_{n=1}^{\infty}$ is defined by

$$\liminf_{n\to\infty} a_n = \lim_{n\to\infty} \left(\inf\{a_\ell | \ell \geq n\}\right).$$

Remark 1.5.12 Let $\{a_n\}_{n=1}^{\infty}$ be a sequence of real numbers. Then we have

(1) $\limsup_{n\to\infty} a_n = a$ for a real number $a \in \mathbb{R}$, if and only if for any $\epsilon > 0$ there exists a $N \in \mathbb{N}$ such that for any $n \geq N$,

$$a - \epsilon < \sup\{a_\ell | \ell \geq n\} < a + \epsilon$$

if and only if for any $\epsilon > 0$, the set $\{\ell \in \mathbb{N} \mid a_\ell > a - \epsilon\}$ is countably infinite and the set $\{\ell \in \mathbb{N} \mid a_\ell > a + \epsilon\}$ is finite.

(2) $\limsup_{n\to\infty} a_n = -\liminf_{n\to\infty}(-a_n)$, in fact we obtain that

$$\begin{aligned}\liminf_{n\to\infty}(-a_n) &= \lim_{n\to\infty}(\inf\{-a_\ell|\ell \geq n\}) \\ &= \lim_{n\to\infty}(-\sup\{a_\ell|\ell \geq n\}) \\ &= -\limsup_{n\to\infty} a_n.\end{aligned}$$

(3) $\{a_n\}_{n=1}^{\infty}$ converges to an extended real number a if and only if

$$\liminf_{n\to\infty} a_n = \limsup_{n\to\infty} a_n = a.$$

[Proof] Note that $\{a_n\}_{n=1}^{\infty}$ converges to a real number a if and only if for any $\epsilon > 0$, there exists a $N \in \mathbb{N}$ such that for any $n \geq N$, $a - \epsilon < a_n < a + \epsilon$ if and only if $\limsup_{n\to\infty} a_n = a$ and $-a = \limsup_{n\to\infty}(-a_n) = -\liminf_{n\to\infty} a_n$ if and only if $\liminf_{n\to\infty} a_n = \limsup_{n\to\infty} a_n = a$. □

(4) For two sequence $\{a_n\}_{n=1}^{\infty}$ and $\{b_n\}_{n=1}^{\infty}$ such that $a_n \leq b_n$ for all $n \in \mathbb{N}$, in general, $\limsup_{n\to\infty} a_n \leq \liminf_{n\to\infty} b_n$ is not true. Consider the following two sequences $\{a_n\}_{n=1}^{\infty}$ and $\{b_n\}_{n=1}^{\infty}$:

$$a_n = \begin{cases} \dfrac{2}{3}, & n : \text{odd} \\ \dfrac{1}{4}, & n : \text{even}, \end{cases} \qquad b_n = \begin{cases} 1, & n : \text{odd} \\ \dfrac{1}{2}, & n : \text{even}. \end{cases}$$

Then for any $n \in \mathbb{N}$, $a_n \leq b_n$ and

$$\sup\{a_\ell|\ell \geq n\} = \frac{2}{3}, \qquad \inf\{b_\ell|\ell \geq n\} = \frac{1}{2}.$$

Therefore, we have

$$\limsup_{n\to\infty} a_n = \frac{2}{3} > \frac{1}{2} = \liminf_{n\to\infty} b_n.$$

Let $\{a_n\}_{n=1}^{\infty}$ be a sequence of real numbers. For each $n \in \mathbb{N}$,

$$s_n = \sum_{k=1}^{n} a_k$$

is called a **partial sum**. If the sequence $\{s_n\}_{n=1}^{\infty}$ of partial sums converges to a real number s, the we say that the series $\sum_{k=1}^{\infty} a_k$ is **summable** to s. In this case, we write

$$\sum_{k=1}^{\infty} a_k.$$

If the series $\sum_{k=1}^{\infty} |a_k|$ is summable, then we say that the series $\sum_{k=1}^{\infty} a_k$ is **absolutely summable**,

Proposition 1.5.13 Let $\{a_n\}_{n=1}^{\infty}$ be a sequence of real numbers.

(i) The series $\sum_{k=1}^{\infty} a_k$ is summable if and only if for any $\epsilon > 0$ there exists a $N \in \mathbb{N}$ such that for any $n \geq N$ and any $m \in \mathbb{N}$,

$$\left|\sum_{k=n}^{n+m} a_k\right| < \epsilon.$$

(ii) Every absolutely summable series is summable.

(iii) If for any $k \in \mathbb{N}$, a_k is nonnegative, then the series $\sum_{k=1}^{\infty} a_k$ is summable if and only if the sequence $\{s_n\}_{n=1}^{\infty}$ of partial sums is bounded.

[Proof] (i) The series $\sum_{k=1}^{\infty} a_k$ is summable to a real number s if and only if the sequence $\{s_n\}_{n=1}^{\infty}$ of partial sums converges to s if and only if for any $\epsilon > 0$ there exists a $N \in \mathbb{N}$ such that for any $n \geq N$ and any $m \in \mathbb{N}$,

$$|s_{n+m} - s_{n-1}| < \epsilon$$

if and only if for any $\epsilon > 0$ there exists a $N \in \mathbb{N}$ such that for any $n \geq N$ and any $m \in \mathbb{N}$,

$$\left|\sum_{k=n}^{n+m} a_k\right| < \epsilon.$$

(ii) Let $\sum_{k=1}^{\infty} a_k$ be an absolutely summable series. Then for any ϵ there exists a $N \in \mathbb{N}$ such that for any $n \geq N$ and any $m \in \mathbb{N}$,

$$\sum_{k=n}^{n+m} |a_k| < \epsilon$$

and so for any $n \geq N$ and any $m \in \mathbb{N}$,

$$\left|\sum_{k=n}^{n+m} a_k\right| \leq \sum_{k=n}^{n+m} |a_k| < \epsilon.$$

Therefore, the series $\sum_{k=1}^{\infty} a_k$ is summable.

(iii) Since a_k is nonnegative for any $k \in \mathbb{N}$, the sequence $\{s_n\}_{n=1}^{\infty}$ is increasing. Therefore, the series $\sum_{k=1}^{\infty} a_k$ is summable if and only if the sequence $\{s_n\}_{n=1}^{\infty}$ converges if and only if $\{s_n\}_{n=1}^{\infty}$ is bounded. □

Theorem 1.5.14 Let $\{a_n\}_{n=1}^{\infty}$ be a sequence of complex numbers such that the series $\sum_{n=1}^{\infty} a_n$ is absolutely summable, i.e. $\sum_{n=1}^{\infty} |a_n| < \infty$. Then for any one-to-one and onto map $\sigma : \mathbb{N} \to \mathbb{N}$, $\sum_{n=1}^{\infty} a_{\sigma(n)}$ is absolutely summable and $\sum_{n=1}^{\infty} a_{\sigma(n)}$ converges to the limit of $\sum_{n=1}^{\infty} a_n$.

[Proof] Let $\epsilon > 0$ be given. Since the series $\sum_{n=1}^{\infty} a_n$ is absolutely summable, there exists $N \in \mathbb{N}$ such that for any $m \geq n \geq N$,

$$\sum_{k=n}^{m} |a_k| < \epsilon.$$

There exists $M \in \mathbb{N}$ with $M \geq N$ such that

$$\{1, 2, \cdots, N\} \in \{\sigma(1), \sigma(2), \cdots, \sigma(M)\}.$$

In fact, $M \geq \max\{k \in \mathbb{N} \,|\, \sigma(k) \in \{1, 2, \cdots, N\}\}$ and it is obvious that $\max\{k \in \mathbb{N} \,|\, \sigma(k) \in \{1, 2, \cdots, N\}\} \geq N$. Then for any $n \geq M$, we have

$$\left| \sum_{k=1}^{n} a_k - \sum_{k=1}^{n} a_{\sigma(k)} \right| \leq \sum_{i \in \{\sigma(1), \cdots, \sigma(n)\} \triangle \{1, \cdots, n\}} |a_i| < \epsilon$$

since $(\{\sigma(1), \cdots, \sigma(n)\} \triangle \{1, \cdots, n\}) \cap \{1, \cdots, n\} = \emptyset$, where $A \triangle B = (A \setminus B) \cup (B \setminus A)$ for two sets A and B. Therefore,

$$\lim_{n\to\infty} \left(\sum_{k=1}^{n} a_k - \sum_{k=1}^{n} a_{\sigma(k)} \right) = 0$$

and $\lim_{n\to\infty} \sum_{k=1}^{n} a_k$ converges and so $\lim_{n\to\infty} \sum_{k=1}^{n} a_{\sigma(k)}$ converges (since $\sum_{n=1}^{\infty} a_n$ is absolutely summable), and

$$\lim_{n\to\infty} \sum_{k=1}^{n} a_{\sigma(k)} = \lim_{n\to\infty} \sum_{k=1}^{n} a_k,$$

which gives the proof. □

1.6 Continuous Functions

Let D be a subset of $\mathbb{R}$ and $f : D \to \mathbb{R}$ a function. Then we say that

(1) f is **continuous** at $x = a \in D$ if for any ϵ, there exists a $\delta > 0$ such that for any $x \in D$ with $|x - a| < \delta$,

$$|f(x) - f(a)| < \epsilon;$$

(2) f is **continuous** (on D) if f is continuous at each point in D;

(3) f is said to be **Lipschitz** if

$$|f(x) - f(y)| \leq c|x - y|, \qquad x, y \in D$$

for some constant $c \geq 0$.

Remark 1.6.1 Every Lipschitz function is continuous. But the converse is not true, that is, there exists a continuous function which is not Lipschitz.

Proposition 1.6.2 Let D be a subset of $\mathbb{R}$ and $f : D \to \mathbb{R}$ a function. Then f is continuous at $x = a \in D$ if and only if for any sequence $\{a_n\}_{n=1}^{\infty} \subset D$ converging to a, $\{f(a_n)\}_{n=1}^{\infty}$ converges to $f(a)$.

[Proof] $\Rightarrow$) Suppose that f is continuous at $x = a \in D$ and $\{a_n\}_{n=1}^{\infty} \subset D$ is a sequence converging to a. Let ϵ be given. Then since f is continuous at $x = a$, there exists a $\delta > 0$ such that for any $x \in D$ with $|x - a| < \delta$,

$$|f(x) - f(a)| < \epsilon.$$

On the other hand, since $\{a_n\}_{n=1}^{\infty}$ converges to a, there exists a $N \in \mathbb{N}$ such that for any $n \geq N$, $|a_n - a| < \delta$ and so $|f(a_n) - f(a)| < \epsilon$. Hence, $\{f(a_n)\}_{n=1}^{\infty}$ converges to $f(a)$.

$\Leftarrow$) For given $\epsilon > 0$, put

$$D_\epsilon = \{x \in D \,|\, |f(x) - f(a)| < \epsilon\}.$$

Suppose that f is not continuous at $x = a$. Then there exists an $\epsilon > 0$ such that for any $\delta > 0$,

$$B_\delta(a) = \{x \in D \,|\, |x - a| < \delta\} \not\subseteq D_\epsilon,$$

that is, for any $n \in \mathbb{N}$, there exists $a_n \in B_{1/n}(a)$ such that $|f(a_n) - f(a)| \geq \epsilon$. Therefore, the sequence $\{a_n\}_{n=1}^{\infty}$ converges to a, but $\{f(a_n)\}_{n=1}^{\infty}$ does not converge to $f(a)$, which is a contradiction. □

Proposition 1.6.3 Let D be a subset of $\mathbb{R}$ and $f : D \to \mathbb{R}$ a function. Then f is continuous on D if and only if for any open subset $\mathcal{O}$ of $\mathbb{R}$, $f^{-1}(\mathcal{O})$ is open in D, i.e., there exists an open subset U of $\mathbb{R}$ such that

$$f^{-1}(\mathcal{O}) = D \cap U.$$

[Proof] $\Rightarrow$) Suppose that f is continuous on D and let $\mathcal{O}$ is an open subset of $\mathbb{R}$. For any $a \in f^{-1}(\mathcal{O})$, there exists an $\epsilon_a > 0$ such that

$$B_{\epsilon_a}(f(a)) = \{y \in \mathbb{R} \,|\, |f(a) - y| < \epsilon_a\} \subset \mathcal{O}.$$

Since f is continuous at $x = a$, for the $\epsilon_a > 0$, there exists a $\delta_a > 0$ such that

$$f(D \cap B_{\delta_a}(a)) = f(\{x \in D \,|\, |x - a| < \delta_a\}) \subset B_{\epsilon_a}(f(a)),$$

where $B_{\delta_a}(a) = \{x \in \mathbb{R} \,|\, |x - a| < \delta_a\}$, which implies that

$$D \cap B_{\delta_a}(a) = \{x \in D \,|\, |x - a| < \delta_a\} \subset f^{-1}(B_{\epsilon_a}(f(a))).$$

Put

$$U = \bigcup_{a \in f^{-1}(\mathcal{O})} B_{\delta_a}(a).$$

Then U is an open set, $f^{-1}(\mathcal{O}) \subset U$ and so $f^{-1}(\mathcal{O}) \subset D \cap U$, and

$$\begin{aligned} D \cap U &= \bigcup_{a \in f^{-1}(\mathcal{O})} (D \cap B_{\delta_a}(a)) \subset \bigcup_{a \in f^{-1}(\mathcal{O})} f^{-1}(B_{\epsilon_a}(f(a))) \\ &\subset f^{-1}(\mathcal{O}). \end{aligned}$$

Therefore, we have

$$f^{-1}(\mathcal{O}) = D \cap U.$$

$\Leftarrow$) Suppose that for any open subset $\mathcal{O}$ of $\mathbb{R}$, $f^{-1}(\mathcal{O})$ is open in D. Let a be a point in D and $\epsilon > 0$ be given. Then since $B_\epsilon(f(a)) = \{y \in \mathbb{R} \,|\, |f(a) - y| < \epsilon\}$ is open, by assumption there exists an open set U such that

$$f^{-1}(B_\epsilon(f(a))) = D \cap U.$$

Also, $a \in U$ and U is open, there exists a $\delta > 0$ such that $B_\delta(a) = \{x \in \mathbb{R} \,|\, |x - a| < \delta\} \subset U$. Therefore, we have

$$f(D \cap B_\delta(a)) \subset f(D \cap U) = f\left(f^{-1}(B_\epsilon(f(a)))\right) \subset B_\epsilon(f(a))$$

and so f is continuous at $x = a$. □

Lemma 1.6.4 Let K be a compact subset of $\mathbb{R}$ and $f : K \to \mathbb{R}$ a continuous function. Then f is bounded.

[Proof] Since f is a continuous function on K, for given $\epsilon = 1$ and any $x \in K$, there exists a $\delta_x > 0$ such that for any $y \in K$ and $y \in B_{\delta_x}(x) = \{y \in \mathbb{R} \,|\, |y - x| < \delta_x\}$, $|f(y) - f(x)| < 1$, which implies that $|f(y)| < |f(x)| + 1$. Since $\{B_{\delta_x}(x) \,|\, x \in K\}$ is an open cover of K and K is compact, there exists a finite set $\{x_1, \cdots, x_n\} \subset K$ such that $K \subset \cup_{i=1}^n B_{\delta_{x_i}}(x_i)$. Then for any $y \in K$, there exists $i \in \{1, 2, \cdots, n\}$ such that $y \in B_{\delta_{x_i}}(x_i)$ and so

$$|f(y)| < |f(x_i)| + 1.$$

Therefore, for any $y \in K$,

$$|f(y)| < M = \max\{|f(x_i)| + 1 \,|\, i = 1, 2, \cdots, n\},$$

which implies that f is bounded. □

Theorem 1.6.5 **[Extreme Value Theorem]** Let K be a compact subset of $\mathbb{R}$ and $f : K \to \mathbb{R}$ a continuous function. There exist $x_0, x_1 \in K$ such that

$$f(x_0) \le f(x) \le f(x_1)$$

for all $x \in K$. In this case, $f(x_0)$ and $f(x_1)$ are called the **minimum value** and **maximum value** of f, respectively.

[Proof] By Lemma 1.6.4, f is bounded. Put

$$M = \sup\{f(x) \,|\, x \in K\} < \infty.$$

If $f(x) \ne M$ for any $x \in K$, then the function $g : K \to \mathbb{R}$ defined by

$$g(x) = \frac{1}{f(x) - M}, \qquad x \in K$$

is a continuous, but it is not bounded, which is a contradiction to Lemma 1.6.4. Therefore, there exists a $x_1 \in K$ such that

$$f(x_1) = M = \sup\{f(x) \,|\, x \in K\} < \infty,$$

which implies that for any $x \in K$, $f(x) \le f(x_1)$. Since f is continuous, $-f$ is continuous and then we put

$$-m = \sup\{-f(x) \,|\, x \in K\}.$$

then there exists a $x_0 \in K$ such that

$$-f(x_0) = -m = \sup\{-f(x) \,|\, x \in K\}$$

and for any $x \in K$, $f(x) \ge m = f(x_0)$. □

Remark 1.6.6 In the proof of Theorem 1.6.5, since $M = \sup(f(E)) < \infty$, there exists a sequence $\{x_n\}_{n=1}^\infty \subset E$ such that $\{f(x_n)\}_{n=1}^\infty$ is an increasing sequence and $\lim_{n\to\infty} f(x_n) = M$, and then since E is bounded, $\{x_n\}_{n=1}^\infty$ is a bounded sequence. Therefore, by the Bolzano-Weierstrass Theorem (Theorem 1.5.7), there exists a subsequence $\{x_{n_k}\}_{k=1}^\infty$ of $\{x_n\}_{n=1}^\infty$ such that x_{n_k} converges to a point $x_1 \in \mathbb{R}$. Since E is closed, $x_1 \in E$ and $f(x_1) = \lim_{k\to\infty} f(x_{n_k}) = M$, which gives another proof of Theorem 1.6.5.

Theorem 1.6.7 **[Intermediate Value Theorem]** Let $a, b \in \mathbb{R}$ and $f : [a, b] \to \mathbb{R}$ a continuous function. Then for any c between $f(a)$ and $f(b)$ ($f(a) < c < f(b)$ or $f(b) < c < f(a)$), there exists a $x_0 \in (a, b)$ such that

$$f(x_0) = c.$$

[Proof] Suppose that $f(a) < c < f(b)$. Put

$$(a_1, b_1) = (a, b), \quad (a_n, b_n) = \begin{cases} (a_{n-1}, m_{n-1}), \text{ if } c < f(m_{n-1}) \\ (m_{n-1}, b_{n-1}), \text{ if } c \geq f(m_{n-1}), \end{cases} \quad n \geq 2,$$

where m_n is the midpoint of $[a_n, b_n]$. then for any $n \in \mathbb{N}$, $f(a_n) \leq c \leq f(b_n)$ and $b_n - a_n = (b-a)/2^{n-1}$. By the Nested Set Theorem (Theorem 1.4.19),

$$F := \bigcap_{n=1}^{\infty} [a_n, b_n] \neq \emptyset$$

and take $x_0 \in F$. Then we obtain that

$$|a_n - x_0| \leq |a_n - b_n| = \frac{b-a}{2^{n-1}}, \quad |b_n - x_0| \leq |b_n - a_n| = \frac{b-a}{2^{n-1}},$$

and so $\lim_{n\to\infty} a_n = \lim_{n\to\infty} b_n = x_0$. Therefore, since f is continuous,

$$\lim_{n\to\infty} f(a_n) = \lim_{n\to\infty} f(b_n) = f(x_0).$$

On the other hand, for any $n \in \mathbb{N}$, $f(a_n) \leq c$ and so

$$f(x_0) = \lim_{n\to\infty} f(a_n) \leq c \leq \lim_{n\to\infty} f(b_n) = f(x_0),$$

which implies that $f(x_0) = c$. □

Definition 1.6.8 Let D be a subset of $\mathbb{R}$ and $f : D \to \mathbb{R}$ a function. Then f is said to be ***uniform continuous*** on D if for any $\epsilon > 0$, there exists a $\delta > 0$ such that for any $x_1, x_2 \in D$ with $|x_1 - x_2| < \delta$,

$$|f(x_1) - f(x_2)| < \epsilon. \tag{1.7}$$

Remark 1.6.9 Let D be a subset of $\mathbb{R}$ and $f : D \to \mathbb{R}$ a function.

(1) If f is continuous on D, then f is continuous at each point $x_0 \in D$ and so for any $\epsilon > 0$, there exists a $\delta > 0$ such that for any $x \in D$ with $|x_0 - x| < \delta$, $|f(x_0) - f(x)| < \epsilon$. In this case, the choice of δ is depend on the point x_0 and also on the given ϵ.

(2) If f is uniform continuous on D, then for any $\epsilon > 0$, there exists a $\delta > 0$ such that for any $x_1, x_2 \in D$ with $|x_1 - x_2| < \delta$, (1.7) holds. In this case, the δ must be chosen independently of x, i.e., we must choose only one δ which works for the whole domain D.

Proposition 1.6.10 If $f : D \to \mathbb{R}$ is uniformly continuous, then f is continuous.

[Proof] Exercise **1.14**. □

Theorem 1.6.11 Let K be a compact subset of $\mathbb{R}$ and $f : K \to \mathbb{R}$ a continuous function. Then f is uniform continuous on K.

[Proof] Let $\epsilon > 0$ be given. Since f is continuous on K, f is continuous at each point $x \in K$ and so there exists a $\delta_\epsilon(x) > 0$ such that for any $w \in K$ with $|x - w| < \delta_\epsilon(x)$,

$$|f(x) - f(w)| < \frac{\epsilon}{2}.$$

For each $r > 0$, put

$$B_r(x) = \{w \in K \,|\, |x - w| < r\}.$$

Then

$$\left\{ B_{\frac{1}{2}\delta_\epsilon(x)}(x) \,\middle|\, x \in K \right\}$$

becomes an open cover of the compact set K. Therefore, there exists a finite subset $\{x_1, x_2, \cdots, x_n\} \subset K$ such that

$$\left\{ B_{\frac{1}{2}\delta_\epsilon(x_i)}(x_i) \,\middle|\, i = 1, 2, \cdots, n \right\}$$

covers K. Put

$$\delta = \frac{1}{2} \min \{\delta_\epsilon(x_i) \,|\, i = 1, 2, \cdots, n\}.$$

Then for any $x, y \in K$ with $|x - y| < \delta$, we have that there exists a $m \in \{1, 2, \cdots, n\}$ such that $x \in B_{\frac{1}{2}\delta_\epsilon(x_m)}(x_m)$ and so

$$|x_m - y| \le |x_m - x| + |x - y| < \frac{1}{2}\delta_\epsilon(x_m) + \delta \le \delta_\epsilon(x_m),$$

and

$$|f(x) - f(y)| \le |f(x) - f(x_m)| + |f(x_m) - f(y)| < \frac{\epsilon}{2} + \frac{\epsilon}{2} = \epsilon.$$

Therefore, f is uniform continuous on K. □

Example 1.6.12 (1) The function $f : \mathbb{R} \to \mathbb{R}$ defined by $f(x) = x^2$ for $x \in \mathbb{R}$ is continuous on $\mathbb{R}$, but f is not uniform continuous on $\mathbb{R}$. However, for any $a < b$, f is uniform continuous on $[a, b]$.

(2) The function $f : (0, 1) \to \mathbb{R}$ defined by $f(x) = 1/x$ for $x \in (0, 1)$ is continuous on $(0, 1)$, but f is not uniform continuous on $(0, 1)$. However, for any $0 < a < 1$, f is uniform continuous on $[a, 1)$, and so f is also uniform continuous on $(a, 1)$,

Definition 1.6.13 Let E be a subset of $\mathbb{R}$ and $f : E \to \mathbb{R}$ a function. Then f is said to be

(i) **increasing** if for any $x, y \in E$ with $x \leq y$, $f(x) \leq f(y)$;

(ii) **decreasing** if for any $x, y \in E$ with $x \leq y$, $f(x) \geq f(y)$;

(iii) **monotone** if it is increasing or decreasing;

(iv) **strictly increasing** if for any $x, y \in E$ with $x < y$, $f(x) < f(y)$;

(v) **strictly decreasing** if for any $x, y \in E$ with $x < y$, $f(x) > f(y)$;

(vi) **strictly monotone** if it is strictly increasing or strictly decreasing.

Proposition 1.6.14 Every strictly monotone continuous function defined on an interval has the continuous inverse.

[Proof] Let I be an interval and $f : I \to \mathbb{R}$ a strictly monotone continuous function defined on I. We may assume that I is an open interval. Let $y_0 \in f(I)$. Then there exists $x_0 \in I$ such that $f(x_0) = y_0$. For any given $\epsilon > 0$ (we can consider only small $\epsilon > 0$ such that $B_\epsilon(x_0) \subset I$), $f(B_\epsilon(x_0))$ is an open interval containing y_0. Therefore, there exists $\delta > 0$ such that

$$B_\delta(y_0) \subset f(B_\epsilon(x_0)),$$

implies that

$$f^{-1}(B_\delta(y_0)) \subset B_\epsilon(x_0),$$

where we used the fact that f is one to one, and so f^{-1} is continuous. □

Remark 1.6.15 In the proof of Proposition 1.6.14, if we suppose that f is strictly increasing, then for any given $\epsilon > 0$ and $x \in B_\epsilon(x_0)$, i.e., $x_0 - \epsilon < x < x_0 + \epsilon$,

$$f(x_0 - \epsilon) < f(x) < f(x_0 + \epsilon)$$

and so we can take δ as

$$\delta = \min\{f(x_0) - f(x_0 - \epsilon), f(x_0 + \epsilon) - f(x_0)\}.$$

If f is strictly decreasing, we can take δ as

$$\delta = \min\{f(x_0) - f(x_0 + \epsilon), f(x_0 - \epsilon) - f(x_0)\}.$$

1.7 Urysohn's Lemma and Tietze Extension Theorem

A pair (X, τ) of a nonempty set X and a collection τ of subsets of X is called a **topological space** if τ satisfies the following properties:

(1) $X, \emptyset \in \tau$,

(2) τ is closed under arbitrary unions, i.e., the union of any family of sets in τ belongs to τ.

(3) τ is closed under finite intersections, i.e., the intersection of any finite family of sets in τ belongs to τ.

Definition 1.7.1 Let (X, τ) be a topological space, E a subset of X and $x \in X$.

(1) Every element of τ is called an **open set**.

(2) E is called a **closed set** if there exists an open set $U \in \tau$ such that $E = U^c$.

(3) A subset $U \subset X$ containing x is called a **neighborhood** of x if there exists an open set O containing x such that $O \subset U$.

Definition 1.7.2 [**Separation Axioms**] Let (X, τ) be a topological space.

(1) (X, τ) is called a **T_1-space** if for each distinct points x and y in X, there exists an open set $O \in \tau$ such that $x \in O$ and $y \notin O$.

(2) (X, τ) is said to be **Hausdorff** (or called a **T_2-space**) if for each distinct points x and y in X, there exist disjoint open sets U and V such that $x \in U$ and $y \in V$.

(3) (X, τ) is said to be **regular** if for each closed subset F of X and each point $x \in X \setminus F$, there exist disjoint open sets U and V such that $F \subset U$ and $x \in V$.

(4) (X, τ) is called a **T_3-space** if (X, τ) is a regular T_1-space.

(5) (X, τ) is said to be **normal** if for each disjoint closed subsets E and F of X, there exist disjoint open sets U and V such that $E \subset U$ and $F \subset V$.

(6) (X, τ) is called a **T_4-space** if (X, τ) is a normal T_1-space.

Proposition 1.7.3 Let (X, τ) be a topological space. Then (X, τ) is a T_1-space if and only if for any $x \in X$, $\{x\}$ is closed.

[Proof] $\Rightarrow$) Let $x \in X$. Then for any $y \in X \setminus \{x\}$, $x \neq y$ and, since (X, τ) is a T_1-space, there exists an open set $O_y \in \tau$ such that $y \in O_y$ and $x \notin O_y$ (which implies that $\{x\} \subset O_y^{\mathrm{c}}$ and so $O_y \subset X \setminus \{x\}$). Put

$$O = \bigcup_{y \in X \setminus \{x\}} O_y.$$

Then O is open and $X \setminus \{x\} = O$, and so $\{x\}$ is closed.

$\Leftarrow$) It is obvious. In fact, for distinct points $x, y \in X$, by the assumption $O = X \setminus \{x\}$ is an open set such that $y \in O$ and $x \notin O$. □

It is easy to see that

$$T_4 \Rightarrow T_3 \Rightarrow T_2 \Rightarrow T_1.$$

Definition 1.7.4 Let (X_1, τ_1) and (X_2, τ_2) be topological spaces, and let $f : X_1 \to X_2$ a function. Then f is said to be **continuous** if for any open set $O \in \tau_2$, $f^{-1}(O) \in \tau_1$.

Let (X, d) be a metric space and F a closed subset of X. For each $x \in X$, put

$$d_F(x) = \inf\{d(x, y) | y \in F\}.$$

Then the function $d_F : X \to \mathbb{R}$ is continuous. In fact, for each given $x, y \in X$, by the triangle inequality, we obtain that

$$d(x, z) \leq d(x, y) + d(y, z), \qquad d(y, z) \leq d(x, y) + d(x, z),$$

which implies that

$$\begin{aligned} d_F(x) &= \inf\{d(x, z) \,|\, z \in F\} \leq d(x, y) + \inf\{d(y, z) \,|\, z \in F\} = d(x, y) + d_F(y), \\ d_F(y) &= \inf\{d(y, z) \,|\, z \in F\} \leq d(y, x) + \inf\{d(x, z) \,|\, z \in F\} = d(y, x) + d_F(x). \end{aligned}$$

Example 1.7.5 Every metric space is a T_1-space.

[Proof] Let (X, d) be a metric space. For given distinct points $x, y \in X$, put

$$O = \left\{ z \in X \,\middle|\, d(x, z) < \frac{1}{2} d(x, y) \right\}.$$

Then it is obvious that O is open and $x \in O$, $y \notin O$. Hence the metric space (X, d) is a T_1-space. □

Example 1.7.6 Every metric space is normal and so every metric space is a T_4-space.

[Proof] Let (X, d) be a metric space. For given disjoint closed subsets E and F of X, put

$$U = \{x \in X | d_E(x) < d_F(x)\} \quad \text{and} \quad V = \{x \in X | d_E(x) > d_F(x)\}.$$

Then it is obvious that U and V are disjoint and $E \subset U$, $F \subset V$. On the other hand,

$$U = (d_E - d_F)^{-1}((-\infty, 0)), \qquad V = (d_E - d_F)^{-1}((0, \infty))$$

are open. Hence the metric space (X, d) is normal. □

Proposition 1.7.7 Let (X, τ) be a topological space. Then (X, τ) is normal if and only if for each closed subset F of X and open set U containing F, there exists an open set $V \in \tau$ such that $F \subset V \subset \overline{V} \subset U$.

[Proof] $\Rightarrow$) Suppose that (X, τ) is normal, and a closed set F and an open set U containing F are given. Then F and U^c are disjoint closed subsets of X. Therefore, there exist disjoint open sets V and W such that $F \subset V$ and $U^c \subset W$. Then $V \subset W^c \subset U$ and so

$$F \subset V \subset \overline{V} \subset W^c \subset U.$$

$\Leftarrow$) Let E and F be disjoint closed subsets of X. Then $E \subset F^c$ and F^c is open. Therefore, by the assumption, there exists an open set V such that $E \subset V \subset \overline{V} \subset F^c$. Put $U = \overline{V}^c$. Then U is open and $U \cap V = \emptyset$, and $F \subset \overline{V}^c = U$. Therefore, (X, τ) is normal. □

Proposition 1.7.8 Let E be a closed subset of a normal topological space (X, τ) and let U be a neighborhood of E, i.e., U is an open subset of X containing E. Then there exists a dense subset Λ of $(0, 1)$ and a family $\{O_\lambda\}_{\lambda \in \Lambda}$ of open subsets of X such that

(i) for any $\lambda \in \Lambda$, $F \subset O_\lambda \subset \overline{O}_\lambda \subset U$,

(ii) for any $\lambda_1, \lambda_2 \in \Lambda$ with $\lambda_1 < \lambda_2$,

$$\overline{O}_{\lambda_1} \subset O_{\lambda_2}.$$

[Proof] For each n, put

$$\Lambda_n = \{m/2^n \,|\, 1 \le m \le 2^n - 1\}$$

and $\{\Lambda_n\}_{n=1}^{\infty}$ be an ascending family of dyadic numbers. Put

$$\Lambda = \bigcup_{n=1}^{\infty} \Lambda_n.$$

Then Λ be a dense subset of $(0,1)$ consisting of dyadic rational numbers in $(0,1)$. Now, we construct inductively a sequence of collections $\{O_\lambda\}_{\lambda\in\Lambda_n}$. Firstly, for $n=1$ and $1/2 \in \Lambda_1$, by Proposition 1.7.7, there exists an open subset $O_{1/2} \in \tau$ such that

$$F \subset O_{1/2} \subset \overline{O}_{1/2} \subset U,$$

and so we have the collection $\{O_\lambda\}_{\lambda\in\Lambda_1}$. Suppose that for $n = k-1$ $(k \geq 2)$, we have the collection $\{O_\lambda\}_{\lambda\in\Lambda_{k-1}}$. Then for any $\lambda \in \Lambda_k \setminus \Lambda_{k-1}$, there exists $m \in \{1, 2, \cdots, 2^{k-1}-1\}$ such that

$$\lambda = \frac{1}{2}\left(\frac{m}{2^{k-1}} + \frac{m+1}{2^{k-1}}\right),$$

and then it holds that

$$\overline{O}_{m/2^{k-1}} \subset O_{(m+1)/2^{k-1}}.$$

Therefore, by applying Proposition 1.7.7 again, there exists an open subset O_λ of X such that

$$\overline{O}_{m/2^{k-1}} \subset O_\lambda \subset \overline{O}_\lambda \subset O_{(m+1)/2^{k-1}},$$

and so we construct the collection $\{O_\lambda\}_{\lambda\in\Lambda_k}$. □

Proposition 1.7.9 Let (X,τ) be a topological space. For a dense subset Λ of the open interval $(0,1)$, let $\{O_\lambda\}_{\lambda\in\Lambda}$ be a family of open subsets of X such that for any $\lambda_1, \lambda_2 \in \Lambda$ with $\lambda_1 < \lambda_2$,

$$\overline{O}_{\lambda_1} \subset O_{\lambda_2}.$$

Define the function $f : X \to \mathbb{R}$ by

$$f(x) = \begin{cases} 0, & x \in \bigcap_{\lambda\in\Lambda} O_\lambda, \\ \sup\{\lambda \in \Lambda \,|\, x \notin O_\lambda\}, & x \in \left(\bigcap_{\lambda\in\Lambda} O_\lambda\right)^{c} = \bigcup_{\lambda\in\Lambda} O_\lambda^{c}. \end{cases} \tag{1.8}$$

Then $f : X \to [0,1]$ is continuous.

[Proof] Let $x \in X$ and $\epsilon > 0$ be given.

(i) Suppose that $0 < f(x) < 1$. There exist $\lambda_1, \lambda_2 \in \Lambda$ such that

$$f(x) - \epsilon < \lambda_1 < f(x) < \lambda_2 < f(x) + \epsilon. \tag{1.9}$$

Then from the definition of the function f, for any $\lambda \in \Lambda$ with $\lambda_1 < \lambda < f(x)$, $x \notin O_\lambda$. On the other hand, it holds that $\overline{O}_{\lambda_1} \subset O_\lambda$ and so $x \notin \overline{O}_{\lambda_1}$. Also, it holds that $x \in O_{\lambda_2}$.

Therefore, for the open set $U = O_{\lambda_2} \setminus \overline{O}_{\lambda_1}$, U is an open neighborhood of x and for any $y \in U$, from the definition of f it holds that $\lambda_1 \leq f(y) \leq \lambda_2$. Hence from (1.9) we have

$$|f(y) - f(x)| < \epsilon$$

for $y \in U$, i.e., $U \subset f^{-1}((f(x) - \epsilon, f(x) + \epsilon))$ so that f is continuous at x.

(ii) Suppose that $f(x) = 1$. Then for any $\lambda \in \Lambda$, $x \notin \overline{O}_\lambda$ and so

$$x \in \left(\bigcap_{\lambda \in \Lambda} \overline{O}_\lambda \right)^{\mathrm{c}} = \bigcup_{\lambda \in \Lambda} \overline{O}_\lambda^{\mathrm{c}}.$$

Then for the open subset $V = \bigcup_{\lambda \in \Lambda} \overline{O}_\lambda^{\mathrm{c}}$, $f(V) = 1$, i.e., $V \subset f^{-1}((1 - \epsilon, 1 + \epsilon))$, and so f is continuous at x.

(iii) Suppose that $f(x) = 0$. Then $x \in \bigcap_{\lambda \in \Lambda} O_\lambda$. On the other hand, there exists $\lambda \in \Lambda$ such that $0 < \lambda < \epsilon$ and

$$f(O_\lambda) \subset [0, \lambda] \subset [0, \epsilon),$$

which implies that $O_\lambda \subset f^{-1}((-\epsilon, \epsilon))$, and so f is continuous at x. □

Theorem 1.7.10 [**Urysohn's Lemma**] Let (X, τ) be a topological space. Then (X, τ) is normal if and only if for any disjoint pair E, F of closed subsets of X, there exists a continuous real-valued function $f : X \to [0, 1]$ such that $f(E) = \{0\}$ and $f(F) = \{1\}$.

[Proof] $\Leftarrow$) Let E, F be disjoint nonempty closed subsets of X. Let $f : X \to [0, 1]$ be a continuous function such that $f(E) = \{0\}$ and $f(F) = \{1\}$. For any $a, b \in \mathbb{R}$ with $0 < a < b < 1$, put

$$U = f^{-1}([0, a)), \qquad V = f^{-1}((b, 1]).$$

Since f is continuous, U and V are open and it holds that $E \subset f^{-1}([0, a)) = U$, $F \subset f^{-1}((b, 1]) = V$ and $U \cap V = \emptyset$. Therefore, (X, τ) is normal.

$\Rightarrow$) Let E, F be a disjoint pair of nonempty closed subsets of X. Then $E \subset F^{\mathrm{c}} \equiv U$ and by Proposition 1.7.8, there exists a dense subset Λ of $(0, 1)$ and a family $\{O_\lambda\}_{\lambda \in \Lambda}$ of open subsets of X such that (i) and (ii) in Proposition 1.7.8 hold. Then by Proposition 1.7.9, the function $f : X \to [0, 1]$ defined as in (1.8) is continuous, and $f(E) = \{0\}$, $f(F) = \{1\}$. □

Lemma 1.7.11 Let (X, τ) be a normal topological space, F a closed subset of X and $f : F \to \mathbb{R}$ a continuous function. Suppose that $|f| \leq a$ for some $a > 0$. Then there exists a continuous function $g : X \to \mathbb{R}$ such that

$$|g| \leq \frac{1}{3}a, \qquad |f - g| \leq \frac{2}{3}a \quad \text{on } F.$$

[Proof] Put

$$C = \{x \in F \mid f(x) \leq -a/3\}, \qquad D = \{x \in F \mid f(x) \geq a/3\}.$$

Then C and D are the inverse images of closed subsets of $\mathbb{R}$ and so closed disjoint subsets of X. Therefore, by Urysohn's lemma (Theorem 1.7.10), there is a continuous function $h : X \to [0, 1]$ such that $h(C) = 0$ and $h(D) = 1$. Put

$$g = \frac{a}{3}(2h - 1).$$

Then $g : X \to [-a/3, a/3]$ is continuous such that $g(C) = -a/3$, $g(D) = a/3$ and $|g| \leq a/3$. In particular, it holds that

$$\begin{aligned} |f - g| &= \left|(f|_C + f|_D + f|_{F\setminus C\cup D}) - (g|_C + g|_D + g|_{F\setminus C\cup D})\right| \\ &\leq \frac{2}{3}a \end{aligned}$$

on F. □

Theorem 1.7.12 [**Tietze Extension Theorem**] Let (X, τ) be a normal topological space, F a closed subset of X and $f : F \to [a, b]$ a continuous function. Then there exists a continuous function $\widetilde{f} : X \to [a, b]$ such that $\widetilde{f}\,|_F = f$.

[Proof] Since the bounded closed interval $[a, b]$ and $[-1, 1]$ are homeomorphic, it is convenient and sufficient to prove the case of $[a, b] = [-1, 1]$. Then from Lemma 1.7.11, there exists a continuous function $g_1 : X \to \mathbb{R}$ such that

$$|g_1| \leq \frac{1}{3}, \qquad |f - g_1| \leq \frac{2}{3} \quad \text{on } F.$$

By applying Lemma 1.7.11 again, we construct a continuous function $g_2 : X \to \mathbb{R}$ such that

$$|g_2| \leq \frac{1}{3}\frac{2}{3}, \qquad |f - g_1 - g_2| \leq \left(\frac{2}{3}\right)^2 \quad \text{on } F.$$

By continuing the above progress, we construction a sequence $\{g_n\}_{n=1}^{\infty}$ of continuous functions $g_n : X \to \mathbb{R}$ such that

$$|g_n| \leq \frac{1}{3}\left(\frac{2}{3}\right)^{n-1}, \qquad |f - [g_1 + g_2 + \cdots + g_n]| \leq \left(\frac{2}{3}\right)^n \quad \text{on } F.$$

Now, put

$$f_n = g_1 + g_2 + \cdots + g_n, \qquad n \geq 1.$$

If $n \geq m$, then

$$|f_n - f_m| = |g_{m+1} + \cdots + g_n|$$

$$\leq \left(\left(\frac{2}{3}\right)^{m+1} + \cdots + \left(\frac{2}{3}\right)^{n}\right)$$
$$= \left(\frac{2}{3}\right)^{m+1} \frac{1-\left(\frac{2}{3}\right)^{n-m}}{1-\frac{2}{3}},$$

which implies that $\{f_n\}_{n=1}^{\infty}$ is Cauchy and converges uniformly to a real-valued function $\widetilde{f}$ on X. Hence $\widetilde{f}$ is continuous. For $x \in F$, it holds that

$$|f(x) - \widetilde{f}(x)| = \lim_{n\to\infty} |f(x) - f_n(x)| \leq \lim_{n\to\infty} \left(\frac{2}{3}\right)^{n} = 0.$$

Hence $\widetilde{f} = f$ on F. Moreover, we obtain that

$$\left|\widetilde{f}(x)\right| = \lim_{n\to\infty} |f_n(x)| = \lim_{n\to\infty} (|g_1(x)| + \cdots + |g_n(x)|) \leq \frac{\frac{1}{3}}{1-\frac{2}{3}}$$
$$= 1,$$

which implies that $f(X) \subset [-1, 1]$. □

Exercises

1.1 Prove that the set $\mathbb{N}$ of all **natural numbers** is the smallest inductive subset of $\mathbb{R}$.

1.2 Prove that the set $\mathbb{Q}$ of rational numbers is countable.

1.3 Prove that the union of a countable collection of countable sets is countable.

1.4 Prove that every compact subset of a Hausdorff space is closed.

1.5 Let $\{F_n\}$ be a descending countable family of nonempty closed subsets of a topological space (X, τ) such that F_1 is a compact, i.e., for any open cover $\{U_\lambda\}_{\lambda\in\Lambda}$ of F_1 has finite subcover. Prove that

$$\bigcap_{n=1}^{\infty} F_n \neq \emptyset.$$

1.6 By applying the Nested Set Theorem (Theorem 1.4.19), prove that for any $a < b$, (a, b) is uncountable.

1.7 Prove that

(1) for any $a, b \in \mathbb{R}$ with $a < b$, (a, b) is an open set,

(2) $\mathbb{R}$ and $\emptyset$ are open in $\mathbb{R}$

(3) the intersection of any finite collection of open subsets of $\mathbb{R}$ is open in $\mathbb{R}$,

(4) the union of any collection of open subsets of $\mathbb{R}$ is open in $\mathbb{R}$.

1.8 Let E be a subset of $\mathbb{R}$. Prove that

$$\overline{E} = \bigcap_{F:\text{closed};E\subset F} F.$$

1.9 Prove that every convergent sequence of real numbers is bounded.

1.10 Let $\{a_n\}_{n=1}^{\infty}$ be a monotone sequence. Then $\{a_n\}_{n=1}^{\infty}$ converges if and only if $\{a_n\}_{n=1}^{\infty}$ is bounded.

1.11 Let E be a subset of $\mathbb{R}$. Then E is said to be **sequentially compact** if every sequence valued in E has a subsequence converging to a point in E. Prove that a subset K of $\mathbb{R}$ is sequentially compact if and only if K is closed and bounded.

1.12 Let E be a subset of $\mathbb{R}$. Prove that E is sequentially compact if and only if K is compact.

1.13 Construct a continuous function which is not Lipschitz.

1.14 Prove that if $f : D \to \mathbb{R}$ is uniformly continuous, then f is continuous.

1.15 Construct a uniform continuous unbounded function defined on an infinite interval.

1.16 Let $f : X \to \mathbb{R}$ be a function and A, B nonempty subsets of X. Prove that if $\inf\{f(x) \,|\, x \in A\} \geq \inf\{f(x) \,|\, x \in B\}$, then

$$\inf\{f(x) \,|\, x \in A \cup B\} = \inf\{f(x) \,|\, x \in B\}.$$

Chapter 2

General Measure Theory

2.1 Extended Real Number System

The set $\overline{\mathbb{R}} = \mathbb{R} \cup \{-\infty, \infty\}$ with the ordering extended from the usual ordering on $\mathbb{R}$ by declaring that $-\infty < x < \infty$ for all $x \in \mathbb{R}$ is called the **extended real number system**. Every subset E of $\overline{\mathbb{R}}$ has the **supremum** (denoted by $\sup(E)$) and the **infimum** (denoted by $\inf(E)$). In fact, for subset E of $\overline{\mathbb{R}}$, if $E \subset \mathbb{R}$ is not bounded above or $\infty \in E$, then $\sup(E) = \infty$. Similarly, for subset E of $\overline{\mathbb{R}}$, if $E \subset \mathbb{R}$ is not bounded below or $-\infty \in E$, then $\inf(E) = -\infty$. The arithmetical operations on $\mathbb{R}$ are partially extended to $\overline{\mathbb{R}}$ by

$$x \pm \infty = \pm\infty, \quad \infty + \infty = \infty, \quad -\infty - \infty = -\infty^{1)},$$

and

$$x \cdot (\pm\infty) = \begin{cases} \pm\infty, & x > 0 \\ 0, & x = 0 \\ \mp\infty, & x < 0 \end{cases}$$

for any real number $x \in \mathbb{R}$.

2.2 Measurable Spaces

Let X be a set and $\mathcal{P}(X)$ be the power set of X. A nonempty collection $\mathcal{F} \subset \mathcal{P}(X)$ of subsets of X is called an **algebra** if it satisfies the following conditions:

(a1) $\mathcal{F}$ is closed under the complements, i.e., for any $E \in \mathcal{F}$, $E^c \in \mathcal{F}$,

(a2) $\mathcal{F}$ is closed under the finite unions, i.e., for any finite family $\{E_k\}_{k=1}^n \subset \mathcal{F}$, $\cup_{k=1}^n E_k \in \mathcal{F}$.

A **σ-algebra** [2] $\mathcal{F}$ is an algebra which is closed under the countable unions, i.e., for any countable family $\{E_k\}_{k=1}^\infty \subset \mathcal{F}$,

$$\bigcup_{k=1}^{\infty} E_k \in \mathcal{F}.$$

---

1) We don't define $\infty - \infty$

2) Some authors use the notions **field** and **σ-field** instead of algebra and σ-algebra, respectively

Remark 2.2.1 Let $\mathcal{F}$ be an algebra of subsets of X. Then $\mathcal{F}$ is nonempty and so there exists a subset E of X such that $E \in \mathcal{F}$ and then $E^c \in \mathcal{F}$ by **(a1)**. Also, by **(a1)** and **(a2)**, $\mathcal{F}$ is closed under finite intersections. Therefore, $\emptyset = E \cap E^c \in \mathcal{F}$ and $X = E \cup E^c \in \mathcal{F}$. Hence, a family $\mathcal{F} \subset \mathcal{P}(X)$ of subsets of X is a σ-algebra if and only if $\mathcal{F}$ satisfies the following conditions hold:

(σ1) $X, \emptyset \in \mathcal{F}$,

(σ2) $\mathcal{F}$ is closed under the complements,

(σ3) $\mathcal{F}$ is closed under the countable unions.

Example 2.2.2 (1) Let X be a set. Then the power set $\mathcal{P}(X)$ of X and $\{X, \emptyset\}$ are σ-algebras.

(2) Let X be a set and E a proper nonempty subset of X. Then $\{X, \emptyset, E, E^c\}$ is a σ-algebra.

Remark 2.2.3 (1) Let $\mathcal{A}_1$ and $\mathcal{A}_2$ be algebras of subsets of disjoint (distinct) sets X_1 and X_2, respectively. Put

$$\mathcal{A}_1 \cup \mathcal{A}_2 = \{A_1 \cup A_2, \,|\, A_1 \in \mathcal{A}_1, \ A_2 \in \mathcal{A}_2\}.$$

Then $\mathcal{A}_1 \cup \mathcal{A}_2$ is an algebra of subsets of the set $X := X_1 \cup X_2$.

(2) Let $\mathcal{A}_1$ and $\mathcal{A}_2$ be σ-algebras of subsets of disjoint (distinct) sets X_1 and X_2, respectively. Then $\mathcal{A}_1 \cup \mathcal{A}_2$ is a σ-algebra of subsets of $X := X_1 \cup X_2$.

(3) Let $\mathcal{A}$ be an algebra of subsets of a set X and X_0 a subset of X. Put

$$X_0 \cap \mathcal{A} = \{X_0 \cap A \,|\, A \in \mathcal{A}\}.$$

Then $X_0 \cap \mathcal{A}$ is an algebra of subsets of X_0.

(4) Let $\mathcal{A}$ be a σ-algebra of subsets of a set X and X_0 a subset of X. Then $X_0 \cap \mathcal{A}$ is a σ-algebra of subsets of X_0. If in addition $X_0 \in \mathcal{A}$, then we have

$$\mathcal{A} = (X_0 \cap \mathcal{A}) \cup (X_0^c \cap \mathcal{A}).$$

Remark 2.2.4 (1) For a family $\{\mathcal{F}_\lambda\}_{\lambda \in \Lambda}$ of σ-algebras $\mathcal{F}_\lambda$ of subsets of X,

$$\mathcal{F} \equiv \bigcap_{\lambda \in \Lambda} \mathcal{F}_\lambda$$

is a σ-algebra. In fact, $X \in \mathcal{F}$ and for any $A \in \mathcal{F}$, $A^c \in \mathcal{F}$. Also, for any countable family $\{A_k\}_{k=1}^\infty \subset \mathcal{F}$, $\{A_k\}_{k=1}^\infty \subset \mathcal{F}_\lambda$ for any $\lambda \in \Lambda$ and so $\cup_{k=1}^\infty A_k \in \mathcal{F}_\lambda$ for any $\lambda \in \Lambda$ since $\mathcal{F}_\lambda$ (for any $\lambda \in \Lambda$) is a σ-algebra, and so $\cup_{k=1}^\infty A_k \in \mathcal{F}$.

(2) For a family $\mathcal{C}$ of subsets of a set X, the **smallest σ-algebra** containing $\mathcal{C}$ is unique and denoted by $\sigma(\mathcal{C})$. In fact, we have

$$\sigma(\mathcal{C}) = \bigcap_{\mathcal{F}:\sigma\text{-algebra containing } \mathcal{C}} \mathcal{F}.$$

(3) For any σ-algebra $\mathcal{F}$ containing $\mathcal{C}$, $\sigma(\mathcal{C}) \subset \mathcal{F}$.

Proposition 2.2.5 For families $\mathcal{C}_1$ and $\mathcal{C}_2$ of subsets of X with $\mathcal{C}_1 \subset \mathcal{C}_2$, $\sigma(\mathcal{C}_1) \subset \sigma(\mathcal{C}_2)$.

[Proof] Since $\sigma(\mathcal{C}_2)$ is the smallest σ-algebra containing $\mathcal{C}_2$, $\mathcal{C}_1 \subset \mathcal{C}_2 \subset \sigma(\mathcal{C}_2)$, i.e., $\sigma(\mathcal{C}_2)$ is a σ-algebra containing $\mathcal{C}_1$. Therefore, by (3) in Remark 2.2.4, $\sigma(\mathcal{C}_1) \subset \sigma(\mathcal{C}_2)$. □

Remark 2.2.6 Let $\mathcal{A}$ be a family of subsets of a set X and X_0 a subset of X. Let $\mathcal{B} = \sigma(\mathcal{A})$ be the smallest σ-algebra containing $\mathcal{A}$. Then $X_0 \cap \mathcal{B}$ is the smallest σ-algebra of subsets of X_0 containing $X_0 \cap \mathcal{A}$.

Definition 2.2.7 The smallest σ-algebra containing all open subsets of $\mathbb{R}$ is called the ***Borel σ-algebra*** and is denoted by $\mathcal{B}(\mathbb{R})$. An elements of $\mathcal{B}(\mathbb{R})$ is called a **Borel set**.

Put

$$\mathcal{I}_1 = \{(a,b) \,|\, a,b \in \mathbb{R}\}, \tag{2.1}$$
$$\mathcal{I}_2 = \{(a,b] \,|\, a,b \in \mathbb{R}\}, \tag{2.2}$$
$$\mathcal{I}_3 = \{[a,b) \,|\, a,b \in \mathbb{R}\}, \tag{2.3}$$
$$\mathcal{I}_4 = \{[a,b] \,|\, a,b \in \mathbb{R}\}, \tag{2.4}$$
$$\mathcal{I}_5 = \{(-\infty,b) \,|\, a,b \in \mathbb{R}\}, \tag{2.5}$$
$$\mathcal{I}_6 = \{(-\infty,b] \,|\, a,b \in \mathbb{R}\}, \tag{2.6}$$
$$\mathcal{I}_7 = \{(a,\infty) \,|\, a,b \in \mathbb{R}\}, \tag{2.7}$$
$$\mathcal{I}_8 = \{[a,\infty) \,|\, a,b \in \mathbb{R}\}. \tag{2.8}$$

Proposition 2.2.8 We have

$$\mathcal{B}(\mathbb{R}) = \sigma(\mathcal{I}_1) = \sigma(\mathcal{I}_2) = \sigma(\mathcal{I}_3) = \sigma(\mathcal{I}_4) = \sigma(\mathcal{I}_5) = \sigma(\mathcal{I}_6) = \sigma(\mathcal{I}_7) = \sigma(\mathcal{I}_8).$$

[Proof] For any $a < b$, we obtain that

$$(a,b] = \bigcap_{n=1}^{\infty} \left(a, b + \frac{1}{n}\right) \in \mathcal{B}(\mathbb{R}),$$
$$[a,b) = \bigcap_{n=1}^{\infty} \left(a - \frac{1}{n}, b\right) \in \mathcal{B}(\mathbb{R}),$$

$$[a,b]=\bigcap_{n=1}^{\infty}\left(a-\frac{1}{n},b+\frac{1}{n}\right)\in\mathcal{B}(\mathbb{R}),$$
$$(-\infty,b]=\bigcap_{n=1}^{\infty}\left(-\infty,b+\frac{1}{n}\right)\in\mathcal{B}(\mathbb{R}),$$
$$[a,\infty)=\bigcap_{n=1}^{\infty}\left(a-\frac{1}{n},\infty\right)\in\mathcal{B}(\mathbb{R}),$$

which implies that for any $i=1,2,\cdots,8$, $\mathcal{I}_i\subset\mathcal{B}(\mathbb{R})$ and so $\sigma(\mathcal{I}_i)\subset\mathcal{B}(\mathbb{R})$. On the other hand, For any $a<b$, we obtain that

$$(a,b)=\bigcup_{n=1}^{\infty}\left(a,b-\frac{1}{n}\right]\in\sigma(\mathcal{I}_2),$$
$$(a,b)=\bigcup_{n=1}^{\infty}\left[a+\frac{1}{n},b\right)\in\sigma(\mathcal{I}_3),$$
$$(a,b)=\bigcup_{n=1}^{\infty}\left[a+\frac{1}{n},b-\frac{1}{n}\right]\in\sigma(\mathcal{I}_4),$$
$$(a,b)=(-\infty,a]^{c}\bigcap\left(\bigcup_{n=1}^{\infty}\left(-\infty,b-\frac{1}{n}\right]\right)\in\sigma(\mathcal{I}_6),$$
$$(a,b)=\left(\bigcup_{n=1}^{\infty}\left[a-\frac{1}{n},\infty\right)\right)\bigcap[b,\infty)^{c}\in\sigma(\mathcal{I}_8),$$

which, by Theorem 1.4.6, implies that all open subsets belong to $\sigma(\mathcal{I}_i)$ for any $i=1,2,\cdots,8$. Therefore, $\mathcal{B}(\mathbb{R})\subset\sigma(\mathcal{I}_i)$ for any $i=1,2,\cdots,8$. □

Remark 2.2.9 Let (X,τ) be a topological space. Then the smallest σ-algebra containing all open subsets of X is called the **Borel σ-algebra** and denoted by $\mathcal{B}(X)$.

Let $\{X_\lambda\}_{\lambda\in\Lambda}$ be a family of nonempty sets and $\{\mathcal{F}_\lambda\}_{\lambda\in\Lambda}$ a family of σ-algebras $\mathcal{F}_\lambda$ of subsets of X_λ. Put

$$X=\prod_{\lambda\in\Lambda}X_\lambda$$

and for each $\lambda\in\Lambda$, the projection from X onto X_λ is denoted by

$$\mathrm{Pr}_\lambda : X=\prod_{\lambda\in\Lambda}X_\lambda\longrightarrow X_\lambda.$$

Then the smallest σ-algebra containing

$$\{\mathrm{Pr}_\lambda^{-1}(F_\lambda)\,|\,F_\lambda\in\mathcal{F}_\lambda,\ \ \lambda\in\Lambda\}$$

is called the product σ-algebra on X and denoted by $\otimes_{\lambda\in\Lambda}\mathcal{F}_\lambda$, i.e.,

$$\bigotimes_{\lambda\in\Lambda}\mathcal{F}_\lambda=\sigma\left(\{\mathrm{Pr}_\lambda^{-1}(F_\lambda)\,|\,F_\lambda\in\mathcal{F}_\lambda,\ \ \lambda\in\Lambda\}\right).$$

Definition 2.2.10 Let X be a set and $\mathcal{M}$ a σ-algebra of subsets of X. Then $(X, \mathcal{M})$ is called a **measurable space** for which a subset E of X is said to be **measurable** if $E \in \mathcal{M}$.

2.3 Measure Spaces

A function defined on a family of sets is called a **set function**.

Definition 2.3.1 Let $\mathcal{S}$ be a collection of subsets of a set X. Let $\rho : \mathcal{S} \to [0, \infty]$ be a set function.

(i) ρ is said to be **monotone** if for any $E, F \in \mathcal{S}$ with $E \subset F$,

$$\rho(E) \leq \rho(F). \tag{2.9}$$

(ii) ρ is said to be **finitely monotone** if for any $E \in \mathcal{S}$ and any finite collection $\{E_k\}_{k=1}^n \subset \mathcal{S}$ such that $E \subset \cup_{k=1}^n E_k$,

$$\rho(E) \leq \sum_{k=1}^n \rho(E_k). \tag{2.10}$$

(iii) ρ is said to be **countably monotone** if for any $E \in \mathcal{S}$ and any countable collection $\{E_k\}_{k=1}^\infty \subset \mathcal{S}$ such that $E \subset \cup_{k=1}^\infty E_k$,

$$\rho(E) \leq \sum_{k=1}^\infty \rho(E_k). \tag{2.11}$$

Remark 2.3.2 Let $\mathcal{S}$ be a collection of subsets of a set X. Let $\rho : \mathcal{S} \to [0, \infty]$ be a set function.

(i) It is obvious that if ρ is finitely monotone, then ρ is monotone.

(ii) Suppose that ρ is countably monotone. If $\emptyset \in \mathcal{S}$ and $\rho(\emptyset) = 0$, then ρ is finitely monotone. In fact, for any $E \in \mathcal{S}$ and finite collection $\{E_k\}_{k=1}^n \subset \mathcal{S}$ such that $E \subset \bigcup_{k=1}^n E_k$, by taking $E_k = \emptyset$ for $k > n$, we have a countable collection $\{E_k\}_{k=1}^\infty \subset \mathcal{S}$ such that $E \subset \bigcup_{k=1}^\infty E_k$, and

$$\rho(E) \leq \sum_{k=1}^\infty \rho(E_k) = \sum_{k=1}^n \rho(E_k).$$

Definition 2.3.3 Let $(X, \mathcal{M})$ be a measurable space. A **measure** $\mu : \mathcal{M} \to [0, \infty]$ is a extended real-valued set function for which $\mu(\emptyset) = 0$ and which is **countable additive**, i.e., for any countable disjoint collection $\{E_n\}_{n=1}^{\infty}$ of measurable sets in $\mathcal{M}$,

$$\mu\left(\bigcup_{n=1}^{\infty} E_n\right) = \sum_{n=1}^{\infty} \mu(E_n). \tag{2.12}$$

A **measure space** $(X, \mathcal{M}, \mu)$ is consisting of a set X, a σ-algebra $\mathcal{M}$ and a measure μ defined on $\mathcal{M}$.

Example 2.3.4 (1) Let X be a set and $\mathcal{P}(X)$ the power set of X. Define a set function c on $\mathcal{P}(X)$ by

$$c(A) = \begin{cases} c(A) = \text{ the number of elements in } A, & A \text{ is finite,} \\ \infty, & A \text{ is infinite.} \end{cases}$$

Then we can easily see that c is a measure on $\mathcal{P}(X)$ (see Exercise **2.11**) and so $(X, \mathcal{P}(X))$ is a measurable space and $(X, \mathcal{P}(X), c)$ is a measure space. In this case, c is called a **counting measure**.

(2) Let $(X, \mathcal{M})$ be a measure space and $x_0 \in X$. Define a set function δ_{x_0} on $\mathcal{M}$ by

$$\delta_{x_0}(A) = \begin{cases} 1, \ x_0 \in A, \\ 0, \ x_0 \notin A. \end{cases}$$

Then we can easily see that δ_{x_0} is a measure on $\mathcal{M}$ (see Exercise **2.12**) and so $(X, \mathcal{M}, \delta_{x_0})$ is a measure space. In this case, δ_{x_0} is called a **Dirac measure**.

(3) Let X be a uncountable set and $\mathcal{C}$ be the family of subsets of X such that every element of $\mathcal{C}$ is countable or the complement of a countable subset of X. Then we can easily see that $\mathcal{C}$ is a σ-algebra (see Exercise **2.13**) and so $(X, \mathcal{C})$ is a measurable space. Define a set function ν on $\mathcal{C}$ by

$$\nu(E) = \begin{cases} 0, \ E \text{ is countable,} \\ 1, \ E \text{ is the complement of a countable set.} \end{cases}$$

Then ν is a measure on $\mathcal{C}$ (see Exercise **2.13**) and so $(X, \mathcal{C}, \nu)$ is a measure space.

(4) Let $(X, \mathcal{M}, \mu)$ be a measure space and let $X_0 \in \mathcal{M}$. Put $\mathcal{M}_0$ be the collect of subsets of X_0 which is contained in $\mathcal{M}$. In fact, we have

$$\mathcal{M}_0 = X_0 \cap \mathcal{M} = \{X_0 \cap A \,|\, A \in \mathcal{M}\}.$$

Then for the restriction μ_0 of μ to $\mathcal{M}_0$, $(X_0, \mathcal{M}_0, \mu_0)$ is a measure space. In this case, μ_0 is sometimes denoted by $\mu|_{X_0}$.

Remark 2.3.5 Let $(X, \mathcal{M}, \mu)$ be a measure space and $\{E_k\}_{k=1}^{m}$ a finite disjoint collection of measurable sets. By considering $E_k = \emptyset$ for $k > m$ and using the countable additivity (see (2.12)) of μ, we have

$$\mu\left(\bigcup_{k=1}^{m} E_k\right) = \mu\left(\bigcup_{k=1}^{\infty} E_k\right) = \sum_{k=1}^{\infty} \mu(E_k) = \sum_{k=1}^{m} \mu(E_k), \tag{2.13}$$

which is called the **finite additivity**.

Definition 2.3.6 Let $(X, \mathcal{M}, \mu)$ be a measure space.

(i) If $\mu(X) < \infty$, then μ is said to be **finite**. In this case, $(X, \mathcal{M}, \mu)$ is called a **finite measure space**.

(ii) If there exists a countable family $\{E_n\}_{n=1}^{\infty} \subset \mathcal{M}$ with $\mu(E_n) < \infty$ for each $n \in \mathbb{N}$ such that $X = \cup_{n=1}^{\infty} E_n$, then μ is said to be **σ-finite**. In this case, $(X, \mathcal{M}, \mu)$ is called a **σ-finite measure space**.

(iii) If $E = \cup_{n=1}^{\infty} E_n$ for a countable family $\{E_n\}_{n=1}^{\infty} \subset \mathcal{M}$ with $\mu(E_n) < \infty$ for each $n \in \mathbb{N}$, then E is said to be **σ-finite** for μ.

Proposition 2.3.7 Let $(X, \mathcal{M}, \mu)$ be a measure space and E a measurable set.

(i) [**Monotonicity**] If $A \subset B$ for $A, B \in \mathcal{M}$, then

$$\mu(A) \leq \mu(B). \tag{2.14}$$

(ii) [**Excision**] If $A \subset B$ for $A, B \in \mathcal{M}$ with $\mu(A) < \infty$, then

$$\mu(B \setminus A) = \mu(B) - \mu(A). \tag{2.15}$$

(iii) [**Countable Monotonicity**] For any countable collection $\{E_k\}_{k=1}^{\infty}$ of measurable sets such that $E \subset \cup_{k=1}^{\infty} E_k$,

$$\mu(E) \leq \sum_{k=1}^{\infty} \mu(E_k). \tag{2.16}$$

[Proof] (i) Since $B = (B \setminus A) \cup A$ and $B \setminus A$, A are disjoint, by the finite additivity (see (2.13)) we have

$$\mu(B) = \mu(B \setminus A) + \mu(A) \geq \mu(A).$$

(ii) We have

$$\mu(B) = \mu(B \setminus A) + \mu(A),$$

which implies (2.15), since $\mu(A) < \infty$.

(iii) Put

$$G_1 = E_1,$$

$$G_k = E_k \setminus \left(\bigcup_{i=1}^{k-1} E_i \right), \qquad k \geq 2.$$

Then $\{G_k\}_{k=1}^{\infty}$ is a countable disjoint collection of measurable sets and

$$\bigcup_{k=1}^{\infty} G_k = \bigcup_{k=1}^{\infty} E_k.$$

Therefore, by the countable additivity of μ, we have

$$\mu(E) \leq \mu\left(\bigcup_{k=1}^{\infty} E_k\right) = \mu\left(\bigcup_{k=1}^{\infty} G_k\right) = \sum_{k=1}^{\infty} \mu(G_k) \leq \sum_{k=1}^{\infty} \mu(E_k),$$

which gives the proof of (2.16). □

Remark 2.3.8 Let $(X, \mathcal{M}, \mu)$ be a finite measure space, i.e., $\mu(X) < \infty$. Then by (i) in Proposition 2.3.7, $\mu(E) < \infty$ for any $E \in \mathcal{M}$.

Definition 2.3.9 Let $(X, \mathcal{M}, \mu)$ be a measure space. If for each $E \in \mathcal{M}$ with $\mu(E) = \infty$ there exists $F \in \mathcal{M}$ with $F \subset E$ and $0 < \mu(F) < \infty$, then μ is said to be **semifinite**.

Example 2.3.10 (1) The measure space $(X, \mathcal{P}(X), c)$ with the counting measure c is σ-finite if X is countable.

(2) The measure space $(X, \mathcal{M}, \delta_{x_0})$ with the Dirac measure δ_{x_0} is finite.

(3) Every σ-finite measure is semifinite (see Exercise **2.14**), but in general the converse is not true (see Exercise **2.15**).

Definition 2.3.11 A measure space $(X, \mathcal{M}, \mu)$ is said to be **complete** if $\mathcal{M}$ contains all subsets of measurable sets of measure zero.

Proposition 2.3.12 Let $(X, \mathcal{M}, \mu)$ be a measure space. Then there exists a complete measure space $(X, \mathcal{M}_0, \mu_0)$ such that

$$\mathcal{M} \subset \mathcal{M}_0 \quad \text{and} \quad \mu_0|_{\mathcal{M}} = \mu.$$

[Proof] Put

$$\mathcal{M}_0 = \{E \cup A \,|\, E \in \mathcal{M}, \quad A \subset N \text{ for some } N \in \mathcal{M} \text{ with } \mu(N) = 0\}.$$

Then we can see that $\mathcal{M}_0$ is a σ-algebra containing $\mathcal{M}$ (see Exercise **2.16**). Define a set function

$$\mu_0 : \mathcal{M}_0 \to [0, \infty]$$

by

$$\mu_0(E \cup A) = \mu(E)$$

for all $E \cup A \in \mathcal{M}_0$ with $E \in \mathcal{M}$ and $A \subset N$ for some $N \in \mathcal{M}$ with $\mu(N) = 0$. Then we can see that μ_0 is a measure on $\mathcal{M}_0$ (see Exercise **2.17**) such that $\mu_0|_{\mathcal{M}} = \mu$. □

Definition 2.3.13 The complete measure space $(X, \mathcal{M}_0, \mu_0)$ given as in Proposition 2.3.12 is called the **completion** of the measure space $(X, \mathcal{M}, \mu)$.

Definition 2.3.14 Let $\{E_k\}_{k=1}^{\infty}$ be a family of subsets of a set X.

(i) $\{E_k\}_{k=1}^{\infty}$ is said to be **ascending** if $E_{k+1} \supseteq E_k$ for each $k \in \mathbb{N}$,

(ii) $\{E_k\}_{k=1}^{\infty}$ is said to be **descending** if $E_{k+1} \subseteq E_k$ for each $k \in \mathbb{N}$.

Theorem 2.3.15 **[Continuity of Measure]** Let $(X, \mathcal{M}, \mu)$ be a measure space. The measure μ satisfies the following properties:

(i) For any ascending family $\{E_k\}_{k=1}^{\infty}$ of measurable sets, we have

$$\mu\left(\bigcup_{k=1}^{\infty} E_k\right) = \lim_{n\to\infty} \mu(E_n). \tag{2.17}$$

(ii) For any descending family $\{E_k\}_{k=1}^{\infty}$ of measurable sets with $\mu(E_1) < \infty$, we have

$$\mu\left(\bigcap_{k=1}^{\infty} E_k\right) = \lim_{n\to\infty} \mu(E_n). \tag{2.18}$$

[Proof] (1) Define

$$C_1 = E_1, \qquad C_k = E_k \setminus E_{k-1}, \qquad k = 2, 3, \cdots.$$

Then $\{C_k\}_{k=1}^{\infty}$ is a countable disjoint family of measurable sets and

$$\bigcup_{k=1}^{\infty} E_k = \bigcup_{k=1}^{\infty} C_k.$$

Therefore, by the countable additivity (see (2.12)), we obtain that

$$\begin{aligned}\mu\left(\bigcup_{k=1}^{\infty} E_k\right) &= \mu\left(\bigcup_{k=1}^{\infty} C_k\right) = \sum_{k=1}^{\infty} \mu(C_k) = \lim_{n\to\infty} \sum_{k=1}^{n} \mu(C_k) = \lim_{n\to\infty} \left(\bigcup_{k=1}^{n} C_k\right) \\ &= \lim_{n\to\infty} \mu(E_n),\end{aligned}$$

which gives the proof of (2.17).

(2) Define

$$D_0 = \bigcap_{k=1}^{\infty} E_k, \qquad D_k = E_k \setminus E_{k+1}, \qquad k = 1, \cdots.$$

Then $\{D_k\}_{k=0}^{\infty}$ is a countable disjoint family of measurable sets. Therefore, by the countable additivity (see (2.12)) and the excision property (see (ii) in Proposition 2.3.7), we obtain that

$$\begin{aligned}\mu(E_1) &= \mu\left(\bigcup_{k=0}^{\infty} D_k\right) = \sum_{k=0}^{\infty} \mu(D_k) = \mu(D_0) + \lim_{n\to\infty} \sum_{k=1}^{n} \mu(D_k) \\ &= \mu(D_0) + \lim_{n\to\infty} \mu\left(\bigcup_{k=1}^{n} D_k\right) = \mu\left(\bigcap_{k=1}^{\infty} E_k\right) + \lim_{n\to\infty} \mu\left(E_1 \setminus E_{n+1}\right) \\ &= \mu\left(\bigcap_{k=1}^{\infty} E_k\right) + \lim_{n\to\infty} \left(\mu\left(E_1\right) - \mu\left(E_{n+1}\right)\right) \quad (\text{since } \mu(E_{n+1}) < \infty) \\ &= \mu\left(\bigcap_{k=1}^{\infty} E_k\right) + \mu\left(E_1\right) - \lim_{n\to\infty} \mu\left(E_{n+1}\right),\end{aligned}$$

which implies (2.18) since $\mu(E_1) < \infty$. □

Example 2.3.16 Let X be an infinite set and $E = \{x_1, x_2, \cdots\}$ a countably infinite subset of X. Let $(X, \mathcal{P}(X), c)$ be a measure space with the counting measure c. For each $n \in \mathbb{N}$, put

$$E_n = \{x_k \in E \mid k \geq n\}.$$

Then $\{E_n\}_{n=1}^{\infty}$ is a descending sequence of measurable sets and

$$c\left(\bigcap_{n=1}^{\infty} E_n\right) = c(\emptyset) = 0,$$

but $c(E_n) = \infty$ for any $n = 1, 2, \cdots$ and $\lim_{n\to\infty} c(E_n) = \infty$.

Definition 2.3.17 Let $(X, \mathcal{M}, \mu)$ be a measure space and $E \in \mathcal{M}$. We say that a property holds **almost everywhere** on E (or **for almost all** x in E) if there exists a measurable subset $E_0 \subset E$ with $\mu(E_0) = 0$ such that the property holds on $E \setminus E_0$.

Definition 2.3.18 For a sequence $\{A_k\}_{k=1}^{\infty}$ of sets, we set

$$\limsup_{n\to\infty} A_n = \bigcap_{n\geq 1}\bigcup_{k\geq n} A_k,$$
$$\liminf_{n\to\infty} A_n = \bigcup_{n\geq 1}\bigcap_{k\geq n} A_k$$

Theorem 2.3.19 [Borel-Cantelli Lemma] Let $(X, \mathcal{M}, \mu)$ be a measure space. Let $\{E_k\}_{k=1}^{\infty}$ be a countable family of measurable sets such that $\sum_{k=1}^{\infty} m(E_k) < \infty$. Then almost all $x \in X$ belong to at most finitely many E_k's.

[Proof] Put

$$E_0 = \limsup_{n\to\infty} E_n = \bigcap_{n\geq 1}\bigcup_{k\geq n} E_k.$$

Then E_0 is measurable and, by the monotonicity of μ and the assumption, for any $n \in \mathbb{N}$, we have

$$\mu\left(\bigcup_{k\geq n} E_k\right) \leq \sum_{k=n}^{\infty} \mu(E_k) < \infty.$$

Therefore, by applying the continuity of measure μ ((ii) in Theorem 2.3.15), we have

$$\mu(E_0) = \lim_{n\to\infty} \mu\left(\bigcup_{k\geq n} E_k\right) \leq \lim_{n\to\infty} \sum_{k=n}^{\infty} \mu(E_k) = 0.$$

Then for any $x \in X \setminus E_0$, we obtain that

$$\begin{aligned} x \in X \setminus E_0 \quad &\Longleftrightarrow \quad x \in \left(\bigcap_{n\geq 1}\bigcup_{k\geq n} E_k\right)^{\mathrm{c}} \\ &\Longleftrightarrow \quad x \in \bigcup_{n\geq 1}\bigcap_{k\geq n} E_k^{\mathrm{c}} \end{aligned}$$

$$\iff \quad \text{there exists } n_x \in \mathbb{N} \text{ such that for any } k \geq n_x \text{ , } x \notin E_k,$$

which gives the proof. □

Remark 2.3.20 Let $(X, \mathcal{M}, \mu)$ be a measure space. We assume that the σ-algebra $\mathcal{M}$ is sufficiently large such that there exists a descending family $\{E_k\}_{k=1}^{\infty}$ of measurable sets satisfying that $\mu(E_k) = 1/k$ for each $k \in \mathbb{N}$ and

$$\limsup_{n\to\infty} E_k = \bigcap_{n\geq 1} \bigcup_{k\geq n} E_k = \emptyset.$$

Then $\mu(\limsup_{n\to\infty} E_k) = 0$, but

$$\sum_{n=1}^{\infty} \mu(E_n) = \sum_{n=1}^{\infty} \frac{1}{n} = \infty.$$

2.4 Signed Measure Spaces

Definition 2.4.1 Let $(X, \mathcal{M})$ be a measurable space. A set function $\nu : \mathcal{M} \to \mathbb{R}^* = [-\infty, \infty]$ is called a **signed measure** if ν satisfies the following conditions:

(i) ν can take at most one of the values $-\infty$ and $+\infty$.

(ii) $\nu(\emptyset) = 0$.

(iii) For any countable disjoint collection $\{E_k\}_{k=1}^{\infty}$ of measurable sets,

$$\nu\left(\bigcup_{k=1}^{\infty} E_k\right) = \sum_{k=1}^{\infty} \nu(E_k),$$

where it is necessary that if $\nu(\cup_{k=1}^{\infty} E_k)$ is finite, then the series $\sum_{k=1}^{\infty} \nu(E_k)$ converges absolutely.

If ν is a signed measure on a measurable space $(X, \mathcal{M})$, then $(X, \mathcal{M}, \nu)$ is called a **signed measure space**.

Remark 2.4.2 (1) Note that every measure on a measurable space $(X, \mathcal{M})$ is a signed measure. To emphasize positivity of measure, we sometimes refer to measures as **positive measures**.

(2) The difference of two measure, one of which is finite, is a signed measure. In fact, every signed measure is a difference of two such measures (Jordan Decomposition theorem).

Definition 2.4.3 Let $(X, \mathcal{M}, \nu)$ be a signed measure space and E a measurable set. Then E is said to be a

(i) **positive set** if for any measurable subset A of E, $\nu(A) \geq 0$,

(ii) **negative set** if for any measurable subset A of E, $\nu(A) \leq 0$,

(iii) **null set** with respect to ν if for any measurable subset A of E, $\nu(A) = 0$.

Remark 2.4.4 Let $(X, \mathcal{M}, \nu)$ be a signed measure space.

(1) Then $(X, \mathcal{M}, -\nu)$ is a signed measure space.

(2) For a negative set $E \in \mathcal{M}$, the restriction $-\nu|_E$ of $-\nu$ to E is a measure.

(3) For a null set $N \in \mathcal{M}$, we have $\nu(N) = 0$. But the converse is not true. In fact, for a measurable set E with $\nu(E) = 0$, E can be decomposed by a partition $\{A, B\}$ of E such that $\nu(A) = -\nu(B) > 0$ and then $\nu(E) = \nu(A) + \nu(B) = 0$.

(4) For a measure space, the notions of null set and measure zero set coincide.

(5) For measurable sets $A, B \in \mathcal{M}$, if $A \subset B$ and $|\nu(B)| < \infty$, then $|\nu(A)| < \infty$. In fact, we have

$$\nu(B) = \nu((B \setminus A) \cup A) = \nu(B \setminus A) + \nu(A).$$

Therefore, if $|\nu(A)| = \infty$, then $\nu(A) = \pm\infty$ and $\nu(B \setminus A) = \pm\infty$ or finite. Therefore, $\nu(B) = \pm\infty$ and so $|\nu(B)| = \infty$, which is a contradiction.

Proposition 2.4.5 Let $(X, \mathcal{M}, \nu)$ be a signed measure space. Then it holds that

(i) every measurable subset of a positive set is also a positive set,

(ii) the union of countable collections of positive sets is also a positive set.

[Proof] (i) Let E be a positive set and A a measurable subset of E. Then for any measurable subset B of A, B is a subset E and, since E is positive, $\nu(B) \geq 0$. Therefore, A is a positive set.

(ii) Let $\{E_k\}_{k=1}^{\infty}$ be a countable family of positive sets and put

$$E = \bigcup_{k=1}^{\infty} E_k.$$

Let A be a measurable subset of E. Put $F_1 = E_1$ and

$$F_k = E_k \setminus \left(\bigcup_{i=1}^{k-1} E_i \right), \qquad k \geq 2.$$

Then one obtain that

$$A = A \cap E = A \cap \left(\bigcup_{k=1}^{\infty} E_k \right) = A \cap \left(\bigcup_{k=1}^{\infty} F_k \right) = \bigcup_{k=1}^{\infty} (A \cap F_k),$$

which implies that

$$\nu(A) = \sum_{k=1}^{\infty} \nu(A \cap F_k) \geq 0,$$

since $\nu(A \cap F_k) \geq 0$ for each $k \in \mathbb{N}$. Therefore, E is a positive set. □

Theorem 2.4.6 **[Hahn's Lemma]** Let $(X, \mathcal{M}, \nu)$ be a signed measure space and E a measurable set with $0 < \nu(E) < \infty$. Then there exists a measurable subset A of E such that A is a positive set with $\nu(A) > 0$.

[Proof] If E is a positive set, then the proof is terminated. Suppose that E is not a positive set. Then there exists a measurable subset of E of which the measure is negative. Let $m_1 \in \mathbb{N}$ be the smallest natural number such that there exists a measurable subset E_1 of E with $\nu(E_1) < -1/m_1$, and so we have

$$\nu(E) = \nu(E \setminus E_1) + \nu(E_1),$$

which implies that

$$\nu(E \setminus E_1) = \nu(E) + (-\nu(E_1)) > \nu(E) + \frac{1}{m_1} > 0.$$

If $E \setminus E_1$ is a positive set, then the proof is terminated. If $E \setminus E_1$ is not a positive set, then there exists a measurable subset of $E \setminus E_1$ of which the measure is negative. Let $m_2 \in \mathbb{N}$ be the smallest natural number such that there exists a measurable subset E_2 of $E \setminus E_1$ with $\nu(E_2) < -1/m_2$, and so we have

$$\nu(E \setminus E_1) = \nu((E \setminus E_1) \setminus E_2) + \nu(E_2),$$

which implies that

$$\nu((E \setminus E_1) \setminus E_2) = \nu(E \setminus E_1) + (-\nu(E_2)) > \nu(E) + \frac{1}{m_1} + \frac{1}{m_2} > 0.$$

By continuing this selection process, we construct a sequence $\{m_i\}_{i=1}^{k}$ of natural numbers and a family $\{E_i\}_{i=1}^{k}$ of measurable sets such that for $i = 1, \cdots, k$, $m_i \in \mathbb{N}$ is the smallest natural number satisfying that there exists a measurable subset F of $E \setminus \left(\cup_{j=0}^{i-1} E_j\right)$ $(E_0 = \emptyset)$ with $\nu(F) < -1/m_i$, and $E_i \subset E \setminus \left(\cup_{j=0}^{i-1} E_j\right)$ with $\nu(E_i) < -1/m_i$, and so we have

$$\nu\left(E \setminus \left(\bigcup_{j=1}^{k} E_j\right)\right) = \nu\left(E \setminus \left(\bigcup_{j=1}^{k-1} E_j\right)\right) + (-\nu(E_k))$$

$$> \nu(E) + \sum_{j=1}^{k} \frac{1}{m_j} > 0.$$

If this selection process is terminated by finite steps, then the proof is completed. Otherwise, put

$$A = E \setminus \left(\bigcup_{k=1}^{\infty} E_k \right), \quad \text{and so} \quad E = A \cup \left(\bigcup_{k=1}^{\infty} E_k \right).$$

Then, since $0 < \nu(E) < \infty$, $|\nu(\cup_{k=1}^{\infty} E_k)| < \infty$ and by the countable additivity of ν,

$$\sum_{k=1}^{\infty} \frac{1}{m_k} < \sum_{k=1}^{\infty} (-\nu(E_k)) < \infty,$$

which implies that $\lim_{k\to\infty} m_k = \infty$ and, also we have

$$\nu(E) = \nu(A) + \sum_{k=1}^{\infty} \nu(E_k) > 0,$$

which implies that $\nu(A) > 0$. Let B be a measurable subset of A. Then for any $n \geq 1$, we have

$$B \subset A = E \setminus \left(\bigcup_{k=1}^{\infty} E_k \right) \subset E \setminus \left(\bigcup_{k=1}^{n} E_k \right),$$

and by the minimal choice of m_{n+1}, $\nu(B) \geq -1/(m_{n+1} - 1)$, which implies that

$$\nu(B) \geq \sup_{n\in\mathbb{N}} \left\{ -\frac{1}{m_{n+1} - 1} \right\} = 0$$

since $\lim_{n\to\infty} m_n = \infty$. Therefore, A is a positive set. □

Theorem 2.4.7 **[Hahn Decomposition]** Let $(X, \mathcal{M}, \nu)$ be a signed measure space. There exists a pair (P, N) of a positive set P and a negative set N such that $X = P \cup N$ and $P \cap N = \emptyset$.

[Proof] Without loss of generality, we assume that $\nu(E) < +\infty$ for any $E \in \mathcal{M}$ (If there exists $E \in \mathcal{M}$ such that $\nu(E) = \infty$, then $\nu(E) > -\infty$ for any $E \in \mathcal{M}$, and so, from which we can prove). Let $\mathcal{P}$ be the collection of all positive sets in $\mathcal{M}$. Put

$$\lambda = \sup\{\nu(E) \mid E \in \mathcal{P}\}.$$

Then $\emptyset \in \mathcal{P}$ and so $\lambda \geq 0$. There exists a sequence $\{E_k\}_{k=1}^{\infty}$ of positive sets such that $\lim_{k\to\infty} \nu(E_k) = \lambda$. Put

$$P = \bigcup_{k=1}^{\infty} E_k.$$

Then P is a positives set, i.e., $P \in \mathcal{P}$ and so $\nu(P) \le \lambda$ and, also, for any $k \in \mathbb{N}$, we have

$$\nu(P) = \nu(P \setminus E_k) + \nu(E_k) \ge \nu(E_k),$$

which implies that $\nu(P) \ge \lambda$. Therefore, $\nu(P) = \lambda$ and by assumption $\lambda = \nu(P) < +\infty$. Put

$$N = X \setminus P.$$

Suppose that N not a negative set. There exists a measurable subset A of N such that $\nu(A) > 0$. By Hahn's Lemma (Theorem 2.4.6) there exists a positive subset B of A such that $\nu(B) > 0$ and then $P \cup B \in \mathcal{P}$, which implies that

$$\infty > \lambda = \sup\{\nu(E) \,|\, E \in \mathcal{P}\} \ge \nu(P \cup B) = \nu(P) + \nu(B) = \lambda + \nu(B),$$

which is a contradiction since $\nu(B) > 0$. Therefore, N is a negative set such that $X = P \cup N$ and $P \cap N = \emptyset$. □

Definition 2.4.8 For a signed measure space $(X, \mathcal{M}, \nu)$, there exists a pair (P, N) of a positive set P and a negative set N such that P, N are disjoint and $X = P \cup N$, which is called a **Hahn Decomposition** of X for ν.

Let X be a set. Then for any subsets A and B of X, the **symmetric difference** $A \triangle B$ of A and B is defined by

$$A \triangle B = (A \setminus B) \cup (B \setminus A).$$

Remark 2.4.9 Let $(X, \mathcal{M}, \nu)$ be a signed measure space.

(i) Let (P, N) be a Hahn Decomposition of X for ν. Let A be a null subset of P. Then $(P \setminus A, N \cup A)$ is a Hahn Decomposition of X for ν. Therefore, Hahn's decomposition is not unique.

(ii) Let (P, N) and (P', N') be Hahn Decompositions of X for ν. Then $P \vartriangle P'$ and $N \vartriangle N'$ are null sets for ν. In fact, since $P \setminus P' \subset N'$ and $P' \setminus P \subset N$, for any $E \subset P \vartriangle P'$ with $E \in \mathcal{M}$,

$$0 \le \nu(E \cap (P \setminus P')) \le \nu(P \setminus P') \le 0,$$
$$0 \le \nu(E \cap (P' \setminus P)) \le \nu(P' \setminus P) \le 0,$$

which implies that $\nu(E) = 0$. Therefore, $P \vartriangle P'$ is a null set for ν.

Definition 2.4.10 Let μ_1 and μ_2 be measures on a measurable space $(X, \mathcal{M})$. Then μ_1 and μ_2 are said to be **mutually singular** if there exist two disjoint measurable sets A and B such that $X = A \cup B$ and $\mu_1(A) = \mu_2(B) = 0$.

Let $(X, \mathcal{M}, \nu)$ be a signed measure space and $X = P \cup N$ be a Hahn decomposition of X for ν, where P is a positive set and N is a negative set. Define set functions ν^+ and ν^- on the measurable space $(X, \mathcal{M})$ by

$$\nu^+(E) = \nu(E \cap P) \quad \text{and} \quad \nu^-(E) = -\nu(E \cap N) \tag{2.19}$$

for any $E \in \mathcal{M}$. Then ν^+ and ν^- are measures on the measurable space $(X, \mathcal{M})$. In fact, $\nu^\pm(\emptyset) = \pm\nu(\emptyset) = 0$ and for any countable disjoint collection $\{E_k\}_{k=1}^\infty$ of measurable sets, for $X^+ = P$ and $X^- = N$, we have

$$\begin{aligned}\nu^\pm\left(\bigcup_{k=1}^\infty E_k\right) = \pm\nu\left(\left(\bigcup_{k=1}^\infty E_k\right) \cap X^\pm\right) &= \pm\nu\left(\bigcup_{k=1}^\infty \left(E_k \cap X^\pm\right)\right) \\ &= \sum_{k=1}^\infty \pm\nu\left(E_k \cap X^\pm\right) \\ &= \sum_{k=1}^\infty \nu^\pm\left(E_k\right),\end{aligned}$$

which implies that ν^+ and ν^- are countable additive. The measures ν^+ and ν^- are called the **positive** and **negative variations** of ν

Theorem 2.4.11 [**Jordan Decomposition**] Let $(X, \mathcal{M}, \nu)$ be a signed measure space. Then there exist only one pair (ν^+, ν^-) of two mutually singular measures ν^+ and ν^- on the measurable space $(X, \mathcal{M})$ such that

$$\nu = \nu^+ - \nu^-,$$

which is called the **Jordan Decomposition** of ν.

[Proof] Let $X = P \cup N$ be a Hahn decomposition of X for ν, where P is a positive set and N is a negative set. Then for the two measures ν^+ and ν^- defined as in (2.19), we obtain that

$$\nu^-(P) = \nu^+(N) = \nu(\emptyset) = 0,$$

which implies that ν^+ and ν^- are mutually singular, and, for any $E \in \mathcal{M}$,

$$\nu(E) = \nu(E \cap (P \cup N)) = \nu(E \cap P) + \nu(E \cap N) = \nu^+(E) - \nu^-(E),$$

which implies that $\nu^+ - \nu^-$. Suppose that there exist two mutually singular measure μ_1 and μ_2 on the measurable space $(X, \mathcal{M})$ such that

$$\nu = \mu_1 - \mu_2.$$

There exist two disjoint measurable sets A and B such that $\mu_1(A) = \mu_2(B) = 0$. Then for any $E \in \mathcal{M}$ with $E \subset B$, we have

$$\nu(E) = \mu_1(E) - \mu_2(E) = \mu_1(E) \geq 0,$$

and so B is a positive set, and, similarly, we see that A is a negative set. Therefore, (B, A) is a Hahn decomposition and so by (ii) in Remark 2.4.9, $P \triangle B$ and $N \triangle A$ are null sets for ν. Therefore, for any measurable set $E \in \mathcal{M}$, we have

$$\begin{aligned}
\mu_1(E) &= \mu_1(E \cap B) + \mu_1(E \cap A) = \mu_1(E \cap B) \quad \text{since } \mu_1(A) = 0 \text{ and so } \mu_1(E \cap A) = 0 \\
&= \mu_1(E \cap B) - \mu_2(E \cap B) = \nu(E \cap B) \quad \text{since } \mu_2(E \cap B) = 0 \\
&= \nu^+(E \cap B \cap P) - \nu^-(E \cap B \cap N) \\
&= \nu^+(E \cap B \cap P) \quad \text{since } B \text{ is a positive set and so } \nu^-(E \cap B \cap N) = 0 \\
&= \nu^+(E \cap P) \\
&= \nu^+(E),
\end{aligned}$$

which implies that $\mu_1 = \nu^+$, and similarly, we see that $\mu_2 = \nu^-$. □

Remark 2.4.12 Let $(X, \mathcal{M}, \nu)$ be a signed measure space. Since ν can take at most one of the values $-\infty$ and $+\infty$, ν^+ or ν^- is finite. If ν^+ and ν^- are finite, then ν is said to be **finite**. Also, ν is said to be **σ-finite** if ν^+ and ν^- are σ-finite,.

Definition 2.4.13 Let $(X, \mathcal{M}, \nu)$ be a signed measure space. The **total variation** of ν is the measure $|\nu|$ defined by

$$|\nu| = \nu^+ + \nu^-.$$

Example 2.4.14 Let $(\mathbb{N}, \mathcal{P}(\mathbb{N}), c)$ be a measure space with the counting measure c. Let $\{a_n\}_{n=1}^{\infty}$ be a sequence of real numbers such that

$$\sum_{n=1}^{\infty} |a_n| < \infty.$$

Define a set function $\nu : \mathcal{P}(\mathbb{N}) \to (-\infty, \infty)$ by

$$\nu(E) = \sum_{\ell \in E} a_\ell, \qquad E \subset \mathbb{N}.$$

Then ν is a signed measure (see Exercise **2.21**). Put

$$P = \{n \in \mathbb{N} \,|\, a_n \geq 0\}, \qquad N = \{n \in \mathbb{N} \,|\, a_n < 0\}.$$

Then (P, N) is a Hahn decomposition (see Exercise **2.21**). For the two measures ν^+ and ν^- defined by

$$\nu^+(E) = \sum_{k \in E \cap P} a_k, \qquad \nu^-(E) = \sum_{k \in E \cap N} a_k,$$

$\nu = \nu^+ - \nu^-$ is the Jordan decomposition of which the total variation $|\nu| = \nu^+ + \nu^-$ is given by

$$|\nu|(E) = \sum_{k \in E} |a_k|.$$

2.5 Outer Measures

Definition 2.5.1 A set function $\mu^* : \mathcal{P}(X) \to [0, \infty]$ is called an **outer measure** if the following properties are satisfied:

(i) $\mu^*(\emptyset) = 0$,

(ii) μ^* is countably monotone, i.e., for any $E \in \mathcal{P}(X)$ and any countable collection $\{E_k\}_{k=1}^{\infty} \subset \mathcal{P}(X)$ such that $E \subset \cup_{k=1}^{\infty} E_k$, it holds that

$$\mu^*(E) \leq \sum_{k=1}^{\infty} \mu^*(E_k),$$

By Remark 2.3.2, every outer measure is monotone and finitely monotone.

Proposition 2.5.2 Let $\mu^* : \mathcal{P}(X) \to [0, \infty]$ be a set function. Then μ^* is an outer measure if and only if μ^* satisfies the following properties:

(i) $\mu^*(\emptyset) = 0$,

(ii) [**Monotonicity**] for $A, B \in \mathcal{P}(X)$ with $A \subset B$,

$$\mu^*(A) \leq \mu^*(B),$$

(iii) [**Countable Subadditivity**] for any countable family $\{A_k\}_{k=1}^{\infty} \subset \mathcal{P}(X)$,

$$\mu^*\left(\bigcup_{k=1}^{\infty} A_k\right) \leq \sum_{k=1}^{\infty} \mu^*(A_k).$$

[Proof] $\Rightarrow$) For any countable family $\{A_k\}_{k=1}^{\infty} \subset \mathcal{P}(X)$, by taking $E = \cup_{k=1}^{\infty} A_k$ and the countably monotonicity,

$$\mu^*\left(\bigcup_{k=1}^{\infty} A_k\right) = \mu^*(E) \leq \sum_{k=1}^{\infty} \mu^*(A_k),$$

which proves the countable subadditivity of μ^*. The monotonicity is obvious by Remark 2.3.2.

$\Leftarrow$) For any $E \in \mathcal{P}(X)$ and any countable collection $\{E_k\}_{k=1}^{\infty} \subset \mathcal{P}(X)$ such that

$E \subset \cup_{k=1}^{\infty} E_k$, by the monotonicity and the countable subadditivity of μ^*, we have

$$\mu^*(E) \leq \mu^*\left(\bigcup_{k=1}^{\infty} E_k\right) \leq \sum_{k=1}^{\infty} \mu^*(E_k),$$

which proves the countably monotonicity of μ^*. □

Proposition 2.5.3 Let $\mathcal{E} \subset \mathcal{P}(X)$ be a collection of subsets of a set X and let $\rho : \mathcal{E} \to [0, \infty]$ a set function. Suppose that $\emptyset, X \in \mathcal{E}$ and $\rho(\emptyset) = 0$. Define a set function $\mu^* : \mathcal{P}(X) \to [0, \infty]$ by

$$\mu^*(A) = \inf\left\{ \sum_{k=1}^{\infty} \rho(E_k) \,\middle|\, \{E_k\}_{k=1}^{\infty} \subset \mathcal{E} \text{ and } A \subset \bigcup_{k=1}^{\infty} E_k \right\}.$$

for any $A \in \mathcal{P}(X)$. Then μ^* is an outer measure on X which is called the **outer measure induced by** ρ.

[Proof] Let $A \subset X$. Since we can take a countable family $\{E_k\}_{k=1}^{\infty} \subset \mathcal{E}$ with $E_1 = X$ and $E_k = \emptyset$ for any $k \geq 2$,

$$\left\{ \{E_k\}_{k=1}^{\infty} \subset \mathcal{E} \,\middle|\, A \subset \bigcup_{k=1}^{\infty} E_k \right\} \neq \emptyset$$

and so $\mu^*(A)$ is well-defined. It is obvious that $\emptyset \subset \cup_{k=1}^{\infty} E_k$ for $\{E_k\}_{k=1}^{\infty} \subset \mathcal{E}$ with $E_k = \emptyset$ for all $k \in \mathbb{N}$ and so

$$0 \leq \mu^*(\emptyset) \leq \sum_{k=1}^{\infty} \rho(E_k) = 0,$$

which implies that $\mu^*(\emptyset) = 0$. We now prove the countably monotonicity of μ^*. Let $E \in \mathcal{P}(X)$ and let $\{E_k\}_{k=1}^{\infty} \subset \mathcal{P}(X)$ be a countable collection such that $E \subset \cup_{k=1}^{\infty} E_k$. Let $\epsilon > 0$ be given. Then for each $k \in \mathbb{N}$, there exists a countable family $\{E_{k,l}\}_{l=1}^{\infty} \subset \mathcal{E}$ such that $E_k \subset \cup_{l=1}^{\infty} E_{k,l}$ and

$$\sum_{l=1}^{\infty} \rho(E_{k,l}) \leq \mu^*(E_k) + \frac{\epsilon}{2^k}.$$

Then $\{E_{k,l}\}_{k,l=1}^{\infty} \subset \mathcal{E}$ is a countable collection such that $E \subset \cup_{k,l=1}^{\infty} E_{k,l}$ and so

$$\mu^*(E) \leq \sum_{k=1}^{\infty}\sum_{l=1}^{\infty} \rho(E_{k,l}) \leq \sum_{k=1}^{\infty} \left(\mu^*(E_k) + \frac{\epsilon}{2^k}\right) = \sum_{k=1}^{\infty} \mu^*(E_k) + \epsilon.$$

Since $\epsilon > 0$ is arbitrary, we prove the countably monotonicity of μ^* and so μ^* is an outer measure. □

Remark 2.5.4 Let $\mathcal{E} \subset \mathcal{P}(X)$ be a collection of subsets of a set X and let $\rho : \mathcal{E} \to [0, \infty]$ a set function. If $\emptyset, X \notin \mathcal{E}$, then by adding $\emptyset$ and X to $\mathcal{E}$, we get the family $\mathcal{E}' = \mathcal{E} \cup \{\emptyset, X\}$ of subsets of X and we extend the set function ρ to $\mathcal{E}'$ by defining $\rho(\emptyset) = 0$ and $\rho(X) = \infty$, of which the extension is denoted by ρ'. Similarly, for all other cases:

- $\emptyset \notin \mathcal{E}$ and $X \in \mathcal{E}$;
- $\emptyset \in \mathcal{E}$ with $\rho(\emptyset) = 0$ and $X \notin \mathcal{E}$;
- $\emptyset \in \mathcal{E}$ with $\rho(\emptyset) \neq 0$ and $X \notin \mathcal{E}$;
- $\emptyset \in \mathcal{E}$ and $X \in \mathcal{E}$, but $\rho(\emptyset) \neq 0$,

we can assign a set function ρ' on $\mathcal{E}' = \mathcal{E} \cup \{\emptyset, X\}$ as an extension of $\rho|_{\mathcal{E}\setminus\{\emptyset\}}$ such that $\rho'(\emptyset) = 0$ and $\rho'(X) = \rho(X)$ or $\rho'(X) = \infty$. Then we can construct the outer measure μ^* induced by ρ' as in Proposition 2.5.3, which is called the **outer measure induced by** ρ.

Definition 2.5.5 Let μ^* be an outer measure on a nonempty set X and $E \subset X$ a set. Then E is said to be **μ^*-measurable** (or **measurable** with respect to μ^*) if for any $A \subset X$,

$$\mu^*(A) = \mu^*(A \cap E) + \mu^*(A \cap E^c). \tag{2.20}$$

Remark 2.5.6 The criterion given as in (2.20) is sometimes called the **Carathéodory criterion**.

Remark 2.5.7 (1) For any subsets $E, A \subset X$, by the countable subadditivity of the outer measure, we have

$$\mu^*(A) \leq \mu^*(A \cap E) + \mu^*\left(A \cap E^{C}\right).$$

Therefore, for the measurability criterion, we need to check only that

$$\mu^*(A) \geq \mu^*(A \cap E) + \mu^*\left(A \cap E^{C}\right), \tag{2.21}$$

i.e., E is measurable if and only if for any set $A \subset X$, (2.21) holds. If $\mu^*(A) = \infty$, then it is obvious that (2.21) holds. Therefore, E is measurable if and only if for any set $A \subset X$ with $\mu^*(A) < \infty$, (2.21) holds.

(2) It is obvious that E is measurable if and only if E^c is measurable.

(3) It is clear that $\emptyset$ and X are measurable.

Proposition 2.5.8 **[Excision Property]** Let A be a measurable set with finite outer measure such that A is contained in a set B, Then we have

$$\mu^*(B \setminus A) = \mu^*(B) - \mu^*(A). \tag{2.22}$$

[Proof] Since A is measurable, we have

$$\mu^*(B) = \mu^*(B \cap A) + \mu^*(B \cap A^c) = \mu^*(A) + \mu^*(B \setminus A),$$

which implies (2.22) since $\mu^*(A) < \infty$. □

Let $\mathfrak{M}$ be the class of all measurable sets with respect to the outer measure μ^*.

Proposition 2.5.9 Every set of outer measure zero belongs to $\mathfrak{M}$.

[Proof] Let $E \in \mathcal{P}(X)$ with $\mu^*(E) = 0$. Then for any $A \subset X$, by the monotonicity of the outer measure μ^*, we have $0 \le \mu^*(A \cap E) \le \mu^*(E) = 0$, which implies that $\mu^*(A \cap E) = 0$ and so, again, by the monotonicity of the outer measure μ^*,

$$\mu^*(A) \ge \mu^*(A \cap E^c) = \mu^*(A \cap E) + \mu^*(A \cap E^c).$$

Therefore, E is measurable, i.e., $E \in \mathfrak{M}$. □

Proposition 2.5.10 The union of two measurable sets is measurable. Moreover, $\mathfrak{M}$ is closed under the finite union.

[Proof] Let E_1 and E_2 be measurable. Then we obtain that

$$\begin{aligned}\mu^*(A \cap (E_1 \cup E_2)) &= \mu^*(A \cap (E_1 \cup E_2) \cap E_1) + \mu^*(A \cap (E_1 \cup E_2) \cap E_1^c)\\ &= \mu^*(A \cap E_1) + \mu^*(A \cap E_1^c \cap E_2),\end{aligned}$$

and so we have

$$\begin{aligned}&\mu^*(A \cap (E_1 \cup E_2)) + \mu^*(A \cap (E_1 \cup E_2)^c)\\ &\quad = \mu^*(A \cap E_1) + \mu^*(A \cap E_1^c \cap E_2) + \mu^*(A \cap E_1^c \cap E_2^c)\\ &\quad = \mu^*(A \cap E_1) + \mu^*(A \cap E_1^c)\\ &\quad = \mu^*(A),\end{aligned}$$

which implies that $E_1 \cup E_2$ is measurable. Let $\{E_k\}_{k=1}^n$ be a finite family of measurable sets. If $n = 1$, it is trivial that $E_1 = \cup_{k=1}^n E_k$ is measurable. Suppose that $\cup_{k=1}^{n-1} E_k$ $(n \ge 2)$ is measurable. Then we see that, since the union of two measurable sets is measurable,

$$\bigcup_{k=1}^{n} E_k = \left(\bigcup_{k=1}^{n-1} E_k\right) \cup E_n$$

is measurable. Therefore, by the induction, for any $n \in \mathbb{N}$, $\cup_{k=1}^n E_k$ is measurable. □

Corollary 2.5.11 $\mathfrak{M}$ is closed under the finite intersection.

[Proof] Since $\mathfrak{M}$ is closed under the finite union and complement, $\mathfrak{M}$ is closed under the finite intersection. In fact, for any finite family $\{E_k\}_{k=1}^n$ of measurable sets,

$$\bigcap_{k=1}^n E_k = \left(\left(\bigcap_{k=1}^n E_k\right)^{\mathrm{c}}\right)^{\mathrm{c}} = \left(\bigcup_{k=1}^n E_k^{\mathrm{c}}\right)^{\mathrm{c}}$$

is measurable. □

Proposition 2.5.12 For any subset $A \subset X$ and any finite disjoint family $\{E_k\}_{k=1}^n$ of measurable sets E_k, we have

$$\mu^*\left(A \cap \bigcup_{k=1}^n E_k\right) = \sum_{k=1}^n \mu^*(A \cap E_k). \tag{2.23}$$

In particular, we have

$$\mu^*\left(\bigcup_{k=1}^n E_k\right) = \sum_{k=1}^n \mu^*(E_k). \tag{2.24}$$

[Proof] If $n = 1$, then it is obvious that (2.23) holds. Suppose that for $n \geq 2$,

$$\mu^*\left(A \cap \bigcup_{k=1}^{n-1} E_k\right) = \sum_{k=1}^{n-1} \mu^*(A \cap E_k).$$

Then by the measurability of E_n, we obtain that

$$\begin{aligned}
\mu^*\left(A \cap \bigcup_{k=1}^n E_k\right) &= \mu^*\left(\left(A \cap \bigcup_{k=1}^n E_k\right) \cap E_n\right) + \mu^*\left(\left(A \cap \bigcup_{k=1}^n E_k\right) \cap E_n^{\mathrm{c}}\right)\\
&= \mu^*(A \cap E_n) + \mu^*\left(A \cap \bigcup_{k=1}^{n-1} E_k\right)\\
&= \sum_{k=1}^n \mu^*(A \cap E_k),
\end{aligned}$$

which implies that (2.23) holds. Therefore, by the induction, for any $n \in \mathbb{N}$, (2.23) holds. By taking $A = X$ in (2.23), we see (2.24). □

Remark 2.5.13 In Proposition 2.5.12, by the measurability of E_k $(k = 1, 2, \cdots, n)$, we obtain that

$$\begin{aligned}
\mu^*\left(A \cap \bigcup_{k=1}^{n} E_k\right) &= \mu^*\left(\left(A \cap \bigcup_{k=1}^{n} E_k\right) \cap E_1\right) + \mu^*\left(\left(A \cap \bigcup_{k=1}^{n} E_k\right) \cap E_1^{\mathrm{c}}\right) \\
&= \mu^*(A \cap E_1) + \mu^*\left(A \cap \bigcup_{k=2}^{n} E_k\right) \\
&= \mu^*(A \cap E_1) + \mu^*\left(\left(A \cap \bigcup_{k=2}^{n} E_k\right) \cap E_2\right) + \mu^*\left(\left(A \cap \bigcup_{k=2}^{n} E_k\right) \cap E_2^{\mathrm{c}}\right) \\
&= \mu^*(A \cap E_1) + \mu^*(A \cap E_2) + \mu^*\left(A \cap \bigcup_{k=3}^{n} E_k\right) \\
&\cdots \\
&= \sum_{k=1}^{n} \mu^*(A \cap E_k).
\end{aligned}$$

Proposition 2.5.14 $\mathfrak{M}$ is closed under the countable union.

[Proof] For any countable disjoint family $\{E_k\}_{k=1}^{\infty}$ of measurable sets and any subset $A \subset X$, we put $E = \cup_{k=1}^{\infty} E_k$ and then we see that

$$\begin{aligned}
\mu^*(A) &= \mu^*\left(A \cap \bigcup_{k=1}^{n} E_k\right) + \mu^*\left(A \cap \left(\bigcup_{k=1}^{n} E_k\right)^{\mathrm{c}}\right) \\
&\geq \mu^*\left(A \cap \bigcup_{k=1}^{n} E_k\right) + \mu^*(A \cap E^{\mathrm{c}}).
\end{aligned}$$

Therefore, by applying (2.23), we obtain that

$$\mu^*(A) \geq \sum_{k=1}^{n} \mu^*(A \cap E_k) + \mu^*(A \cap E^{\mathrm{c}}),$$

which, by the countable subadditivity of μ^*, implies that

$$\begin{aligned}
\mu^*(A) &\geq \sum_{k=1}^{\infty} \mu^*(A \cap E_k) + \mu^*(A \cap E^{\mathrm{c}}) \\
&\geq \mu^*(A \cap E) + \mu^*(A \cap E^{\mathrm{c}}),
\end{aligned}$$

which implies that $E = \cup_{k=1}^{\infty} E_k \in \mathfrak{M}$. For any countable family $\{E_k\}_{k=1}^{\infty}$ of measurable sets E_k, put

$$F_1 = E_1, \qquad F_k = E_k \setminus \bigcup_{i=1}^{k-1} E_i \in \mathfrak{M}, \qquad k = 2, 3, \cdots.$$

Then for $i \neq j$, $F_i \cap F_j = \emptyset$, i.e., $\{F_k\}_{k=1}^{\infty}$ is a countable disjoint family. Therefore, we have

$$\bigcup_{k=1}^{\infty} E_k = \bigcup_{k=1}^{\infty} F_k \in \mathfrak{M}.$$

Hence, $\mathfrak{M}$ is closed under the countable union. □

Theorem 2.5.15 $\mathfrak{M}$ is a σ-algebra.

[Proof] (σ1) By (3) in Remark 2.5.7. $\emptyset, X \in \mathfrak{M}$.

(σ2) If $E \in \mathfrak{M}$, i.e., E is measurable, then E^c is measurable, i.e., $E^c \in \mathfrak{M}$. Therefore, $\mathfrak{M}$ is closed under complements.

(σ3) By Proposition 2.5.14, $\mathfrak{M}$ is closed under countable unions. □

Theorem 2.5.16 [**Carathéodory's Theorem**] Let μ be the restriction of μ^* to $\mathfrak{M}$, i.e., $\mu = \mu^*|_{\mathfrak{M}}$. Then $(X, \mathfrak{M}, \mu)$ is a complete measure space.

[Proof] It is obvious that $\mu(\emptyset) = \mu^*(\emptyset) = 0$. For any countable disjoint family $\{E_k\}_{k=1}^{\infty}$ of measurable sets, from the countable subadditivity (see (iii) in Proposition 2.5.2) of the outer measure μ^*, it is obvious that

$$\begin{aligned} \mu\left(\bigcup_{k=1}^{\infty} E_k\right) &= \mu^*\left(\bigcup_{k=1}^{\infty} E_k\right) \\ &\leq \sum_{k=1}^{\infty} \mu^*(E_k) \\ &= \sum_{k=1}^{\infty} \mu(E_k). \end{aligned}$$

On the other hand, for any $n \in \mathbb{N}$, by the monotonicity (see (ii) in Proposition 2.5.2) of the outer measure μ^* and (2.24), we have

$$\mu\left(\bigcup_{k=1}^{\infty} E_k\right) = \mu^*\left(\bigcup_{k=1}^{\infty} E_k\right) \geq \mu^*\left(\bigcup_{k=1}^{n} E_k\right) = \sum_{k=1}^{n} \mu^*(E_k) = \sum_{k=1}^{n} \mu(E_k),$$

which implies that, by taking supremum in n,

$$\mu\left(\bigcup_{k=1}^{\infty} E_k\right) \geq \sum_{k=1}^{\infty} \mu(E_k).$$

Therefore, μ satisfies the countable additivity (2.12). If N is a μ^*-measurable with $\mu(N) = 0$ and $A \subset N$, then by the monotonicity $\mu^*(A) \leq \mu^*(N) = \mu(N) = 0$ and so, by Proposition 2.5.9, A is μ^*-measurable, i.e., $A \in \mathfrak{M}$. Therefore, $(X, \mathfrak{M}, \mu)$ is a complete measure space. □

Definition 2.5.17 Let $\mathcal{E} \subset \mathcal{P}(X)$ be a collection of subsets of a set X such that $\emptyset, X \in \mathcal{E}$, and let $\rho : \mathcal{E} \to [0, \infty]$ a set function with $\rho(\emptyset) = 0$. Let μ^* be the outer measure induced by ρ. The measure μ which is the restriction of the outer measure μ^* to the σ-algebra $\mathfrak{M}$ of μ^*-measurable sets is called the **Carathéodory measure induced by** ρ.

For a collection $\mathcal{E}$ of subsets of X, we denote $\mathcal{E}_\sigma$ the family of countable unions of sets of $\mathcal{E}$ and we denote $\mathcal{E}_{\sigma\delta}$ the family of countable intersections of sets of $\mathcal{E}_\sigma$. If $\mathcal{E}$ is the collection of all open intervals of real numbers, then $\mathcal{E}_\sigma$ is the family of all open subsets of $\mathbb{R}$ and $\mathcal{E}_{\sigma\delta}$ coincides with the family of all G_δ subsets of $\mathbb{R}$.

Theorem 2.5.18 Let $\rho : \mathcal{E} \to [0, \infty]$ be a set function defined on a collection $\mathcal{E}$ of subsets of a set X. Suppose that $\emptyset, X \in \mathcal{E}$ and $\rho(\emptyset) = 0$. Let μ^* be the outer measure induced by ρ. Let E be a subset of X with $\mu^*(E) < \infty$. Then there exists a subset A of X such that $A \in \mathcal{E}_{\sigma\delta}$, $E \subset A$ and $\mu^*(E) = \mu^*(A)$. Furthermore, if E is μ^*-measurable, then

$$\mu^*(A \setminus E) = 0.$$

[Proof] Let $n \in \mathbb{N}$ be given. Then there exists a countable family $\{E_{n,k}\}_{k=1}^\infty \subset \mathcal{E}$ such that $E \subset \cup_{k=1}^\infty E_{n,k}$ and

$$\sum_{k=1}^\infty \rho(E_{n,k}) \le \mu^*(E) + \frac{1}{n}.$$

Put

$$A_n = \bigcup_{k=1}^\infty E_{n,k},$$

and then we construct a countable family $\{A_n\}_{n=1}^\infty$ of subsets of X in $\mathcal{E}_\sigma$. Put

$$A = \bigcap_{n=1}^\infty A_n \in \mathcal{E}_{\sigma\delta}.$$

Then $E \subset A$ and, for any $n \in \mathbb{N}$, we have

$$\mu^*(E) \le \mu^*(A) \le \mu^*(A_n) \le \sum_{k=1}^\infty \rho(E_{n,k}) \le \mu^*(E) + \frac{1}{n},$$

which implies that $\mu^*(E) = \mu^*(A)$.

If E is μ^*-measurable with $\mu(E) = \mu^*(E) < \infty$, then by the excision property of μ^* (see Proposition 2.5.8), we have

$$\mu^*(A \setminus E) = \mu^*(A) - \mu^*(E) = 0,$$

which gives the proof. □

A set function $\rho : \mathcal{E} \to [0, \infty]$ defined on a nonempty collection $\mathcal{E}$ of subsets of X is said to be

(1) **[finitely additive]** if for any finite disjoint collection $\{E_k\}_{k=1}^n \subset \mathcal{E}$ such that $\cup_{k=1}^n E_k \in \mathcal{E}$,

$$\rho\left(\bigcup_{k=1}^n E_k\right) = \sum_{k=1}^n \rho(E_k); \tag{2.25}$$

(2) **[countably additive]** if for any countable disjoint collection $\{E_k\}_{k=1}^\infty \subset \mathcal{E}$ such that $\cup_{k=1}^\infty E_k \in \mathcal{E}$,

$$\rho\left(\bigcup_{k=1}^\infty E_k\right) = \sum_{k=1}^\infty \rho(E_k). \tag{2.26}$$

Proposition 2.5.19 Let $\mathcal{E}$ be a collection of subsets of X such that $\emptyset, X \in \mathcal{E}$ and $\mu_* : \mathcal{E} \to [0, \infty]$ a set function with $\mu_*(\emptyset) = 0$. If the Carathéodory measure μ induced by μ_* is an extension of μ_*, then μ_* is finitely additive and countably monotone.

[Proof] Since μ is a measure, μ is countably additive and so finitely additive. Therefore, for any finite disjoint collection $\{E_k\}_{k=1}^n \subset \mathcal{E}$ such that $\cup_{k=1}^n E_k \in \mathcal{E}$,

$$\mu_*\left(\bigcup_{k=1}^n E_k\right) = \mu\left(\bigcup_{k=1}^n E_k\right) = \sum_{k=1}^n \mu(E_k) = \sum_{k=1}^n \mu_*(E_k),$$

which implies the finitely additivity of μ_*. For any $E \in \mathcal{E}$ and countable family $\{E_k\}_{k=1}^\infty \subset \mathcal{E}$ such that $E \subset \cup_{k=1}^\infty E_k$, since every outer measure is countably monotone,

$$\mu_*(E) = \mu(E) = \mu^*(E) \le \sum_{k=1}^\infty \mu^*(E_k) = \sum_{k=1}^\infty \mu(E_k) = \sum_{k=1}^\infty \mu_*(E_k),$$

which implies the countable monotonicity of μ_*. □

Remark 2.5.20 In Proposition 2.5.19, let $\mathcal{E}$ be a collection of subsets of X such that $\emptyset, X \in \mathcal{E}$ and $\mu_* : \mathcal{E} \to [0, \infty]$ a set function with $\mu_*(\emptyset) = 0$. Suppose that the Carathéodory measure μ induced by μ_* is an extension of μ_*. Then since μ is a measure, μ is countably additive. Therefore, for any countable disjoint collection $\{E_k\}_{k=1}^\infty \subset \mathcal{E}$ such that $\cup_{k=1}^\infty E_k \in \mathcal{E}$,

$$\mu_*\left(\bigcup_{k=1}^\infty E_k\right) = \mu\left(\bigcup_{k=1}^\infty E_k\right) = \sum_{k=1}^\infty \mu(E_k) = \sum_{k=1}^\infty \mu_*(E_k),$$

which implies the countably additivity of μ_*.

Definition 2.5.21 Let $\mathcal{E}$ be a collection of subsets of X such that $\emptyset, X \in \mathcal{E}$ and $\mu_* : \mathcal{E} \to [0, \infty]$ a set function with $\mu_*(\emptyset) = 0$. Then μ_* is called a **premeasure** if μ_* is finitely additive and countably monotone.

Proposition 2.5.22 Let $\mathcal{E}$ be an algebra of subsets of X and $\mu_* : \mathcal{E} \to [0, \infty]$ a set function with $\mu_*(\emptyset) = 0$. Suppose that μ_* is countably additive. Then for any countable ascending family $\{E_k\}_{k=1}^{\infty} \subset E$ with $\cup_{k=1}^{\infty} E_k \in \mathcal{E}$,

$$\mu_*\left(\bigcup_{k=1}^{\infty} E_k\right) = \lim_{n\to\infty} \mu_*(E_n).$$

[Proof] Let $\{E_k\}_{k=1}^{\infty} \subset E$ be a countable ascending family such that $\cup_{k=1}^{\infty} E_k \in \mathcal{E}$. For each $k \in \mathbb{N}$, put

$$F_k = E_k \setminus E_{k-1}, \qquad E_0 = \emptyset.$$

Then since $\mathcal{E}$ is an algebra, $\{F_k\}_{k=1}^{\infty}$ is a countable disjoint family of sets in $\mathcal{E}$ such that

$$\bigcup_{k=1}^{\infty} F_k = \bigcup_{k=1}^{\infty} E_k \in \mathcal{E}.$$

Since μ_* is countably additive by assumption,

$$\begin{aligned}\mu_*\left(\bigcup_{k=1}^{\infty} E_k\right) &= \mu_*\left(\bigcup_{k=1}^{\infty} F_k\right) = \sum_{k=1}^{\infty} \mu_*(F_k) = \lim_{n\to\infty} \sum_{k=1}^{n} \mu_*\left(\bigcup_{k=1}^{n} F_k\right) \\ &= \lim_{n\to\infty} \mu_*(E_n),\end{aligned}$$

which gives the proof. □

Proposition 2.5.23 Let $\mathcal{E}$ be an algebra of subsets of X and $\mu_* : \mathcal{E} \to [0, \infty]$ a set function with $\mu_*(\emptyset) = 0$. Then μ_* is a premeasure if and only if μ_* is countably additive.

[Proof] $\Rightarrow$) Let $\{E_k\}_{k=1}^{\infty} \subset \mathcal{E}$ be a countable disjoint collection such that $\cup_{k=1}^{\infty} E_k \in \mathcal{E}$. Since μ_* is a premeasure, μ_* is countably monotone and and so we have

$$\mu_*\left(\bigcup_{k=1}^{\infty} E_k\right) \leq \sum_{k=1}^{\infty} \mu_*(E_k).$$

Also, since μ_* is countably monotone and $\mu_*(\emptyset) = 0$, μ_* is monotone. Therefore, for any $n \in \mathbb{N}$, we have

$$\mu_*\left(\bigcup_{k=1}^{\infty} E_k\right) \geq \mu_*\left(\bigcup_{k=1}^{n} E_k\right) = \sum_{k=1}^{n} \mu_*(E_k),$$

where we used the finitely additivity of μ_* for the last identity, which implies that

$$\mu_*\left(\bigcup_{k=1}^{\infty} E_k\right) \geq \sum_{k=1}^{\infty} \mu_*(E_k).$$

Hence μ_* is countably additive.

$\Leftarrow$) Suppose that μ_* is countably additive. Then since $\mu_*(\emptyset) = 0$, μ_* is finitely additive. We now prove that μ_* is countably monotone. Let $E \in \mathcal{E}$ and $\{E_k\}_{k=1}^{\infty} \in \mathcal{E}$ such that $E \subset \cup_{k=1}^{\infty} E_k$. For each $k \in \mathbb{N}$, put

$$F_k = E_k \setminus \cup_{i=1}^{k-1} E_i, \qquad E_0 = \emptyset.$$

Then $\{F_k\}_{k=1}^{\infty}$ is a countable disjoint family such that

$$\bigcup_{k=1}^{\infty} F_k = \bigcup_{k=1}^{\infty} E_k.$$

For each $n \in \mathbb{N}$, put

$$G_n = \bigcup_{k=1}^{n} F_k = \bigcup_{k=1}^{n} E_k.$$

Then $\{E \cap G_n\}_{k=1}^{\infty} \subset \mathcal{E}$ be a countable ascending family of sets in $\mathcal{E}$. Therefore, by Proposition 2.5.22, we obtain that

$$\begin{aligned}
\mu_*(E) = \mu_*\left(E \cap \bigcup_{n=1}^{\infty} G_n\right) &= \lim_{n\to\infty} \mu_*\left(E \cap G_n\right) \\
&\leq \lim_{n\to\infty} \mu_*\left(G_n\right) = \lim_{n\to\infty} \mu_*\left(\bigcup_{k=1}^{n} F_k\right) = \lim_{n\to\infty} \sum_{k=1}^{n} \mu_*\left(F_k\right) \\
&\leq \sum_{k=1}^{\infty} \mu_*\left(E_k\right).
\end{aligned}$$

Hence μ_* is countably monotone. □

Proposition 2.5.24 Let $\mathcal{A}$ be an algebra of subsets of X and $\mu_* : \mathcal{A} \to [0, \infty]$ a premeasure. Then $\sigma(\mathcal{A}) \subset \mathfrak{M}$, where $\mathfrak{M}$ is the σ-algebra of all μ^*-measurable sets and μ^* is the outer measure induced by μ_*.

[Proof] Let $E \in \mathcal{A}$. Then for any $A \subset X$ with finite outer measure and $\epsilon > 0$, there exists a countable family $\{E_k\}_{k=1}^{\infty} \subset \mathcal{A}$ such that $A \subset \cup_{k=1}^{\infty} E_k$ and

$$\mu^*(A) \geq \sum_{k=1}^{\infty} \mu_*(E_k) - \epsilon.$$

For each $k \in \mathbb{N}$, since the premeasure μ_* is finitely additive,

$$\mu_*(E_k) = \mu_*((E_k \cap E) \cup (E_k \cap E^c)) = \mu_*(E_k \cap E) + \mu_*(E_k \cap E^c).$$

Therefore, by the definition of outer measure, we have

$$\begin{aligned} \mu^*(A) &\geq \sum_{k=1}^{\infty} \mu_*(E_k \cap E) + \sum_{k=1}^{\infty} \mu_*(E_k \cap E^c) - \epsilon \\ &\geq \mu^*(A \cap E) + \mu^*(A \cap E^c) - \epsilon. \end{aligned}$$

Since $\epsilon > 0$ is arbitrary, we have

$$\mu^*(A) \geq \mu^*(A \cap E) + \mu^*(A \cap E^c),$$

which implies that E is μ^*-measurable, i.e., $\mathcal{A} \subset \mathfrak{M}$. Therefore, $\sigma(\mathcal{A}) \subset \mathfrak{M}$. □

Proposition 2.5.25 Let $\mathcal{A}$ be an algebra of subsets of X and $\mu_* : \mathcal{A} \to [0, \infty]$ a premeasure. Then $\mu^*|_{\mathcal{A}} = \mu_*$.

[Proof] Let $E \in \mathcal{A}$. Let $\{E_n\}_{n=1}^{\infty} \subset \mathcal{A}$ be a countable sequence given by $E_1 = E$ and $E_n = \emptyset$ for $n \geq 2$. Then

$$E \subset \bigcup_{n=1}^{\infty} E_n$$

and

$$\mu^*(E) \leq \sum_{n=1}^{\infty} \mu_*(E_n) = \mu_*(E_1) = \mu_*(E).$$

Also, for any countable sequence $\{E_n\}_{n=1}^{\infty} \subset \mathcal{A}$ with $E \subset \cup_{n=1}^{\infty} E_n$, by the countably monotonicity of μ_*, we have

$$\mu_*(E) \leq \sum_{n=1}^{\infty} \mu_*(E_n).$$

Therefore, $\mu_*(E) \leq \mu^*(E)$. Finally, $\mu_*(E) = \mu^*(E)$. □

Theorem 2.5.26 [**Carathéodory-Hahn Theorem**] Let $\mathcal{A}$ be an algebra of subsets of X and $\mu_* : \mathcal{A} \to [0, \infty]$ a premeasure. Then the Carathéodory measure $\mu : \mathfrak{M} \to [0, \infty]$ induced by μ_* is an extension of μ_*, which is called the **Carathéodory extension** of μ_*.

[Proof] By Proposition 2.5.24, $\sigma(\mathcal{A}) \subset \mathfrak{M}$ and so for the measure $\mu = \mu^*|_{\mathfrak{M}}$,

$$\mu|_{\mathcal{A}} = (\mu^*|_{\mathfrak{M}})|_{\mathcal{A}} = \mu^*|_{\mathcal{A}} = \mu_*,$$

where for the last identity, we applied Proposition 2.5.25. Therefore, the Carathéodory measure μ induced by μ_* is an extension of μ_*. □

Theorem 2.5.27 Let $\mathcal{A}$ be an algebra of subsets of X and $\mu_* : \mathcal{A} \to [0, \infty]$ a premeasure. Let $\mu : \mathfrak{M} \to [0, \infty]$ be the Carathéodory extension of μ_*.

(i) If ν is a measure on $\mathfrak{M}$ which extends μ_*, then $\nu(E) \le \mu(E)$ for any $E \in \mathfrak{M}$, and $\nu(E) = \mu(E)$ whenever $\mu(E) < \infty$.

(ii) If μ_* is σ-finite, then μ is σ-finite.

(ii) If μ_* is σ-finite, then μ is the unique measure on $\mathfrak{M}$ which extends μ_*.

[Proof] (i) Let $E \in \mathfrak{M}$. Then for any countable family $\{E_k\}_{k=1}^{\infty} \subset \mathcal{A}$ with $E \subset \bigcup_{k=1}^{\infty} E_k$, we have

$$\nu(E) \le \sum_{k=1}^{\infty} \nu(E_k) = \sum_{k=1}^{\infty} \mu_*(E_k)$$

and so $\nu(E) \le \mu^*(E) = \mu(E)$. If $\mu(E) < \infty$, then for any $\epsilon > 0$ there exists a countable family $\{F_k\}_{k=1}^{\infty} \subset \mathcal{A}$ such that $E \subset \bigcup_{k=1}^{\infty} F_k$ and

$$\mu(E) = \mu^*(E) > \sum_{k=1}^{\infty} \mu_*(F_k) - \epsilon = \sum_{k=1}^{\infty} \mu(F_k) - \epsilon,$$

which implies that

$$\begin{aligned} \mu\left(\bigcup_{k=1}^{\infty} F_k \setminus E\right) &= \mu\left(\bigcup_{k=1}^{\infty} F_k\right) - \mu(E) \\ &\le \sum_{k=1}^{\infty} \mu(F_k) - \mu(E) \\ &< \epsilon. \end{aligned}$$

Therefore, we have

$$\begin{aligned} \mu(E) &\le \mu\left(\bigcup_{k=1}^{\infty} F_k\right) = \lim_{n\to\infty} \mu\left(\bigcup_{k=1}^{n} F_k\right) = \lim_{n\to\infty} \nu\left(\bigcup_{k=1}^{n} F_k\right) \\ &= \nu\left(\bigcup_{k=1}^{\infty} F_k\right) = \nu(E) + \nu\left(\bigcup_{k=1}^{\infty} F_k \setminus E\right) \\ &\le \nu(E) + \mu\left(\bigcup_{k=1}^{\infty} F_k \setminus E\right) \end{aligned}$$

$$< \nu(E) + \epsilon.$$

Since $\epsilon > 0$ is arbitrary, $\mu(E) \leq \nu(E)$.

(ii) It is obvious. In fact, if μ_* is σ-finite, then there exists a countable disjoint family $\{X_k\}_{k=1}^{\infty} \subset \mathcal{A}$ such that $\mu_*(X_k) < \infty$ for each $k \in \mathbb{N}$ and $X = \bigcup_{k=1}^{\infty} X_k$.

(iii) Suppose $X = \bigcup_{k=1}^{\infty} X_k$ with $\mu_*(X_k) < \infty$ for each $k \in \mathbb{N}$. We may assume that $\{X_k\}_{k=1}^{\infty}$ is a disjoint family. Suppose ν is another extension of μ_* to $\mathcal{M}$. Let $E \in \mathcal{M}$. Then by (i), we have

$$\begin{aligned} \mu(E) &= \sum_{k=1}^{\infty} \mu(E \cap X_k) = \sum_{k=1}^{\infty} \nu(E \cap X_k) \\ &= \nu(E), \end{aligned}$$

which implies that $\nu = \mu$. □

Corollary 2.5.28 Let $\mathcal{A}$ be an algebra of subsets of X. Let μ and ν be two measures on $\sigma(\mathcal{A})$ which are σ-finite on $\mathcal{A}$. Then $\mu = \nu$ on $\sigma(\mathcal{A})$ if and only if $\mu = \nu$ on $\mathcal{A}$.

[Proof] $\Rightarrow$) The proof is obvious.

$\Leftarrow$) Suppose that μ and ν are finite. Put

$$\mathcal{S} = \{A \in \sigma(\mathcal{A}) \mid \mu(A) = \nu(A)\}.$$

Then by the assumption and $\emptyset, X \in \mathcal{A} \subset \mathcal{S}$. If $A \in \mathcal{S}$, then since $\mu(A) = \nu(A) < \infty$, by the excision property of measure, we see that

$$\mu(A^c) = \mu(X) - \mu(A) = \nu(X) - \nu(A) = \nu(A^c),$$

which implies that $A \in \mathcal{S}$. Let $\{A_k\}_{k=1}^{\infty} \subset \mathcal{S}$ be a countable family of sets in $\mathcal{S}$. Without loss of generality, we may assume that $\{A_k\}_{k=1}^{\infty}$ is a disjoint family. Then we have

$$\begin{aligned} \mu\left(\bigcup_{k=1}^{\infty} A_k\right) &= \sum_{k=1}^{\infty} \mu(A_k) \\ &= \sum_{k=1}^{\infty} \nu(A_k) \\ &= \nu\left(\bigcup_{k=1}^{\infty} A_k\right), \end{aligned}$$

which implies that $\mathcal{S}$ is closed under the countable unions. Therefore, $\mathcal{S}$ is a σ-algebra containing $\mathcal{S}$ and $\sigma(\mathcal{A})$ is the smallest σ-algebra containing $\mathcal{A}$, and so $\sigma(\mathcal{A}) \subset \mathcal{S}$. Therefore, $\mu = \nu$ on $\sigma(\mathcal{A})$. Let $X_0 \in \mathcal{A}$ with $\mu(X_0) = \nu(X_0) < \infty$. Then by Remark 2.2.3, $X_0 \cap \mathcal{A}$ is an algebra of subsets of X_0 and $X_0 \cap \sigma(\mathcal{A})$ is the smallest σ-algebra of subsets of X_0 containing $X_0 \cap \mathcal{A}$. Therefore, $\mu = \nu$ on $X_0 \cap \mathcal{A}$. Since μ and ν are σ-finite, there exists a countable disjoint family $\{X_i\}_{i=1}^{\infty} \subset \mathcal{A}$ such that $\mu(X_i) = \nu(X_i) < \infty$ for all $i \in \mathbb{N}$ and $X = \cup_{i=1}^{\infty} X_i$. Then for any $E \in \sigma(\mathcal{A})$, we have

$$\begin{aligned}\mu(E) &= \sum_{i=1}^{\infty} \mu(X_i \cap E) \\ &= \sum_{i=1}^{\infty} \nu(X_i \cap E) \\ &= \nu(E).\end{aligned}$$

Therefore, $\mu = \nu$ on $\sigma(\mathcal{A})$. □

Example 2.5.29 Let $\mathbb{Q}$ be the set of rational numbers and $\mathcal{A}$ an algebra generated by $\{(a, b] \cap \mathbb{Q} \,|\, a \leq b\}$. Let $\mu_* : \mathcal{A} \to [0, \infty]$ be the premaesure defined by

$$\mu_*(A) = \begin{cases} \infty, & A \neq \emptyset, \\ 0, & A = \emptyset, \end{cases} \qquad A \in \mathcal{A}.$$

Let c be the counting measure on $\sigma(\mathcal{A})$ and let μ be the measure on $\sigma(\mathcal{A})$ defined by

$$\mu(B) = \begin{cases} \infty, & B \neq \emptyset, \\ 0, & B = \emptyset, \end{cases} \qquad B \in \sigma(\mathcal{A}).$$

Then c and μ are extension of μ_* to $\sigma(\mathcal{A})$, but $c \neq \mu$. In fact, or any $a \in \mathbb{Q}$,

$$\begin{aligned}[a, a] \cap \mathbb{Q} &= \bigcap_{n=1}^{\infty} \left(\left(a - \frac{1}{n}, a \right] \cap \mathbb{Q} \right) \\ &\in \sigma(\mathcal{A})\end{aligned}$$

and $c(\{a\}) = 1 \neq \infty = \mu(\{a\})$.

Exercises

2.1 Prove that every algebra of subsets of a set is closed under finite intersections.

2.2 Prove that every σ-algebra of subsets of a set is closed under countable intersections.

2.3 Let $\mathcal{A}_1$ and $\mathcal{A}_2$ be algebras of subsets of disjoint (distinct) sets X_1 and X_2, respectively. Prove that

$$\mathcal{A}_1 \cup \mathcal{A}_2 = \{A_1 \cup A_2, \,|\, A_1 \in \mathcal{A}_1, \;\; A_2 \in \mathcal{A}_2\}$$

is an algebra of subsets of the set $X := X_1 \cup X_2$.

2.4 Let $\mathcal{A}$ be an algebra of subsets of a set X and X_0 a subset of X with $X_0 \in \mathcal{A}$. Prove that

$$X_0 \cap \mathcal{A} = \{X_0 \cap A \,|\, A \in \mathcal{A}\}.$$

is an algebra of subsets of X_0.

2.5 Let $\mathcal{A}$ be a family of subsets of a set X and X_0 a subset of X. Let $\mathcal{B}$ be the smallest σ-algebra containing $\mathcal{A}$. Prove that $X_0 \cap \mathcal{B}$ is the smallest σ-algebra of subsets of X_0 containing $X_0 \cap \mathcal{A}$.

2.6 For a family $\mathcal{C}$ of subsets of a set X, prove that

$$\sigma(\mathcal{C}) = \bigcap_{\mathcal{F}:\sigma\text{-algebra containing } \mathcal{C}} \mathcal{F}.$$

2.7 Let X be an uncountable set and

$$\mathcal{F} = \{E \subset X \mid E \text{ is countable or } E^{c} \text{ is countable}\}.$$

Prove that $\mathcal{F}$ is a σ-algebra, which is called the **σ-algebra of countable or co-countable sets**.

2.8 Let $\{X_\lambda\}_{\lambda\in\Lambda}$ be a family of nonempty sets and $\{\mathcal{F}_\lambda\}_{\lambda\in\Lambda}$ a family of σ-algebras $\mathcal{F}_\lambda$ of subsets of X_λ. Prove that if Λ is a countable set, then

$$\bigotimes_{\lambda\in\Lambda} \mathcal{F}_\lambda = \sigma\left(\left\{\prod_{\lambda\in\Lambda} F_\lambda \,\middle|\, F_\lambda \in \mathcal{F}_\lambda\right\}\right).$$

2.9 Let $\{X_\lambda\}_{\lambda\in\Lambda}$ be a family of nonempty sets and $\{\mathcal{C}_\lambda\}_{\lambda\in\Lambda}$ a family of sets of subsets of X_λ. Prove that

$$\bigotimes_{\lambda\in\Lambda} \sigma\left(\mathcal{C}_\lambda\right) = \sigma\left(\left\{\mathrm{Pr}_\lambda^{-1}(F_\lambda) \mid F_\lambda \in \mathcal{C}_\lambda,\ \ \lambda \in \Lambda\right\}\right)$$

2.10 Prove that

$$\mathcal{B}(\mathbb{R}^n) = \bigotimes_{i=1}^{n} \mathcal{B}\left(\mathbb{R}\right).$$

2.11 Let X be a set and $\mathcal{P}(X)$ the power set of X. Define a set function c on $\mathcal{P}(X)$ by

$$c(A) = \begin{cases} c(A) = \text{ the number of elements in } A, & A \text{ is finite,} \\ \infty, & A \text{ is infinite.} \end{cases}$$

Prove that c is a measure on $\mathcal{P}(X)$.

2.12 Let $(X, \mathcal{M})$ be a measure space and $x_0 \in X$. Define a set function δ_{x_0} on $\mathcal{M}$ by

$$\delta_{x_0}(A) = \begin{cases} 1, \ x_0 \in A, \\ 0, \ x_0 \notin A. \end{cases}$$

Prove that δ_{x_0} is a measure on $\mathcal{M}$.

2.13 Let X be an uncountable set and

$$\mathcal{C} = \{E \subset X \mid E \text{ is countable or } E^{c} \text{ is countable}\}.$$

Define a set function ν on $\mathcal{C}$ by

$$\nu(E) = \begin{cases} 0, \ E \text{ is countable,} \\ 1, \ E \text{ is the complement of a countable set.} \end{cases}$$

Prove that ν is a measure on $\mathcal{C}$.

2.14 Prove that every σ-finite maesure is semifinite.

2.15 Construct a measure space $(X, \mathcal{M}, \mu)$ for which μ is semifinite, but μ is not σ-finite.

2.16 Let $(X, \mathcal{M}, \mu)$ be a measure space. Put

$$\mathcal{M}_0 = \{E \cup A \,|\, E \in \mathcal{M}, \quad A \subset N \text{ for some } N \in \mathcal{M} \text{ with } \mu(N) = 0\}.$$

Prove that $\mathcal{M}_0$ is a σ-algebra.

2.17 For a set function $\mu_0 : \mathcal{M}_0 \to [0, \infty]$ on $\mathcal{M}_0$, given as in Exercise **2.16**, defined by

$$\mu_0(E \cup A) = \mu(E)$$

for all $E \cup A \in \mathcal{M}_0$ with $E \in \mathcal{M}$ and $A \subset N$ for some $N \in \mathcal{M}$ with $\mu(N) = 0$, prove that μ_0 is a measure.

2.18 Let $(X, \mathcal{M}, \mu)$ be a measure space and $\{E_n\}_{n=1}^{\infty} \subset \mathcal{M}$.

(1) Prove that if $\mu(E) < \infty$ or $\mu(F) < \infty$, then

$$\mu(E \cup F) = \mu(E) + \mu(F) - \mu(E \cap F).$$

(2) Prove that $\mu(\liminf_{n\to\infty} E_n) \leq \liminf_{n\to\infty} \mu(E_n)$.

(3) Prove that if $\mu(\cup_{n=1}^{\infty} E_n) < \infty$, then $\mu(\limsup_{n\to\infty} E_n) \geq \limsup_{n\to\infty} \mu(E_n)$.

2.19 Let μ, ν be measures on a measurable space $(X, \mathcal{M})$ and $\alpha > 0$.

(1) Prove that $\alpha\mu$ is a measure on the measurable space $(X, \mathcal{M})$.

(2) Prove that $\mu + \nu$ is a measure on the measurable space $(X, \mathcal{M})$.

2.20 Let $(X, \mathcal{M}, \mu)$ be a measure space.

(1) Prove that if $E, F \in \mathcal{M}$ and $\mu(E \triangle F) = 0$, then $\mu(E) = \mu(E \cap F) = \mu(F)$.

(2) Define a relation $\sim$ on $\mathcal{M}$ by

$$E \sim F \quad \Longleftrightarrow \quad \mu(E \triangle F) = 0, \qquad E, F \in \mathcal{M}.$$

Provc that $\sim$ ia an equivalence relation.

2.21 Let $(\mathbb{N}, \mathcal{P}(\mathbb{N}), c)$ be a measure space with the counting measure c. Let $\{a_n\}_{n=1}^{\infty}$ be an absolutely summable sequence of real numbers, i.e.,

$$\sum_{n=1}^{\infty} |a_n| < \infty.$$

Define a set function $\nu : \mathcal{P}(\mathbb{N}) \to (-\infty, \infty)$ by

$$\nu(E) = \sum_{\ell \in E} a_\ell, \qquad E \subset \mathbb{N}.$$

(1) Prove that ν is a signed measure.

(2) Find all Hahn decomposition of $\mathbb{N}$ for ν.

(3) Prove that ν is finite.

(4) For a subset E of $\mathbb{N}$ with $0 < \nu(E) < \infty$, find a positive (measurable) subset A of E such that $\nu(A) > 0$.

2.22 Let $(X, \mathcal{M}, \nu)$ be a signed measure space. Prove that

$$|\nu(E)| \leq |\nu|(E)$$

for all $E \in \mathcal{M}$.

2.23 Let $\mathfrak{S}$ be the set of all finite signed measures on a measurable space $(X, \mathcal{M})$. Prove that $\mathfrak{S}$ is a vector space over $\mathbb{R}$.

2.24 Let $\mathcal{A} \subset \mathcal{P}(X)$ be an algebra for a set X and μ_* be a premeasure defined on $\mathcal{A}$. Let $E \subset X$. Prove that E is μ^*-measurable if and only if for any $F \in \mathcal{A}$,

$$\mu^*(F) = \mu^*(F \cap E) + \mu^*(F \cap E^c),$$

where μ^* is the outer measure induced by μ_*.

2.25 Let $\mathcal{A}$ be an algebra of subsets of X and μ_* be a premeasure defined on $\mathcal{A}$. Let $A \subset X$. Prove that if A is not μ^*-measurable, then the restriction $\mu^*|_{\sigma(\mathfrak{M} \cup \{A\})}$ of μ^* to the σ-algebra $\sigma(\mathfrak{M} \cup \{A\})$ does not satisfy the countable additivity.

Chapter 3

Lebesgue Measurable Sets and Lebesgue Measure

In this chapter, as an application of the general construction of a measure studied in the previous chapter, we study the measure space $(\mathbb{R}, \mathfrak{M}_L, m)$, called the **Lebesgue measure space**, consisting of **Lebesgue σ-algebra** $\mathfrak{M}_L$ and the **Lebesgue measure** m. Here an element of $\mathfrak{M}_L$ is called a **Lebesgue measurable set**.

Then the Lebesgue measure space $(\mathbb{R}, \mathfrak{M}_L, m)$ satisfies the following important properties:

(m1) $\mathcal{B}(\mathbb{R}) \subset \mathfrak{M}_{\mathrm{L}}$,

(m2) $m(I) = \ell(I)$ any interval I, where $\ell(I)$ denotes the **length of the interval** I defined by

$$\ell(I) = \begin{cases} 0, & I = \emptyset, \\ \text{the difference of its endpoints}, & I \text{ is bounded }, \\ \infty, & I \text{ is unbounded.} \end{cases}$$

(m3) m is translation invariant, i.e., for any $E \in \mathfrak{M}_{\mathrm{L}}$ and $x \in \mathbb{R}$, $m(E + x) = m(E)$.

Our strategy to construct $\mathfrak{M}_{\mathrm{L}}$ and m is the following:

(1) we first construct a premeasure m_* defined on the algebra $\mathcal{A}_{\mathrm{oc}}$ containing all finite disjoint unions of open-closed intervals,

(2) we study the **Lebesgue outer measure** $m^* := (m_*)^*$ defined on the power set $\mathcal{P}(\mathbb{R})$ of $\mathbb{R}$, which is induced by the premeasure m_*,

(3) we construct the σ-algebra $\mathfrak{M}_{\mathrm{L}}$ of all (Lebesgue) measurable sets, i.e.,$\mathfrak{M}_{\mathrm{L}}$ is the σ-algebra of all m^*-measurable subsets of $\mathbb{R}$,

(4) finally m is constructed as the unique Carathéothory extension of m_* i.e., m is the restriction of m^* to $\mathfrak{M}_{\mathrm{L}}$. In this case, m is called the **Lebesgue measure**.

3.1 Lebesgue Outer Measure

The set (interval) of the form $(a,b]$ or (a,∞) or $\emptyset$ for $-\infty \le a < b < \infty$ is called an **open-closed interval**. Let $\mathcal{I}_{\rm oc}$ be the set of all open-closed intervals in $\mathbb{R}$. Let $\mathcal{A}_{\rm oc}$ be the set of all finite disjoint unions of open-closed intervals. Then $\mathcal{A}_{\rm oc}$ is an algebra (see Exercise **3.1**).

The length of intervals is a set function defined on $\mathcal{I}_{\rm oc}$ which is denoted by ℓ again. Then by using the set function ℓ, we can define an outer measure of subsets of $\mathbb{R}$ which is called the Lebesgue outer measure.

Define a set function $m_* : \mathcal{A}_{\rm oc} \to [0,\infty]$ by

$$m_*(E) = \sum_{i=1}^{k} \ell(I_i), \qquad E = \cup_{i=1}^{k} I_i \in \mathcal{A}_{\rm oc} \tag{3.1}$$

for a finite disjoint family $\{I_i\}_{i=1}^k$ of open-closed intervals. In fact, the premeasure defined as in (3.1) is well-defined, i.e., if $E \in \mathcal{A}_{\rm oc}$ and there exist finite disjoint families $\{I_i\}_{i=1}^k$ and $\{J_j\}_{j=1}^n$ of open-closed intervals such that

$$E = \bigcup_{i=1}^{k} I_i = \bigcup_{j=1}^{n} J_j,$$

then it holds that

$$\sum_{i=1}^{k} \ell(I_i) = \sum_{i=1}^{k}\sum_{j=1}^{n} \ell(I_i \cap J_j) = \sum_{j=1}^{n}\sum_{i=1}^{k} \ell(I_i \cap J_j) = \sum_{j=1}^{n} \ell(J_j).$$

Proposition 3.1.1 The set function m_* on $\mathcal{A}_{\rm oc}$ is a premeasure.

[Proof] It is obvious that $\ell(\emptyset) = 0$ by definition. Let $\{E_k\}_{k=1}^n \subset \mathcal{A}_{\rm oc}$ be a finite disjoint family. Then for each $1 \le k \le n$, there exists a finite disjoint family $\{I_{ki}\}_{i=1}^{n_k}$ of open-closed intervals such that

$$E_k = \bigcup_{i=1}^{n_k} I_{ki}, \qquad \bigcup_{k=1}^{n} E_k = \bigcup_{k=1}^{n}\bigcup_{i=1}^{n_k} I_{ki},$$

where all unions are disjoint unions. Therefore, we obtain that

$$m_*\left(\bigcup_{k=1}^{n} E_k\right) = \sum_{k=1}^{n}\sum_{i=1}^{n_k} \ell(I_{ki}) = \sum_{k=1}^{n} m_*(E_k),$$

and so m_* is finitely additive.

Let $E \in \mathcal{A}_{\rm oc}$ and $\{E_k\}_{k=1}^\infty \subset \mathcal{A}_{\rm oc}$ with $E \subset \cup_{k=1}^\infty E_k$. If $\sum_{k=1}^\infty m_*(E_k) = \infty$, then it is obvious that $m_*(E) \le \sum_{k=1}^\infty m_*(E_k)$. Suppose that $\sum_{k=1}^\infty m_*(E_k) < \infty$ and $E =$

$\cup_{i=1}^{n_0}(a_i, b_i]$ a disjoint union. Since there exist a sequence $\{n_k\}_{k=1}^{\infty} \subset \mathbb{N}$ and a family $\{I_i\}_{i=n_0+1}^{\infty}$ of open-closed intervals such that

$$E_k = \bigcup_{i=n_{k-1}+1}^{n_k} I_i \quad \text{(disjoint union)}, \quad k = 1, 2, \cdots,$$

where $I_i = (c_i, d_i]$ for some real numbers c_i and d_i,

$$E \subset \bigcup_{k=1}^{\infty} E_k = \bigcup_{i=n_0+1}^{\infty} I_i \quad \text{and} \quad \sum_{k=1}^{\infty} m_*(E_k) = \sum_{i=n_0+1}^{\infty} \ell(I_i),$$

which implies that E is bounded and so $\overline{E} = \cup_{i=1}^{n_0}[a_i, b_i]$ is closed and bounded. Let $\epsilon > 0$ be given. For each $i = 1, 2, \cdots, n_0$, put

$$\widetilde{I_i} = \left(a_i - \frac{\epsilon}{2n_0}, a_i + \frac{\epsilon}{2n_0}\right),$$

and for each $i = n_0 + 1, n_0 + 2, \cdots$, put

$$\widetilde{I_i} = \left(c_i, d_i + \frac{\epsilon}{2^i}\right).$$

Then $\{\widetilde{I_i}\}_{i=1}^{\infty}$ is an open cover of $\overline{E}$ and so there exists a finite family $\{I_i\}_{i=1}^{n}$ such that

$$E \subset \overline{E} \subset \bigcup_{i=1}^{n} \widetilde{I_i},$$

which implies that

$$\begin{aligned}
m_*(E) &\leq \sum_{i=1}^{n} \ell(\widetilde{I_i}) \leq \sum_{i=1}^{n_0} \ell(\widetilde{I_i}) + \sum_{i=n_0+1}^{n} \ell(\widetilde{I_i}) \\
&= \epsilon + \sum_{i=n_0+1}^{n} \ell(I_i) + \sum_{i=n_0+1}^{n} \frac{\epsilon}{2^i} \\
&\leq 2\epsilon + \sum_{i=n_0+1}^{n_1+\cdots+n_k} \ell(I_i)
\end{aligned}$$

for some $k \in \mathbb{N}$ with $n_1 + \cdots + n_k \geq n$. Therefore, we obtain that

$$\begin{aligned}
m_*(E) &\leq 2\epsilon + \sum_{i=n_0+1}^{n_1+\cdots+n_k} \ell(I_i) \leq 2\epsilon + \sum_{i=n_0+1}^{\infty} \ell(I_i) \\
&= 2\epsilon + \sum_{k=1}^{\infty} \sum_{i=n_{k-1}+1}^{n_k} \ell(I_i) = 2\epsilon + \sum_{k=1}^{\infty} m_*(E_k),
\end{aligned}$$

Since $\epsilon > 0$ is arbitrary,

$$m_*(E) \leq \sum_{k=1}^{\infty} m_*(E_k),$$

which implies that m_* is countably monotone. $\square$

Definition 3.1.2 Let $E \subset \mathbb{R}$. The **Lebesgue outer measure** $m^*(E)$ of E is defined as the outer measure induced by the premeasure m_* on $\mathcal{A}_{oc}$, i.e.,

$$m^*(E) = \inf \left\{ \sum_{k=1}^{\infty} m_*(E_k) \,\middle|\, \{E_k\}_{k=1}^{\infty} \subset \mathcal{A}_{oc} \quad \text{and} \quad E \subset \bigcup_{k=1}^{\infty} E_k \right\}. \tag{3.2}$$

Proposition 3.1.3 Let $E \subset \mathbb{R}$. Then it holds that

$$m^*(E) = \inf \left\{ \sum_{k=1}^{\infty} \ell(I_k) \,\middle|\, \{I_k\}_{k=1}^{\infty} \subset \mathcal{I}_{oc} \quad \text{and} \quad E \subset \bigcup_{k=1}^{\infty} I_k \right\}. \tag{3.3}$$

[Proof] Let $E \subset \mathbb{R}$ be given. Put

$$\lambda(E) = \inf \left\{ \sum_{k=1}^{\infty} \ell(I_k) \,\middle|\, \{I_k\}_{k=1}^{\infty} \subset \mathcal{I}_{oc} \quad \text{and} \quad E \subset \bigcup_{k=1}^{\infty} I_k \right\}.$$

Then by definition, it is obvious that $m^*(E) \leq \lambda(E)$. Let $\{E_k\}_{k=1}^{\infty} \subset \mathcal{A}_{oc}$ such that $E \subset \cup_{k=1}^{\infty} E_k$. Then there exist a sequence $\{n_k\}_{k=1}^{\infty} \subset \mathbb{N}$ and a countable family $\{I_i\}_{i=1}^{\infty}$ of open-closed intervals such that for each $k = 1, 2, \cdots$, $\{I_i\}_{i=n_{k-1}+1}^{n_k}$ is a disjoint family and

$$E_k = \bigcup_{i=n_{k-1}+1}^{n_k} I_i,$$

where $n_0 = 0$. Therefore,

$$E \subset \bigcup_{k=1}^{\infty} E_k = \bigcup_{i=1}^{\infty} I_i,$$

and

$$\lambda(E) \leq \sum_{i=1}^{\infty} \ell(I_i) = \sum_{k=1}^{\infty} m_*(E_k),$$

which implies that

$$\lambda(E) \leq m^*(E).$$

Hence $m^*(E) = \lambda(E)$. □

Remark 3.1.4 (1) For any $\epsilon > 0$ and $a \in \mathbb{R}$, $\emptyset \subset (a, a+\epsilon]$ and so

$$m^*(\emptyset) \leq \ell((a, a+\epsilon]) = \epsilon.$$

Since $\epsilon > 0$ is arbitrary, $m^*(\emptyset) = 0$.

(2) The outer measure is **monotone** in the sense that

$$\text{if } A \subset B, \text{ then } m^*(A) \le m^*(B),$$

since every cover $\{I_k\}_{k=1}^\infty$ of B is also a cover of A.

(3) Every countable set C has outer measure zero. In fact, if $C = \{x_i \,|\, i = 1, 2, \cdots\}$, then for any $\epsilon > 0$, we may take a countable collection $\{I_k = \left(x_k - \frac{\epsilon}{2^{k+1}}, x_k + \frac{\epsilon}{2^{k+1}}\right]\}$ and then

$$C \subset \bigcup_{k=1}^\infty I_k, \qquad 0 \le m^*(C) \le \sum_{k=1}^\infty \ell(I_k) = \sum_{k=1}^\infty \frac{\epsilon}{2^k} = \epsilon,$$

which implies that $m^*(C) = 0$.

Proposition 3.1.5 The outer measure of any interval is its length.

[Proof] (i) Suppose $I = [a, b]$, i.e., I is closed and bounded. Let $\epsilon > 0$ be given. Then there exists a countable collection $\{I_k\}_{k=1}^\infty$ of open-closed intervals such that $I \subset \cup_{k=1}^\infty I_k$ and

$$m^*([a, b]) + \frac{\epsilon}{2} \ge \sum_{k=1}^\infty \ell(I_k).$$

For the countable collection $\{I_k\}_{k=1}^\infty$, there exists a countable collection $\{J_k\}_{k=1}^\infty$ of open intervals J_k such that $I_k \subset J_k$ and $\ell(I_k) + \epsilon/2^{k+1} \ge \ell(J_k)$. Then we have $I = [a, b] \subset \cup_{k=1}^\infty J_k$ and

$$\sum_{k=1}^\infty \ell(I_k) + \frac{\epsilon}{2} \ge \sum_{k=1}^\infty \ell(J_k).$$

Since I is compact (by Heine-Borel theorem), there exists a finite subcollection $\{J_{k_i}\}_{i=1}^n$ covering I. Suppose that there exists no $1 \le i \ne j \le n$ such that $J_{k_i} \subset J_{k_j}$ or $J_{k_j} \subset J_{k_i}$, and

$$J_{k_i} = (a_i, b_i), \qquad a_1 < a_2 < \cdots < a_n, \qquad b_{n-1} < b,$$

which means that $a_1 < a$, $a_{i+1} < b_i$ for $i = 1, 2, \cdots, n-1$ and $b_n > b$. Therefore, we have

$$\begin{aligned} m^*([a, b]) + \epsilon &\ge \sum_{k=1}^\infty \ell(I_k) + \frac{\epsilon}{2} \ge \sum_{k=1}^\infty \ell(J_k) \ge \sum_{i=1}^n \ell(J_{k_i}) = \sum_{i=1}^n (b_i - a_i) \\ &\ge b - a = \ell(I), \end{aligned}$$

which implies that $m^*(I) \ge \ell(I)$. For any given $\epsilon > 0$, $[a, b] \subset (a - \epsilon, b]$ and so $m^*([a, b]) \le b - a + \epsilon = \ell([a, b]) + \epsilon$. Since $\epsilon > 0$ is arbitrary, $m^*(I) \le b - a = \ell(I)$. Hence $m^*(I) = \ell(I)$.

(ii) I is an arbitrary bounded interval. Put $a = \inf\{x \,|\, x \in I\}$ and $b = \sup\{x \,|\, x \in I\}$. Then $I \subset [a, b]$ implies $m^*(I) \le \ell([a, b]) = \ell(I)$ and for any $\epsilon > 0$, by the monotonicity of the outer measure,

$$b - a - \epsilon = m^*([a + \epsilon/2, b - \epsilon/2]) \le m^*(I),$$

which implies that $\ell(I) = b - a \le m^*(I)$. Therefore, $m^*(I) = \ell(I)$.

(iii) Suppose that I is an unbounded interval. Let $x_0 \in I$. Then for any $M > 0$,

$$[-M, x_0] \subset I \text{ or } [x_0, M] \subset I,$$

which means that, by (i), $m^*(I) \ge x_0 + M$ or $m^*(I) \ge M - x_0$ and so $m^*(I) = \infty$. Therefore, $m^*(I) = \ell(I)$. □

Example 3.1.6 For $a < b$, suppose that (a, b) is countable and the by Proposition 3.1.5 and (3) in Remark 3.1.4, $b - a = m^*((a, b)) = 0$, which is in contradiction to $a < b$. This example gives another proof of Theorem 1.3.8.

Remark 3.1.7 We now consider another proof of Proposition 3.1.5. In Section 3.3, the Carathéodory measure is denoted by m and called the Lebesgue measure, i.e., for the σ-algebra $\mathfrak{M}_{\mathrm{L}}$ of Lebesgue measurable (m^*-measurable) sets,

$$m = m^*|_{\mathfrak{M}_{\mathrm{L}}}.$$

Since m_* is σ-finite, by Theorem 2.5.27 the Lebesgue measure m is the unique extension of m_* to $\mathfrak{M}_{\mathrm{L}}$, i.e.,

$$m|_{\mathcal{A}_{\mathrm{oc}}} = m_*.$$

Therefore, for any open-closed interval I, $m^*(I) = \ell(I)$. For any arbitrary bounded interval I. Put $a = \inf\{x \,|\, x \in I\}$ and $b = \sup\{x \,|\, x \in I\}$. Then for any $\epsilon > 0$, $I \subset (a - \epsilon, b]$ implies $m^*(I) \le \ell(a - \epsilon, b]) = \ell(I) + \epsilon$ and by the monotonicity of the outer measure,

$$\ell(I) - \epsilon = b - a - \epsilon = m^*((a + \epsilon/2, b - \epsilon/2]) \le m^*(I),$$

which implies that $m^*(I) = \ell(I)$. Suppose that I is an unbounded interval. Let $x_0 \in I$. Then for any $M > 0$,

$$(-M, x_0] \subset I \text{ or } (x_0, M] \subset I,$$

which means that $m^*(I) \ge x_0 + M$ or $m^*(I) \ge M - x_0$ and so $m^*(I) = \infty$. Therefore, $m^*(I) = \ell(I)$.

Proposition 3.1.8 The outer measure is **translation invariant**.

[Proof] Let $E \subset \mathbb{R}$ and $x \in \mathbb{R}$. Note that a countable collection $\{I_k\}_{k=1}^{\infty}$ of open-closed intervals, $\{I_k\}_{k=1}^{\infty}$ covers E if and only if $\{I_k + x\}_{k=1}^{\infty}$ covers $E + x$. Also, for any interval I, $\ell(I) = \ell(I + x)$. Therefore, the proof is immediate. □

Proposition 3.1.9 [**Countable Subadditivity**] The outer measure satisfies the countable subadditivity, i.e., for any countable collection $\{E_k\}_{k=1}^{\infty}$ of subsets of $\mathbb{R}$,

$$m^*\left(\bigcup_{k=1}^{\infty} E_k\right) \le \sum_{k=1}^{\infty} m^*(E_k). \tag{3.4}$$

[Proof] Exercise **3.3**. □

3.2 Lebesgue Measurable Sets

The outer measure does not satisfy the countable additivity. In fact, the outer measure does not satisfy even the finite additivity, i.e., there exist disjoint subsets A and B of $\mathbb{R}$ such that

$$m^*(A \cup B) < m^*(A) + m^*(B),$$

see Theorem 3.4.4, of which the reason is that the domain $\mathcal{P}(\mathbb{R})$ of the Lebesgue outer measure m^* is too large. Therefore, we need to consider a restricted class, called the Lebesgue measurable sets, of subsets of $\mathbb{R}$ satisfying the (Lebesgue) measurability condition.

Definition 3.2.1 A set $E \subset \mathbb{R}$ is said to be **Lebesgue measurable** or simply **measurable** if for any set $A \subset \mathbb{R}$,

$$m^*(A) = m^*(A \cap E) + m^*\left(A \cap E^{\mathrm{C}}\right). \tag{3.5}$$

Remark 3.2.2 (1) For any set $E, A \subset \mathbb{R}$, by the countable subadditivity of the outer measure, we have

$$m^*(A) \leq m^*(A \cap E) + m^*\left(A \cap E^{\mathrm{C}}\right).$$

Therefore, for the measurability criterion (3.5), we need to check only that

$$m^*(A) \geq m^*(A \cap E) + m^*\left(A \cap E^{\mathrm{C}}\right), \tag{3.6}$$

i.e., E is measurable if and only if for any set $A \subset \mathbb{R}$, (3.6) holds. If $m^*(A) = \infty$, then (3.6) holds obviously. Therefore, E is measurable if and only if for any set $A \subset \mathbb{R}$ with $m^*(A) < \infty$, (3.6) holds.

(2) It is obvious that E is measurable if and only if E^{c} is measurable.

(3) It is clear that $\emptyset$ and $\mathbb{R}$ are measurable.

Let $\mathfrak{M}_{\mathrm{L}}$ be the class of all Lebesgue measurable sets.

Proposition 3.2.3 Every set of Lebesgue outer measure zero is Lebesgue measurable. In particular, every countable set belongs to $\mathfrak{M}_{\mathrm{L}}$.

[Proof] Exercise **3.4**. □

Lemma 3.2.4 The union of two Lebesgue measurable sets is measurable.

[Proof] Exercise **3.5**. □

Corollary 3.2.5 $\mathfrak{M}_{\mathrm{L}}$ is closed under finite unions.

[Proof] The proof is immediate from Lemma 3.2.4. □

Corollary 3.2.6 $\mathfrak{M}_{\mathrm{L}}$ is closed under finite intersections.

[Proof] The proof is immediate from Corollary 3.2.5 and (2) in Remark 3.2.2. □

Theorem 3.2.7 Every interval is measurable. Moreover, every Borel set belongs to $\mathfrak{M}_{\mathrm{L}}$, i.e., $\mathcal{B}(\mathbb{R}) \subset \mathfrak{M}_{\mathrm{L}}$.

[Proof] By Proposition 2.5.24, it holds that $\sigma(\mathcal{A}_{\mathrm{oc}}) \subset \mathfrak{M}_{\mathrm{L}}$. For the class $\mathcal{I}_2 = \mathcal{I}_{\mathrm{oc}}$ (see (2.2)) of all open-closed intervals,

$$\mathcal{B}(\mathbb{R}) = \sigma(\mathcal{I}_2) = \sigma(\mathcal{A}_{\mathrm{oc}}) \subset \mathfrak{M}_{\mathrm{L}}.$$

Therefore, $\mathcal{B}(\mathbb{R}) = \sigma(\mathcal{I}_2) \subset \mathfrak{M}_{\mathrm{L}}$. □

Proposition 3.2.8 For any subset $A \subset \mathbb{R}$ and any finite disjoint family $\{E_k\}_{k=1}^{n}$ of measurable sets E_k, we have

$$m^*\left(A \cap \bigcup_{k=1}^{n} E_k\right) = \sum_{k=1}^{n} m^*\left(A \cap E_k\right). \tag{3.7}$$

In particular, we have

$$m^*\left(\bigcup_{k=1}^{n} E_k\right) = \sum_{k=1}^{n} m^*\left(E_k\right). \tag{3.8}$$

[Proof] Exercise **3.6**. □

Proposition 3.2.9 $\mathfrak{M}_{\mathrm{L}}$ is closed under countable unions.

[Proof] Exercise **3.7**. □

Theorem 3.2.10 $\mathfrak{M}_{\mathrm{L}}$ is a σ-algebra.

[Proof] The proof is immediate from Corollary 3.2.9 and (2)-(3) in Remark 3.2.2. □

Definition 3.2.11 The pair $(\mathbb{R}, \mathfrak{M}_{\mathrm{L}})$ is called the **Lebesgue measurable space**.

Theorem 3.2.12 $\mathfrak{M}_{\mathrm{L}}$ is translation invariant, i.e., for any $E \in \mathfrak{M}_{\mathrm{L}}$ and $x \in \mathbb{R}$, $E + x \in \mathfrak{M}_{\mathrm{L}}$.

[Proof] Let $E \in \mathfrak{M}_{\mathrm{L}}$ and $x \in \mathbb{R}$. Since m^* is translation invariant, for any $A \subset \mathbb{R}$ we obtain that

$$\begin{aligned} m^*(A) &= m^*(A - x) = m^*((A - x) \cap E) + m^*((A - x) \cap E^{\mathrm{c}}) \\ &= m^*(A \cap (E + x)) + m^*(A \cap (E + x)^{\mathrm{c}}), \end{aligned}$$

which implies that $E + x$ is measurable. □

Remark 3.2.13 By modifying the solution of Exercise **3.2**, we see that the Lebesgue outer measure $m^*(E)$ of E can be defined as in (3.3) by taking the infimum over all countable collections $\{I_k\}_{k=1}^{\infty}$ of any type (open or closed or closed-open or open-closed) of intervals I_k. However, by Exercise **3.1**, $\mathcal{A}_{\mathrm{oc}}$ is an algebra . Therefore, by Carathéodory-Hahn Theorem (Theorem 2.5.26) and Theorem 2.5.27, the Carathéodory measure $\mu : \mathfrak{M}_{\mathrm{L}} \to [0, \infty]$ induced by m_* is the unique extension of m_*, which is called the Lebesgue measure (see Section 3.3).

Proposition 3.2.14 [**Excision Property**] Let E be a measurable set with finite outer measure such that E is contained in a set A. Then we have

$$m^*(A \setminus E) = m^*(A) - m^*(E). \tag{3.9}$$

[Proof] Exercise **3.8**. □

Remark 3.2.15 In Exercise **3.2**, we see that the outer measures m^* and m^{**} coincide. Therefore, for the definition of the Lebesgue outer measure m^*, we sometimes use the definition of m^{**}.

Definition 3.2.16 A countable intersection of open sets is called a G_δ **set**. A countable union of closed sets is called a F_σ **set**.

Remark 3.2.17 (1) Every G_δ set is measurable.
(2) Every F_σ set is measurable.

Theorem 3.2.18 [**Regularity of Outer Measure**] Let E be a subset of $\mathbb{R}$. Then the following are equivalent:

(m) E is measurable,

(i) For each $\epsilon > 0$, there exists an open set O containing E such that $m^*(O \setminus E) < \epsilon$,

(ii) There exists a G_δ set G containing E such that $m^*(G \setminus E) = 0$,

(iii) For each $\epsilon > 0$, there exists a closed set F contained in E such that $m^*(E \setminus F) < \epsilon$,

(iv) There exists a F_σ set F contained E such that $m^*(E \setminus F) = 0$.

[Proof] (m) $\Rightarrow$ (i)) (1) Suppose that $m^*(E) < \infty$. Then by the definition of the Lebesgue outer measure, for each $\epsilon > 0$, there exists a countable family $\{I_k = (a_k, b_k]\}_{k=1}^{\infty}$ of open-closed intervals I_k such that

$$E \subset \bigcup_{k=1}^{\infty} I_k \quad \text{and} \quad \sum_{k=1}^{\infty} \ell(I_k) \leq m^*(E) + \frac{\epsilon}{2}.$$

For each $k \in \mathbb{N}$, put

$$J_k = \left(a_k, b_k + \frac{\epsilon}{2^{k+1}}\right).$$

Then $\{J_k\}_{k=1}^{\infty}$ is a countable family of open intervals such that

$$E \subset \bigcup_{k=1}^{\infty} J_k \quad \text{and} \quad \sum_{k=1}^{\infty} \ell(J_k) \leq m^*(E) + \epsilon.$$

Put

$$O = \bigcup_{k=1}^{\infty} J_k.$$

Then O is an open set and containing E. Also, by the monotonicity, we have

$$m^*(O) \leq \sum_{k=1}^{\infty} \ell(J_k) \leq m^*(E) + \epsilon,$$

which, by the excision property since E is measurable and $m^*(E) < \infty$, we have

$$m^*(O \setminus E) = m^*(O) - m^*(E) < \epsilon.$$

(2) For an arbitrary measurable set E, there exists a countable family $\{E_k\}_{k=1}^{\infty}$ of measurable sets E_k with $m^*(E_k) < \infty$ such that $E = \cup_{k=1}^{\infty} E_k$. Then for each $k \in \mathbb{N}$, by (1), there exists an open set O_k containing E_k such that

$$m^*(O_k \setminus E_k) < \frac{\epsilon}{2^k}.$$

Put

$$O = \bigcup_{k=1}^{\infty} O_k.$$

Then O is an open set and $E \subset O$. Moreover, we have

$$m^*(O \setminus E) \leq m^*\left(\bigcup_{k=1}^{\infty} (O_k \setminus E_k)\right) \leq \sum_{k=1}^{\infty} m^*(O_k \setminus E_k) < \sum_{k=1}^{\infty} \frac{\epsilon}{2^k} = \epsilon,$$

which gives the proof.

(i) $\Rightarrow$ (ii)) For any $n \in \mathbb{N}$, there exists an open set O_n containing E such that $m^*(O_n \setminus E) < \frac{1}{n}$. Put

$$G = \bigcap_{n=1}^{\infty} O_n.$$

Then G is a G_δ set containing E and, by the monotonicity of the Lebesgue outer measure m^*, we have

$$m^*(G \setminus E) \leq m^*(O_n \setminus E) < \frac{1}{n}$$

for any $n \in \mathbb{N}$. Therefore, $m^*(G \setminus E) = 0$.

(ii) $\Rightarrow$ (m)) By assumption, there exists a G_δ set G containing E such that $m^*(G \setminus E) = 0$. Since every Lebesgue outer measure zero set is measurable, $G \setminus E$ is measurable. Also, every G_δ set is measurable. Therefore,

$$E = G \cap (G \setminus E)^c$$

is measurable.

(m) $\Rightarrow$ (iii)) Since (m) $\Leftrightarrow$ (i) $\Leftrightarrow$ (ii) and E^c is measurable, for each $\epsilon > 0$, there exists an open set O containing E^c such that $m^*(O \setminus E^c) < \epsilon$. Then $F = O^c$ is a closed set contained in E since $E^c \subset O$ if and only if $F = O^c \subset E$ and

$$O \setminus E^c = O \cap E = E \cap (O^c)^c = E \setminus F$$

and so $m^*(E \setminus F) = m^*(O \setminus E^c) < \epsilon$.

(iii) $\Rightarrow$ (iv)) For each $n \in \mathbb{N}$, there exists a closed set F_n contained in E such that $m^*(E \setminus F_n) < \frac{1}{n}$. Put $F = \cup_{n=1}^{\infty} F_n$. Then F is a F_σ set contained in E and

$$m^*(E \setminus F) \leq m^*(E \setminus F_n) < \frac{1}{n}$$

for any $n \in \mathbb{N}$, which implies that $m^*(E \setminus F) = 0$.

(iv) $\Rightarrow$ (m) By assumption, there exists a F_σ set F contained in E such that $m^*(E \setminus F) = 0$. Since every Lebesgue outer measure zero set is measurable, $E \setminus F$ is measurable. Also, every F_σ set is measurable. Therefore,

$$E = F \bigcup (E \setminus F)$$

is measurable. □

Theorem 3.2.19 **[Littlewood's First Principle]** Let E be a measurable set with finite Lebesgue (outer) measure. Then for each $\epsilon > 0$, there exists a finite disjoint family $\{I_k\}_{k=1}^{n_0}$ of open intervals such that for $O = \cup_{k=1}^{n_0} I_k$,

$$m^*(E \bigtriangleup O) < \epsilon,$$

where $E \bigtriangleup O = (E \setminus O) \cup (O \setminus E)$.

[Proof] For each $\epsilon > 0$, by (i) in Theorem 3.2.18, there exists an open set U containing E such that $m^*(U \setminus E) < \frac{\epsilon}{2}$. Then since $m^*(E) < \infty$, by the excision property (see Proposition 3.2.14), we have

$$m^*(U \setminus E) = m^*(U) - m^*(E) < \frac{\epsilon}{2} \quad \Longrightarrow \quad m^*(U) < m^*(E) + \frac{\epsilon}{2} < \infty.$$

Since every open set is a countable union of disjoint open intervals (see Theorem 1.4.6), there exists a countable disjoint family $\{I_k\}_{k=1}^{\infty}$ of open intervals such that $U = \cup_{k=1}^{\infty} I_k$.

Then for any $n \in \mathbb{N}$, by (3.8) and the monotonicity of the Lebesgue outer measure, we have

$$\sum_{k=1}^{n} \ell(I_k) = m^*\left(\bigcup_{k=1}^{n} I_k\right) \le m^*(U) < \infty,$$

which implies that

$$\sum_{k=1}^{\infty} \ell(I_k) < \infty,$$

and there exists $n_0 \in \mathbb{N}$ such that $\sum_{k=n_0+1}^{\infty} \ell(I_k) < \frac{\epsilon}{2}$. Therefore, by taking the finite disjoint family $\{I_k\}_{k=1}^{n_0}$ and $O = \cup_{k=1}^{n_0} I_k$, we obtain that

$$m^*(O \setminus E) \le m^*(U \setminus E) < \frac{\epsilon}{2},$$
$$m^*(E \setminus O) \le m^*(U \setminus O) = m^*\left(\bigcup_{k=n_0+1}^{\infty} I_k\right) \le \sum_{k=n_0+1}^{\infty} \ell(I_k) < \frac{\epsilon}{2},$$

which implies that

$$m^*(E \bigtriangleup O) = m^*((E \setminus O) \cup (O \setminus E)) \le m^*(E \setminus O) + m^*(O \setminus E) < \frac{\epsilon}{2} + \frac{\epsilon}{2} = \epsilon,$$

which gives the proof. □

Remark 3.2.20 **[Littlewood's Three Principles]** The Littlewood's three principles (see Section 4.2) for the Lebesgue measure on $\mathbb{R}$ are roughly expressed as the following:

(1) Every Lebesgue measurable set of finite Lebesgue measure is nearly a finite union of intervals,

(2) **[Lusin's Theorem]** Every Lebesgue measurable function is nearly continuous (see, Theorem 4.2.5),

(3) **[Egorov's Theorem]** Every convergent sequence of (Lebesgue measurable) functions is nearly uniformly convergent (see Theorem 4.2.3).

Remark 3.2.21 Let E be a measurable set of finite outer measure. Then by the regularity of the Lebesgue outer measure (Theorem 3.2.18), There exist G_δ-set G and F_σ-set F such that

$$F \subset E \subset G$$

and

$$m^*(G \setminus E) = m^*(E \setminus F) = 0.$$

Since F is also measurable with finite outer measure, by the excision property, we have

$$0 = m^*(G \setminus E) = m^*(G) - m^*(E), \qquad 0 = m^*(E \setminus F) = m^*(E) - m^*(F),$$

which implies that

$$m^*(F) = m^*(E) = m^*(G),$$

see Remark 3.4.5.

Remark 3.2.22 [Lebesgue] Let $E \subset \mathbb{R}$ be of finite Lebesgue outer measure. Suppose that for any bounded open interval $I = (a, b)$, the following equality holds:

$$m^*(I) = m^*(I \cap E) + m^*(I \cap E^c).$$

Since $E \subset \mathbb{R}$ is of finite Lebesgue outer measure, for each given $\epsilon > 0$, there exists a countable family $\{I_n\}_{n=1}^{\infty}$ of open intervals such that

$$O := \bigcup_{n=1}^{\infty} I_n \supset E \quad \text{and} \quad \sum_{n=1}^{\infty} \ell(I_n) < m^*(E) + \epsilon.$$

Then we obtain that

$$\begin{aligned} m^*(O \setminus E) + m^*(E) = m^*(O \cap E^c) + m^*(O \cap E) &\leq \sum_{n=1}^{\infty} m^*(I_n \cap E^c) + \sum_{n=1}^{\infty} m^*(I_n \cap E) \\ &= \sum_{n=1}^{\infty} (m^*(I_n \cap E^c) + m^*(I_n \cap E)) = \sum_{n=1}^{\infty} \ell(I_n) < m^*(E) + \epsilon, \end{aligned}$$

which implies that $m^*(O \setminus E) < \epsilon$ and so by the regularity of outer measure (Theorem 3.2.18), E is measurable.

Remark 3.2.23 Let $\mathcal{A}$ be a family of subsets of $\mathbb{R}$. Put

$$\mathfrak{M}_{\mathrm{L}}(\mathcal{A}) = \{E \subset \mathbb{R} \,|\, m^*(A) = m^*(A \cap E) + m^*(A \cap E^c) \text{ for all } A \in \mathcal{A}\}.$$

Then we can prove that $\mathfrak{M}_{\mathrm{L}}(\mathcal{A})$ is a σ-algebra (see Exercise **3.9**). Therefore, for the descending sequence

$$\mathcal{P}(\mathbb{R}) = \mathcal{A}_0 \supset \mathcal{A}_1 \supset \mathcal{A}_2 \cdots \supset \mathcal{A}_\infty = \{\mathbb{R}, \emptyset\}$$

of families of subsets of $\mathbb{R}$, we construct ascending sequence of σ-algebras:

$$\mathfrak{M}_{\mathrm{L}} = \mathfrak{M}_{\mathrm{L}}(\mathcal{A}_0) \subset \mathfrak{M}_{\mathrm{L}}(\mathcal{A}_1) \subset \mathfrak{M}_{\mathrm{L}}(\mathcal{A}_2) \subset \cdots \subset \mathfrak{M}_{\mathrm{L}}(\{\mathbb{R}, \emptyset\}) = \mathcal{P}(\mathbb{R}).$$

3.3 Lebesgue Measure

Definition 3.3.1 The restriction of the Lebesgue outer measure m^* to the σ-algebra $\mathfrak{M}_{\mathrm{L}}$ is called the **Lebesgue measure** and denoted by m, i.e. $m = m^* |_{\mathfrak{M}_{\mathrm{L}}}$, and the triple $(\mathbb{R}, \mathfrak{M}_{\mathrm{L}}, m)$ is called the **Lebesgue measure space**.

Remark 3.3.2 (1) By Proposition 3.1.5, the Lebesgue measure of any interval is its length.

(2) The Lebesgue measure is translation invariant. For any measurable set $E \in \mathfrak{M}_{\mathrm{L}}$ and $x \in \mathbb{R}$, since $\mathfrak{M}_{\mathrm{L}}$ is translation invariant (see Theorem 3.2.12), $m(E+x)$ is well-defined. Therefore, since the Lebesgue outer measure is translation invariant (see Proposition 3.1.8), we have

$$m(E+x) = m^*(E+x) = m^*(E) = m(E).$$

(3) The measure space $(\mathbb{R}, \mathfrak{M}_{\mathrm{L}}, m)$ with the σ-algebra $\mathfrak{M}_{\mathrm{L}}$ of all Lebesgue measurable subsets of $\mathbb{R}$ and the Lebesgue measure m on $\mathfrak{M}_{\mathrm{L}}$ is σ-finite.

(4) The measure space $(\mathbb{R}, \mathcal{B}(\mathbb{R}), m)$ with the Borel σ-algebra $\mathcal{B}(\mathbb{R})$ and the Lebesgue measure m on $\mathcal{B}(\mathbb{R})$ is σ-finite. The measure space $([a,b], \mathcal{B}([a,b]), m)$ for a finite interval $[a,b]$ is finite

Theorem 3.3.3 **[Countable Additivity]** For any countable disjoint family $\{E_k\}_{k=1}^{\infty}$ of measurable sets E_k, we have

$$m\left(\bigcup_{k=1}^{\infty} E_k\right) = \sum_{k=1}^{\infty} m(E_k). \tag{3.10}$$

[Proof] Exercise **3.10**. □

The following theorem gives a continuity of the Lebesgue measure m.

Theorem 3.3.4 The Lebesgue measure m satisfies the following properties:

(1) For any ascending family $\{E_k\}_{k=1}^{\infty}$ of measurable sets, we have

$$m\left(\bigcup_{k=1}^{\infty} E_k\right) = \lim_{n\to\infty} m(E_n). \tag{3.11}$$

(2) For any descending family $\{E_k\}_{k=1}^{\infty}$ of measurable sets with $m(E_1) < \infty$, we have

$$m\left(\bigcap_{k=1}^{\infty} E_k\right) = \lim_{n\to\infty} m(E_n). \tag{3.12}$$

[Proof] Exercise **3.11**. □

Remark 3.3.5 In (2) of Theorem 3.3.4, the condition $m(E_1) < \infty$ is necessary. If we consider the decreasing family $\{E_k = [k, \infty)\}_{k=1}^{\infty}$, then

$$\bigcap_{k=1}^{\infty} E_k = \emptyset, \qquad m(E_k) = \infty, \qquad k = 1, 2, \cdots.$$

Therefore, we have

$$m\left(\bigcap_{k=1}^{\infty} E_k\right) = m(\emptyset) = 0 \neq \infty = \lim_{n\to\infty} m(E_k).$$

However, the condition $m(E_1) < \infty$ can be replaced by the condition that there exists $n_0 \in \mathbb{N}$ such that $m(E_{n_0}) < \infty$.

Theorem 3.3.6 [**Borel-Cantelli Lemma**] Let $\{E_k\}_{k=1}^{\infty}$ be a countable family of measurable sets such that $\sum_{k=1}^{\infty} m(E_k) < \infty$. Then almost all $x \in \mathbb{R}$ belong to at most finitely many E_k's.

[Proof] Exercise **3.12**. □

3.4 Nonmeasurable Sets

In this section, we consider the following natural question:

is there any nonmeasurable set?

The answer of the question is yes (see Vitali Theorem (Theorem 3.4.3)), i.e., there exists a subset A of $\mathbb{R}$ such that E is not Lebesgue measurable.

Lemma 3.4.1 Let $E \subset \mathbb{R}$ be a bounded measurable set and $\Lambda \subset \mathbb{R}$ a bounded countably infinite set such that $\{E+\lambda\}_{\lambda\in\Lambda}$ is a countably infinite disjoint family of measurable sets. Then $m(E) = 0$.

[Proof] By the countable additivity of m (see Theorem 3.3.3), we have

$$m\left(\bigcup_{\lambda\in\Lambda}(E+\lambda)\right) = \sum_{\lambda\in\Lambda} m(E+\lambda) = \sum_{\lambda\in\Lambda} m(E).$$

On the other hand, since E and Λ are bounded, $\cup_{\lambda\in\Lambda}(E+\lambda)$ is a bounded set and so

$$m\left(\cup_{\lambda\in\Lambda}(E+\lambda)\right) < \infty,$$

which implies that $\sum_{\lambda\in\Lambda} m(E) < \infty$. Therefore, $m(E) = 0$. □

Let E be a nonempty set of real numbers. Define a relation $\sim$ by

$$x \sim y \quad \text{for } x, y \in E \quad \Longleftrightarrow \quad x - y \in \mathbb{Q}.$$

Then $\sim$ becomes an equivalence relation: (i) (reflexive) for any $x \in E$, $x - x = 0 \in \mathbb{Q}$ if and only if $x \sim x$, (ii) (symmetric) if $x \sim y$, then $x - y \in \mathbb{Q}$ if and only if $y - x = -(x-y) \in \mathbb{Q}$, which implies that $y \sim x$, (iii) (transitive) if $x \sim y$ and $y \sim z$, then $x - z = (x-y) + (y-z) \in \mathbb{Q}$, which implies that $x \sim z$. If $x \sim y$, then we say that x and y are ***rationally equivalent***. Then the set E is decomposed into the disjoint family of equivalence classes. The ***choice set*** C_E is consisting of exactly one element of each equivalence class.

Remark 3.4.2 (1) For two different points $c_x, c_y \in C_E$, $c_x - c_y \notin \mathbb{Q}$.

(2) For each point $x \in E$, there exists a unique $c_x \in C_E$ such that $x - c_x = q \in \mathbb{Q}$.

(3) For any subset $\Lambda \subset \mathbb{Q}$, $\{C_E + \lambda\}_{\lambda \in \Lambda}$ is a disjoint family. In fact, if

$$(C_E + \lambda_1) \cap (C_E + \lambda_2) \neq \emptyset \quad \text{for } \lambda_1 \neq \lambda_2,$$

then there exists two different points $c_x, c_y \in C_E$ such that $c_x + \lambda_1 = c_y + \lambda_2$, which implies that $c_x - c_y = \lambda_2 - \lambda_1 \in \mathbb{Q}$ which is in contradiction.

Theorem 3.4.3 [**Vitali**] For each subset E of $\mathbb{R}$ with positive Lebesgue outer measure contains a nonmeasurable set.

[Proof] We may assume that E is bounded. Then we claim that C_E is not measurable. The claim is verified as following. Since E is bounded, there exists $M \geq 0$ such that $E \subset [-M, M]$. Put

$$\Lambda = [-2M, 2M] \cap \mathbb{Q},$$

and then we have

$$E \subset \bigcup_{\lambda \in \Lambda} (C_E + \lambda),$$

in fact, for any $x \in E$, by (2) in Remark 3.4.2, there exists a unique $c_x \in C_E$ such that $x - c_x = q \in \mathbb{Q}$ and $-2M \leq q \leq 2M$, which implies that $x \in C_E + q$. Therefore, if C_E is Lebesgue measurable, then

$$0 < m^*(E) \leq m\left(\bigcup_{\lambda \in \Lambda} (C_E + \lambda)\right) = \sum_{\lambda \in \Lambda} m(C_E + \lambda) = 0,$$

where the last equality is from Lemma 3.4.1, which is in contradiction. □

Theorem 3.4.4 There are disjoint sets A and B of real numbers such that

$$m^*(A \cup B) < m^*(A) + m^*(B).$$

[Proof] For any sets A and B, by the subadditivity of the Lebesgue outer measure, we have

$$m^*(A \cup B) \leq m^*(A) + m^*(B).$$

If $m^*(A \cup B) = m^*(A) + m^*(B)$ for any disjoint sets A and B, then for any subset C and D of $\mathbb{R}$,

$$m^*(C) = m^*(C \cap (D \cup D^c)) = m^*(C \cap D) + m^*(C \cap D^c)$$

and so every subset D of $\mathbb{R}$ is Lebesgue measurable, which is in contradiction to Theorem 3.4.3. □

Remark 3.4.5 Let E be a nonmeasurable set of finite outer measure. Then for any $n \in \mathbb{N}$, there exists a countable family $\{I_{nk}\}_{k=1}^{\infty}$ of open intervals such that $E \subset \cup_{k=1}^{\infty} I_{nk}$ and

$$\sum_{k=1}^{\infty} \ell(I_{nk}) < m^*(E) + \frac{1}{n}.$$

Put

$$G = \bigcap_{n=1}^{\infty} \left(\bigcup_{k=1}^{\infty} I_{nk} \right).$$

Then G is a G_δ-set and $E \subset G$. Moreover, for any $n \in \mathbb{N}$,

$$m^*(G) \leq m^* \left(\bigcup_{k=1}^{\infty} I_{nk} \right) \leq \sum_{k=1}^{\infty} \ell(I_{nk}) < m^*(E) + \frac{1}{n},$$

which implies that

$$m^*(G) \leq m^*(E)$$

and so $m^*(G) = m^*(E)$ by the monotonicity of the Lebesgue outer measure. However, $m^*(G \setminus E) > 0$, in fact, if $m^*(G \setminus E) = 0$, then by the regularity of the Lebesgue outer measure (Theorem 3.2.18), E should be measurable.

3.5 Cantor Set

From Proposition 3.2.3, we see that every countable subset of $\mathbb{R}$ is Lebesgue measurable and its Lebesgue is zero. Then it is a question:

is every Lebesgue measurable set of Lebesgue measure zero countable?

The answer of the question is no (see Proposition 3.5.1). We now construct the Cantor set which gives an answer of the above question.

Let $C_0 = [0, 1]$ and

$$C_1 = \left[0, \frac{1}{3}\right] \cup \left[\frac{2}{3}, 1\right], \qquad C_2 = \left[0, \frac{1}{3^2}\right] \cup \left[\frac{2}{3^2}, \frac{3}{3^2}\right] \cup \left[\frac{6}{3^2}, \frac{7}{3^2}\right] \cup \left[\frac{8}{3^2}, 1\right].$$

By continuing this way, we construct a descending sequence of $\{C_n\}$ of compact sets C_n such that

C_n is the disjoint union of 2^n closed intervals (**Cantor intervals**) of length 3^{-n}.

Let $\{I_{n,k} \mid k = 1, 2, \cdots, 2^n - 1\}$ be the family of $2^n - 1$ intervals deleted to construct C_n, listing them in order from left to right. For example,

$$I_{1,1} = \left(\frac{1}{3}, \frac{2}{3}\right), I_{2,1} = \left(\frac{1}{9}, \frac{2}{9}\right), \quad I_{2,2} = \left(\frac{3}{9}, \frac{6}{9}\right), \quad I_{2,3} = \left(\frac{7}{9}, \frac{8}{9}\right),$$

The set

$$\mathbf{C} = \bigcap_{n=1}^{\infty} C_n$$

is called the **Cantor set**. Then we have

(1) $\mathbf{C}$ is a nonempty set (by the Nested Set Theorem, see Theorem 1.4.19).

(2) $\mathbf{C}$ is compact (by the Heine-Borel Theorem, see Theorem 1.4.16).

(3) $\mathbf{C}$ is a measurable set.

Proposition 3.5.1 We have

(1) $\mathbf{C}$ is an uncountable set,

(2) $m(\mathbf{C}) = 0$.

[Proof] (1) Suppose that $\mathbf{C}$ is countable and let $\{c_k\}_{k=1}^{\infty}$ be an enumeration of $\mathbf{C}$.

- One of Cantor intervals in C_1 which does not contain c_1 is denoted by F_1,
- One of Cantor intervals in C_2 which is a subset of F_1 and does not contain c_2 is denoted by F_2.

By continuing this way, for $n \geq 2$,

- One of Cantor intervals in C_n which is a subset of F_{n-1} and does not contain c_n is denoted by F_n.

Then $\{F_n\}_{n=1}^{\infty}$ is a countable descending family of nonempty closed sets. Therefore, by the Nested Set Theorem (Theorem 1.4.19)

$$\emptyset \neq \bigcap_{n=1}^{\infty} F_n \subset \bigcap_{n=1}^{\infty} C_n = \mathbf{C}$$

and so there exists $c \in \mathbf{C}$, but $c \notin \{c_k\}_{k=1}^{\infty}$, which is in contradiction.

(2) For any $n \in \mathbb{N}$, we have

$$m(\mathbf{C}) \leq m(C_n) = 2^n 3^{-n},$$

where 2^n is the number of Cantor intervals in C_n and 3^{-n} is the length of each Cantor interval in C_n, and so $m(\mathbf{C}) = 0$. □

3.6 Cantor-Lebesgue Function

From Theorem 3.2.7, we see that every Borel subset $\mathbb{R}$ is Lebesgue measurable. Then it is a question:

is every measurable set a Borel set?

of which the answer is negative (see Proposition 3.6.5). We now construct the Cantor-Lebesgue function from which we give an answer of the above question.

For each $n \in \mathbb{N}$, put

$$O_n = [0,1] \setminus C_n = \bigcup_{k=1}^{2^n-1} I_{n,k}$$

and put

$$\mathbf{O} = \bigcup_{n=1}^{\infty} O_n.$$

Then we note that

(1) for each $n \in \mathbb{N}$, O_n is a finite disjoint union of $2^n - 1$ open intervals,

(2) $\{O_n\}_{n=1}^{\infty}$ is an ascending countable family of open sets O_n,

(3) $\mathbf{O}$ is also an open set with $m(\mathbf{O}) = 1$ since

$$1 = m([0,1]) = m(\mathbf{O}) + m(\mathbf{C}) = m(\mathbf{O}),$$

(4) $[0,1] = \mathbf{C} \cup \mathbf{O}$ is the disjoint union.

The **Cantor-Lebesgue function** (or **Cantor ternary function**) ϕ on $[0,1]$ is defined as following: define ϕ on O_n to be an increasing function which is constant on each of $2^n - 1$ open intervals in O_n and takes $2^n - 1$ values

$$\left\{\frac{1}{2^n}, \frac{2}{2^n}, \frac{3}{2^n}, \ldots, \frac{2^n-1}{2^n}\right\},$$

i.e., $\phi(x) = \frac{k}{2^n}$ for all $x \in I_{n,k}$, and

$$\phi(x) = \begin{cases} 0, & x = 0 \\ \sup\{\phi(t) \,|\, t \in \mathbf{O} \cap [0,x)\}, & x \in \mathbf{C} \setminus \{0\}. \end{cases}$$

We note that for $m, n \in \mathbb{N}$ with $m \le n$, $(\phi|_{O_n})|_{O_m} = \phi|_{O_m}$, and

$$\begin{aligned}\phi(1) &= \sup\{\phi(t) \,|\, t \in \mathbf{O} \cap [0,1)\} = \sup\left\{\phi(t) \,\middle|\, t \in \bigcup_n O_n\right\} = \lim_{n\to\infty} \frac{2^n-1}{2^n} \\ &= 1,\end{aligned}$$

we also note that, if $x \in \mathbf{O}$, then

$$\phi(x) = \sup\{\phi(t) \,|\, t \in \mathbf{O} \cap [0,x)\}. \tag{3.13}$$

Proposition 3.6.1 (i) The Cantor-Lebesgue function ϕ is an increasing continuous function.

(ii) The Cantor-Lebesgue function ϕ maps $[0,1]$ onto $[0,1]$.

(iii) The Cantor-Lebesgue function ϕ is differentiable on $\mathbf{O}$, i.e., ϕ is almost everywhere differentiable, and $\phi' = 0$ on $\mathbf{O}$.

[Proof] (i) For each $x, y \in [0,1]$ with $x \le y$, by (3.13), we have

$$\phi(x) = \sup\{\phi(t) \,|\, t \in \mathbf{O} \cap [0,x)\} \le \sup\{\phi(t) \,|\, t \in \mathbf{O} \cap [0,x)\} = \phi(y),$$

which implies that ϕ is an increasing function.

For each

$$x_0 \in \mathbf{O} = \bigcup_{n=1}^{\infty} O_n = \bigcup_{n=1}^{\infty} \left(\bigcup_{k=1}^{2^n-1} I_{n,k} \right),$$

there exist $n \in \mathbb{N}$ and $k \in \{1, 2, \cdots, 2^n - 1\}$ such that $x_0 \in I_{n,k}$ and ϕ is constant on $I_{n,k}$ and so ϕ is continuous at $x = x_0$. If $x_0 \in \mathbf{C}$, then

$$x_0 \in [0,1] \cap \mathbf{O}^c = [0,1] \cap \left(\bigcup_{n=1}^{\infty} O_n \right)^c = [0,1] \cap \left(\bigcap_{n=1}^{\infty} O_n^c \right) = [0,1] \cap \left(\bigcap_{n=1}^{\infty} \left(\bigcap_{k=1}^{2^n-1} I_{n,k}^c \right) \right)$$

and so for any $n \in \mathbb{N}$ and $k \in \{1, 2, \cdots, 2^n - 1\}$, $x_0 \in [0,1] \cap I_{n,k}^c$. Therefore, if x_0 is not one of the end points of $[0,1]$, i.e., $x_0 \neq 0, 1$, then for any $n \in \mathbb{N}$, there exists $k \in \{0, 1, 2, \cdots, 2^n - 1, 2^n\}$ such that x_0 lies between the intervals $I_{n,k}$ and $I_{n,k+1}$, where $I_{n,0} = \{0\}$ and $I_{n,2^n} = \{1\}$. Let $\epsilon > 0$ be given. Then there exists $n \in \mathbb{N}$ such that $0 < \frac{1}{2^n} < \epsilon$ and x_0 lies between the intervals $I_{n,k}$ and $I_{n,k+1}$ for some $k \in \{0, 1, 2, \cdots, 2^n - 1, 2^n\}$. Take $\delta > 0$ by

$$\delta = \min\left\{ |x_0 - \text{ the right end point of } I_{n,k}|, |\text{ the left end point of } I_{n,k+1} - x_0| \right\}.$$

Then for any $x \in [0,1]$ with $|x - x_0| < \delta$,

$$|\phi(x) - \phi(x_0)| < 2^{-n} < \epsilon$$

and so ϕ is continuous at $x = x_0$. The continuity of ϕ at 0 and 1 can be proved similarly.

(ii) By (i), ϕ is increasing and continuous, and $\phi(0) = 0$ and $\phi(1) = 1$. Therefore, by the Intermediate Value Theorem (Theorem 1.6.7), for any $y \in [0,1]$, there exists $x \in [0,1]$ such that $\phi(x) = y$, which implies that ϕ maps $[0,1]$ onto $[0,1]$.

(iii) Since ϕ is constant on $I_{n,k}$ for any $n \in \mathbb{N}$ and $k \in \{1, 2, \cdots, 2^n - 1\}$ and so ϕ is differentiable on $I_{n,k}$ with $\phi' = 0$ and so ϕ is differentiable on $\mathbf{O}$ with $\phi' = 0$ and $m(\mathbf{O}) = 1$. □

Proposition 3.6.2 Let ϕ be the Cantor-Lebesgue function and let ψ be a function defined by

$$\psi(x) = \phi(x) + x, \qquad x \in [0,1].$$

Then we have that

(i) ψ is a strictly increasing continuous function and maps $[0,1]$ onto $[0,2]$,

(ii) $\psi(\mathbf{C})$ is measurable and $m(\psi(\mathbf{C})) = 1$,

(iii) ψ maps a measurable subset of the Cantor set onto a nonmeasurable set,

[Proof] (i) Since the sum of two continuous functions is continuous, ψ is continuous. Also, since the sum of an increasing function and a strictly increasing function is strictly increasing, and the identity map is strictly increasing, ψ is strictly increasing. Since ϕ is (strictly) increasing and continuous, and $\phi(0) = 0$ and $\phi(1) = 2$. By the Intermediate Value Theorem (Theorem 1.6.7), for any $y \in [0,2]$, there exists $x \in [0,1]$ such that $\phi(x) = y$, which implies that ϕ maps $[0,1]$ onto $[0,2]$.

(ii) Since $[0,1] = \mathbf{C} \cup \mathbf{O}$ is the disjoint union,

$$[0,2] = \psi(\mathbf{C}) \cup \psi(\mathbf{O})$$

is the disjoint union. Since, by Proposition 1.6.14, every strictly monotone continuous function defined on an interval has the continuous inverse, $\psi(\mathbf{C}) = (\psi^{-1})^{-1}(\mathbf{C})$ is closed and $\psi(\mathbf{O}) = (\psi^{-1})^{-1}(\mathbf{O})$ is open and so both of $\psi(\mathbf{C})$ and $\psi(\mathbf{O})$ are measurable. Put

$$U_1 = O_1 = J_{1,1}, \qquad U_n = O_n \setminus O_{n-1} = \bigcup_{k=1}^{2^{n-1}} J_{n,k}, \qquad n \geq 2,$$

where

$$\{J_{n,1}, J_{n,2}, \cdots, J_{n,2^{n-1}}\} = \{I_{n,1}, I_{n,2}, \cdots, I_{n,2^n-1}\} \setminus \{I_{n-1,1}, I_{n-1,2}, \cdots, I_{n-1,2^{n-1}-1}\}$$

and then

$$\mathbf{O} = \bigcup_{n=1}^{\infty} U_n = \bigcup_{n=1}^{\infty}\left(\bigcup_{k=1}^{2^{n-1}} J_{n,k}\right)$$

is a disjoint union of open intervals $J_{n,k}$. Therefore, for a sequence $\{x_{n,k}\}_{n,k}$ of $x_{n,k} \in J_{n,k}$, we have

$$\begin{aligned} m(\psi(\mathbf{O})) &= m\left(\bigcup_{n=1}^{\infty}\left(\bigcup_{k=1}^{2^{n-1}} \psi(J_{n,k})\right)\right) = m\left(\bigcup_{n=1}^{\infty}\left(\bigcup_{k=1}^{2^{n-1}} (\phi(x_{n,k}) + J_{n,k})\right)\right) \\ &= \sum_{n=1}^{\infty}\left(\sum_{k=1}^{2^{n-1}} m\left((\phi(x_{n,k}) + J_{n,k})\right)\right) = \sum_{n=1}^{\infty}\left(\sum_{k=1}^{2^{n-1}} m\left(J_{n,k}\right)\right) = m(\mathbf{O}) = 1, \end{aligned}$$

and hence $m(\psi(\mathbf{C})) = m([0,2]) - m(\psi(\mathbf{O})) = 1$.

(iii) Since by (ii) $m(\psi(\mathbf{C})) = 1 > 0$, by the Vitali's theorem (Theorem 3.4.3), there exists a nonmeasurable subset A of $\psi(\mathbf{C})$ and then $\psi^{-1}(A) \subset \mathbf{C}$ and $m(\psi^{-1}(A)) \leq m(\mathbf{C}) = 0$ and so $\psi^{-1}(A))$ is a measurable set and then ψ maps $\psi^{-1}(A)$ onto the nonmeasurable set A. □

Remark 3.6.3 Let f be a continuous function defined on an interval I. Put

$$\mathcal{E} = \{E \subset \mathbb{R} \,|\, f^{-1}(E) \text{ is a Borel set in } I\}.$$

Then $f^{-1}(\mathbb{R}) = I$ is a Borel set and so $\mathbb{R} \in \mathcal{E}$. If $E \in \mathcal{E}$, then $f^{-1}(E^{\mathrm{c}}) = (f^{-1}(E))^{\mathrm{c}}$ is a Borel set and so $E^{\mathrm{c}} \in \mathcal{E}$. Let $\{E_n\}_{n=1}^{\infty}$ be a countable family of sets in $\mathcal{E}$. Then we have that

$$f^{-1}\left(\bigcup_{n=1}^{\infty} E_n\right) = \bigcup_{n=1}^{\infty} f^{-1}(E_n)$$

is a Borel set. Therefore, $\mathcal{E}$ is a σ-algebra. Moreover, since $f^{-1}(U)$ is open in I for any open set $U \subset \mathbb{R}$, $\mathcal{E}$ contains all open sets and so $\mathcal{E}$ also contains all Borel sets. Therefore, for a Borel set B, $f^{-1}(B)$ is a Borel set.

Proposition 3.6.4 The range of a Borel set by every strictly increasing continuous function defined on an interval is a Borel set.

[Proof] Let f be a strictly increasing continuous function defined on an interval I. Since, by Proposition 1.6.14, every strictly monotone continuous function defined on an interval has the continuous inverse, i.e., f^{-1} is continuous. Put

$$\mathcal{E} = \left\{E \subset I \,|\, f(E) = (f^{-1})^{-1}(E) \text{ is a Borel set}\right\}.$$

Then by Remark 3.6.3, $\mathcal{E}$ is a σ-algebra containing all Borel sets in I. Therefore, for any Borel set B in I, $f(B)$ is a Borel set. □

Proposition 3.6.5 There is a measurable subset of the Cantor set which is not a Borel set.

[Proof] Consider the strictly increasing continuous function ψ which maps $[0,1]$ onto $[0,2]$ given as in Proposition 3.6.2. Then by the proof of (iii) in Proposition 3.6.2, ψ maps the measurable subset $\psi^{-1}(A)$ of the Cantor set $\mathbf{C}$ onto the nonmeasurable set A. If $\psi^{-1}(A)$ is a Borel set, then by Proposition 3.6.4, A should be a Borel set, but A is a nonmeasurable set and so A is not a Borel set. Therefore, $\psi^{-1}(A)$ is not a Borel set. □

Exercises

3.1 Let $\mathcal{A}_{\mathrm{oc}}$ be the set of all finite disjoint unions of open-closed intervals. Prove that $\mathcal{A}_{\mathrm{oc}}$ is an algebra.

3.2 Let $E \subset \mathbb{R}$. An outer measure[1] $m^{**}(E)$ of E is defined by

$$m^{**}(E) = \inf\left\{ \sum_{k=1}^{\infty} \ell(J_k) \,\middle|\, \{J_k\}_{k=1}^{\infty} \subset \mathcal{I}_1 \quad \text{and} \quad E \subset \bigcup_{k=1}^{\infty} J_k \right\}, \tag{3.14}$$

where $\mathcal{I}_1$ is the collection of all open intervals including $\emptyset$ and $\mathbb{R}$. Prove that for any subset $E \subset \mathbb{R}$,

$$m^{**}(E) = m^*(E),$$

where $m^*(E)$ is the Lebesgue outer measure given as in Definition 3.1.2.

3.3 Prove Proposition 3.1.9.

3.4 Prove Proposition 3.2.3.

3.5 Prove Lemma 3.2.4.

3.6 Prove Proposition 3.2.8.

3.7 Prove Proposition 3.2.9.

3.8 Prove Proposition 3.2.14.

3.9 Let $\mathcal{A}$ be a family of subsets of $\mathbb{R}$. Put

$$\mathfrak{M}_{\mathrm{L}}(\mathcal{A}) = \{E \subset \mathbb{R} \,|\, m^*(A) = m^*(A \cap E) + m^*(A \cap E^{\mathrm{c}}) \text{ for all } A \in \mathcal{A}\}.$$

Prove that $\mathfrak{M}_{\mathrm{L}}(\mathcal{A})$ is a σ-algebra.

3.10 Prove Theorem 3.3.3.

3.11 Prove Theorem 3.3.4.

3.12 Prove Theorem 3.3.6.

3.13 Let $F : \mathbb{R} \to \mathbb{R}$ be an increasing and right continuous function. Define a set function $\mu_* : \mathcal{A}_{\mathrm{oc}} \to [0, \infty]$ by

$$\mu_*(E) = \sum_{k=1}^{n} (F(b_k) - F(a_k)),$$

where $E = \cup_{k=1}^{n}(a_k, b_k]$ is a finite disjoint union of open-closed intervals, and let $\mu_*(\emptyset) = 0$. Prove that

(1) μ_* is a premeasure defined on the algebra $\mathcal{A}_{\mathrm{oc}}$,

(2) μ_* is σ-finite,

(3) the Carathéodory measure μ_F on $\mathfrak{M}_F$ (the set of all μ^*-measurable sets, where μ^* is the outer measure induced by the premeasure μ_*) induced by the premeasure is a unique extension of μ_*,

(4) for any $a \in \mathbb{R}$, $\mu_F(\{a\}) = F(a) - F(a-)$, where $F(a-) = \lim_{x \to a-} F(x)$.

The measure μ_F is called the **Lebesgue-Stieltjes measure** associated with F.

---

1) The outer measure m^{**} gives another definition of the ***Legesgue outer measure***.

Chapter 4

Measurable Functions

4.1 Measurable Functions

We start with the general definition of measurable functions. Let $f : X \to Y$ be a function from a set X into Y. Then for a subset $E \subset Y$,

$$f^{-1}(E) = \{x \in X \mid f(x) \in E\}.$$

Definition 4.1.1 Let $(X, \mathcal{M})$ and $(Y, \mathcal{N})$ be measurable spaces and $f : X \to Y$ a function. If for any $E \in \mathcal{N}$, $f^{-1}(E) \in \mathcal{M}$, then f is called a **$(\mathcal{M}, \mathcal{N})$-measurable function** or simply a **measurable function** if there is no confusion.

Proposition 4.1.2 Let $(X, \mathcal{M})$ and $(Y, \mathcal{N})$ be measurable spaces and $f : X \to Y$ a function. Suppose that $\mathcal{N} = \sigma(\mathcal{E})$ for a family $\mathcal{E}$ of subsets of Y. Then f is $(\mathcal{M}, \mathcal{N})$-measurable if and only if for any $E \in \mathcal{E}$, $f^{-1}(E) \in \mathcal{M}$.

[Proof] $\Rightarrow$) It is obvious from the definition.

$\Leftarrow$) Put

$$\mathcal{F} = \{F \subset Y \mid f^{-1}(F) \in \mathcal{M}\}.$$

Then $\mathcal{F}$ is a σ-algebra and by assumption $\mathcal{E} \subset \mathcal{F}$. Therefore, $\sigma(\mathcal{E}) \subset \mathcal{F}$ and so for any $E \in \sigma(\mathcal{E})$, $E \in \mathcal{F}$, which $f^{-1}(E) \in \mathcal{M}$. Hence f is $(\mathcal{M}, \mathcal{N})$-measurable. □

Definition 4.1.3 Let $(X, \mathcal{M})$ and $(Y, \mathcal{N})$ be measurable spaces and $f : X \to Y$ a function. Let $E \in \mathcal{M}$ be a subset of X, i.e., E is a measurable subset of X. Then f is said to be **measurable** on E if for any $F \in \mathcal{N}$, $f^{-1}(F) \cap E \in \mathcal{M}$.

Remark 4.1.4 Let $(X, \mathcal{M})$ and $(Y, \mathcal{N})$ be measurable spaces and $f : X \to Y$ a function. Let $E \in \mathcal{M}$ be a subset of X, i.e., E is a measurable subset of X.

(1) f is measurable on E if and only if $f|_E : E \to Y$ is $(\mathcal{M}_E, \mathcal{N})$-measurable, where $\mathcal{M}_E = \mathcal{M} \cap E$.

(2) f is measurable (on X) if and only if $f|_E$ is $(\mathcal{M}_E, \mathcal{N})$-measurable and $f|_{E^c}$ is $(\mathcal{M}_{E^c}, \mathcal{N})$-measurable.

Proposition 4.1.5 Let $(X, \mathcal{M})$, $(Y, \mathcal{N})$ and $(Z, \mathcal{W})$ be measurable spaces. Let $f : X \to Y$ be a $(\mathcal{M}, \mathcal{N})$-measurable function and $g : Y \to Z$ a $(\mathcal{N}, \mathcal{W})$-measurable function. Then $g \circ f : X \to Z$ is $(\mathcal{M}, \mathcal{W})$-measurable.

[Proof] For any $E \in \mathcal{W}$,

$$(g \circ f)^{-1}(E) = f^{-1}(g^{-1}(E)) \in \mathcal{M},$$

which implies that $g \circ f : X \to Z$ is $(\mathcal{M}, \mathcal{W})$-measurable. □

Definition 4.1.6 Let $(X, \mathcal{M})$ be a measurable space and $f : X \to \mathbb{R}$ (or $\mathbb{C}$) a function. If f is $(\mathcal{M}, \mathcal{B}(\mathbb{R}))$-measurable (or $(\mathcal{M}, \mathcal{B}(\mathbb{C}))$-measurable), then f is called **$\mathcal{M}$-measurable** or simple **measurable**.

Proposition 4.1.7 Let $(X, \mathcal{M})$ be a measurable space and $f : X \to \mathbb{R}$ a function defined on X. Then the following statements are equivalent:

(i) f is measurable,

(ii) for each $c \in \mathbb{R}$, $f^{-1}((c, \infty))$ is measurable,

(iii) for each $c \in \mathbb{R}$, $f^{-1}([c, \infty))$ is measurable,

(iv) for each $c \in \mathbb{R}$, $f^{-1}((-\infty, c))$ is measurable,

(v) for each $c \in \mathbb{R}$, $f^{-1}((-\infty, c])$ is measurable.

[Proof] (i) ⇒ (ii) It is obvious from the definition.
(ii) ⇒ (iii) For each $c \in \mathbb{R}$, we obtain that

$$f^{-1}([c, \infty)) = f^{-1}\left(\bigcap_{n=1}^{\infty}\left(c - \frac{1}{n}, \infty\right)\right) = \bigcap_{n=1}^{\infty} f^{-1}\left(\left(c - \frac{1}{n}, \infty\right)\right) \in \mathcal{M}.$$

(iii) ⇒ (iv) For each $c \in \mathbb{R}$, we obtain that

$$f^{-1}((-\infty, c)) = f^{-1}([c, \infty)^c) = \left(f^{-1}([c, \infty))\right)^c \in \mathcal{M}.$$

(iv) ⇒ (v) For each $c \in \mathbb{R}$, we obtain that

$$f^{-1}((-\infty, c]) = f^{-1}\left(\bigcap_{n=1}^{\infty}\left(-\infty, c + \frac{1}{n}\right)\right) = \bigcap_{n=1}^{\infty} f^{-1}\left(\left(-\infty, c + \frac{1}{n}\right)\right) \in \mathcal{M}.$$

(v) ⇒ (i) Since

$$\mathcal{B}(\mathbb{R}) = \sigma\left(\{(-\infty, c] \mid c \in \mathbb{R}\}\right),$$

the proof is immediate from Proposition 4.1.2. □

Proposition 4.1.8 Let $(X, \mathcal{M})$ be a measurable space and $f : X \to \mathbb{R}$ a function defined on X. Then f is measurable if and only if for any open subset O of $\mathbb{R}$, $f^{-1}(O) \in \mathcal{M}$.

[Proof] The proof is immediate from Proposition 4.1.2. □

Definition 4.1.9 Let $f : \mathbb{R} \to \mathbb{R}$ be a real-valued function. Then

(i) f is called a **Lebesgue measurable function** if f is $\mathfrak{M}_{\mathrm{L}}$-measurable.

(ii) f is called a **Borel measurable function** if f is $\mathcal{B}(\mathbb{R})$-measurable.

Remark 4.1.10 (1) Since $\mathcal{B}(\mathbb{R}) \subset \mathfrak{M}_{\mathrm{L}}$, every Borel measurable function is Lebesgue measurable.

(2) Every continuous function is Borel measurable (see Proposition 4.1.2).

Proposition 4.1.11 Let $f : \mathbb{R} \to \mathbb{R}$ be a function. Then the following statements are equivalent:

(i) f is Lebesgue measurable (Borel resp.),

(ii) for each $c \in \mathbb{R}$, $f^{-1}((c, \infty))$ is Lebesgue measurable (a Borel set resp.),

(iii) for each $c \in \mathbb{R}$, $f^{-1}([c, \infty))$ is Lebesgue measurable (a Borel set resp.),

(iv) for each $c \in \mathbb{R}$, $f^{-1}((-\infty, c))$ is Lebesgue measurable (a Borel set resp.),

(v) for each $c \in \mathbb{R}$, $f^{-1}((-\infty, c])$ is Lebesgue measurable (a Borel set resp.).

[Proof] The proof is immediate from Proposition 4.1.2 or by similar arguments used in the proof of Proposition 4.1.7. □

Proposition 4.1.12 Let $f : \mathbb{R} \to \mathbb{R}$ be a function. Then

(i) f is Lebesgue measurable if and only if for any open subset O of $\mathbb{R}$, $f^{-1}(O)$ is Lebesgue measurable.

(ii) f is Borel measurable if and only if for any open subset O of $\mathbb{R}$, $f^{-1}(O)$ is a Borel set.

[Proof] Since $\mathcal{B}(\mathbb{R})$ be the smallest σ-algebra containing all open subsets of $\mathbb{R}$, by Proposition 4.1.2 the proof is immediate (see Proposition 4.1.8). □

Proposition 4.1.13 Every real valued continuous function defined on $\mathbb{R}$ is Lebesgue (Borel resp.) measurable on E for any Lebesgue measurable set (Bore set resp.) E.

[Proof] Let $f : \mathbb{R} \to \mathbb{R}$ be a continuous function and E a Lebesgue measurable set (Borel set resp.). By (2) in Remark 4.1.10, for any Borel set $B \in \mathcal{B}(\mathbb{R})$, $f^{-1}(B)$ is a Borel set and so Lebesgue measurable set. Therefore, $f^{-1}(B) \cap E$ is Lebesgue (measurable (a Borel set resp.). Hence f is Lebesgue (Borel resp.) measurable on E. □

A real valued function which is increasing or decreasing is called a **monotone function**.

Proposition 4.1.14 Every monotone function defined on an interval is Borel measurable.

[Proof] Let $f : I \to \mathbb{R}$ be an increasing function defined on an interval I. For each $c \in \mathbb{R}$, $f^{-1}([c, \infty)) = \emptyset$ or $f^{-1}([c, \infty)) \neq \emptyset$. If $f^{-1}([c, \infty)) = \emptyset$, then $f^{-1}([c, \infty))$ is a Borel set. If $f^{-1}([c, \infty)) \neq \emptyset$, then put $x_c = \inf\{x \in I | f(x) \geq c\}$ and then

$$f^{-1}([c, \infty)) = I \cap [x_c, \infty) \in \mathcal{B}(I) = \mathcal{B}(\mathbb{R}) \cap I$$

or

$$f^{-1}([c, \infty)) = I \cap (x_c, \infty) \in \mathcal{B}(I).$$

Therefore, by Proposition 4.1.2, f is Borel measurable. The proof for the decreasing function is similar. □

Definition 4.1.15 Let $(X, \mathcal{M})$ be a measurable space and $f : X \to \overline{\mathbb{R}} = \mathbb{R} \cup \{-\infty, \infty\}$ an extended real-valued function defined on X. Put

$$X_{\pm\infty} = \{x \in X \mid f(x) = \pm\infty\}, \qquad X_{\mathrm{f}} = \{x \in \mid f(x) \in \mathbb{R}\}.$$

If $f|_{X_{\mathrm{f}}}$ is $\mathcal{M}$-measurable and $X_{\pm\infty} \in \mathcal{M}$, then f is called **$\mathcal{M}$-measurable** or simple **measurable**.

Remark 4.1.16 Let $(X, \mathcal{M})$ be a measurable space and $f : X \to \overline{\mathbb{R}}$ an extended real-valued function defined on X. Then we can prove that f is measurable if and only if f is $(\mathcal{M}, \mathcal{N})$-measurable, where

$$\mathcal{N} = \sigma\left(\mathcal{B}(\mathbb{R}) \cup \{\{-\infty\}, \{\infty\}\}\right).$$

Example 4.1.17 Let X be a set.

(1) If $\mathcal{M} = \mathcal{P}(X)$ the power set of X, then every function $f : X \to \mathbb{R} \cup \{\pm\infty\}$ is measurable.

(2) If $\mathcal{M} = \{X, \emptyset\}$, then constant function on X is measurable.

(3) Let (X, τ) be a topological space. If X contains τ, then every continuous function defined on X is measurable.

Definition 4.1.18 Let $f : \mathbb{R} \to \overline{\mathbb{R}}$ be an extended real-valued function. Then f is called a **Lebesgue measurable function** if f is $\mathfrak{M}_{\mathrm{L}}$-measurable.

Remark 4.1.19 (1) Let $f : \mathbb{R} \to \overline{\mathbb{R}}$ be an extended real-valued function. Then f is Lebesgue measurable if and only if $f|_{\mathbb{R}_{\mathrm{f}}}$ is Lebesgue measurable and $\mathbb{R}_{\pm\infty} \in \mathfrak{M}_{\mathrm{L}}$, where

$$\mathbb{R}_{\mathrm{f}} = \{x \in \mathbb{R} \mid f(x) \in \mathbb{R}\}, \qquad \mathbb{R}_{\pm\infty} = \{x \in \mathbb{R} \mid f(x) = \pm\infty\}.$$

(2) Since $\mathcal{B}(\mathbb{R}) \subset \mathfrak{M}_{\mathrm{L}}$, every Borel measurable function is Lebesgue measurable.

(3) Every continuous function is Boreal measurable.

Proposition 4.1.20 Let $(X, \mathcal{M}, \mu)$ be a complete measure space. Let f and g be extended real valued functions defined on X.

(i) f is measurable if and only if there exists $X_0 \in \mathcal{M}$ with $\mu(X \setminus X_0) = 0$ such that $f_0 := f|_{X_0}$ is measurable.

(ii) If $f = g$ a.e. on X, then f is measurable if and only if g is measurable.

[Proof] (i) The proof is obvious by (1) in Example 4.1.17 and Remark 4.1.4. In fact, since $\mu(X \setminus X_0) = 0$ and μ is complete, by (1) in Example 4.1.17, every function on $X \setminus X_0$ is measurable.

(ii) There exists a measurable subset $N \subset X$ with $\mu(N) = 0$ such that $f = g$ on $X_0 := X \setminus N$. Then by (i), f is measurable if and only if $f|_{X_0}$ is measurable if and only if $g|_{X_0}$ is measurable if and only if g is measurable. □

Remark 4.1.21 In Proposition 4.1.20, the completeness of the measure space $(X, \mathcal{M}, \mu)$ is necessary (see Exercise **4.7**).

Corollary 4.1.22 Let f and g be extended real valued functions on $\mathbb{R}$. If f is Lebesgue measurable and $g = f$ a.e. on $\mathbb{R}$, then g is Lebesgue measurable.

[Proof] Since the Lebesgue measure space $(\mathbb{R}, \mathfrak{M}_{\mathrm{L}}, m)$ is complete, the proof is immediate from Proposition 4.1.20. □

Proposition 4.1.23 Let $(X, \mathcal{M})$ be a measurable space. Let f and g be real valued measurable functions defined on X. Then

(i) $f + g$ is measurable,

(ii) for any $\alpha \in \mathbb{R}$, αf is measurable,

(iii) fg is measurable.

[Proof] (i) For any $c \in \mathbb{R}$, we have

$$\begin{aligned}\{x \in X \mid f(x) + g(x) > c\} &= \{x \in X \mid f(x) > c - g(x)\} \\ &= \bigcup_{r \in \mathbb{Q}} \Big(\{x \in X \mid f(x) > r\} \bigcap \{x \in X \mid r > c - g(x)\}\Big) \\ &= \bigcup_{r \in \mathbb{Q}} \Big(\{x \in X \mid f(x) > r\} \bigcap \{x \in X \mid g(x) > c - r\}\Big),\end{aligned}$$

which is a countable union of measurable sets and so it is measurable. Therefore, $f + g$ is measurable.

(ii) If $\alpha = 0$, then it is clear that $\alpha f = 0$ is measurable. If $\alpha \neq 0$, then for any $c \in \mathbb{R}$, we have

$$\{x \in X \mid \alpha f(x) > c\} = \begin{cases} \{x \in X \mid \alpha f(x) > c/\alpha\}, \ \alpha > 0, \\ \{x \in X \mid \alpha f(x) < c/\alpha\}, \ \alpha < 0 \end{cases}$$

is measurable and so αf is measurable.

(iii) For any $c \geq 0$, we have

$$\{x \in X \mid f(x)^2 > c\} = \{x \in X \mid f(x) > \sqrt{c}\} \bigcup \{x \in X \mid f(x) < -\sqrt{c}\},$$

which is measurable. If $c < 0$, then $\{x \in X \mid f(x)^2 > c\} = X$ is measurable. Therefore, f^2 is measurable and by (i) and (ii)

$$fg = \frac{1}{4}\left((f+g)^2 - (f-g)^2\right)$$

is measurable. □

Remark 4.1.24 Let $(X, \mathcal{M}, \mu)$ be a complete measure space. Let f and g be extended real valued measurable functions defined on X. If f and g are finite a.e. on X, then there exists a measurable subset $N \subset X$ with $\mu(N) = 0$ such that f and g are finite on $X_0 := X \setminus N$. Therefore, for any real numbers α and β, $\alpha f + \beta g$ is well-defined on X_0. Moreover, by Propositions 4.1.20 and 4.1.23, $\alpha f + \beta g$ is measurable with respect to the measurable space $(X_0, \mathcal{M}_{X_0})$. Therefore, if $(X, \mathcal{M}, \mu)$ is a complete measure space, then by Proposition 4.1.20, any extension of $\alpha f + \beta g$ to X is measurable. Even though $(X, \mathcal{M}, \mu)$ is not a complete measure space, we can extend $\alpha f + \beta g$ to X, of which the extension is denoted by same symbol, by

$$(\alpha f + \beta g)(x) = \begin{cases} \alpha f(x) + \beta g(x), & x \in X_0, \\ 0, & x \in X \setminus X_0, \end{cases}$$

and then $\alpha f + \beta g$ is measurable with respect to the measurable space $(X, \mathcal{M})$. If we study integration theory on then measure space $(X, \mathcal{M}, \mu)$, then the function $\alpha f + \beta g$ defined on $X \setminus X_0$ is negligible since $\mu(X \setminus X_0) = 0$. Therefore, for two extended real valued measurable functions f and g defined on X which are finite a.e. on X, $\alpha f + \beta g$ is considered as any measurable extension which agrees with $\alpha f + \beta g$ on $X_0 = X \setminus N$ for some measurable set with $\mu(N) = 0$. Similarly, we can consider the product of two extended real valued measurable functions f and g defined on X which are finite a.e. on X. With this consideration, Proposition 4.1.23 holds for extended real valued measurable functions f and g defined on X which are finite a.e. on X.

Let $\{f_n\}$ be a sequence of real valued functions defined on X. Then we define

$$\left(\sup_n\{f_n\}\right)(x) = \sup_n\{f_n(x)\}, \qquad \left(\inf_n\{f_n\}\right)(x) = \inf_n\{f_n(x)\}, \qquad x \in X.$$

Proposition 4.1.25 Let $(X, \mathcal{M})$ be a measurable space. Let $\{f_n\}$ be a sequence of real valued measurable functions defined on X. Then $\sup_n\{f_n\}$ and $\inf_n\{f_n\}$ are measurable.

[Proof] For any $c \in \mathbb{R}$,

$$\left\{x \in X \,\middle|\, \left(\sup_n\{f_n\}\right)(x) < c\right\} = \bigcap_{n=1}^{\infty}\{x \in X \mid f_n(x) < c\}$$

is measurable and so $\sup_n\{f_n\}$ is measurable. Similarly, for any $c \in \mathbb{R}$,

$$\left\{x \in X \,\middle|\, \left(\inf_n\{f_n\}\right)(x) > c\right\} = \bigcap_{n=1}^{\infty}\{x \in X \mid f_n(x) > c\}$$

is measurable and so $\inf_n\{f_n\}$ is measurable. □

Proposition 4.1.26 Let $(X, \mathcal{M})$ be a measurable space. Let $\{f_k\}_{k=1}^{n}$ be a finite sequence of real valued measurable functions defined on X. Then $\max\{f_k \mid k = 1, 2, \cdots, n\}$ and $\min\{f_k \mid k = 1, 2, \cdots, n\}$ are measurable, where

$$\max\{f_k \mid k = 1, 2, \cdots, n\}(x) = \max\{f_k(x) \mid k = 1, 2, \cdots, n\},$$
$$\min\{f_k \mid k = 1, 2, \cdots, n\}(x) = \min\{f_k(x) \mid k = 1, 2, \cdots, n\}$$

for $x \in X$.

[Proof] The proof is immediate from Proposition 4.1.25. □

For a real valued function defined on X, we define

$$f^+(x) = \max\{f(x), 0\}, \qquad f^-(x) = \max\{-f(x), 0\}, \qquad |f| = f^+ + f^-.$$

Then we note that

$$f = f^+ - f^-.$$

Proposition 4.1.27 Let $(X, \mathcal{M})$ be a measurable space. Let f be a real valued measurable function defined on X. Then f^+, f^- and $|f|$ are measurable.

[Proof] The proof is immediate from Propositions 4.1.26 and 4.1.23. □

Let $\{f_n\}$ be a sequence of real valued functions defined on X. Then we define

$$\limsup_{n\to\infty} f_n = \inf_{n\geq 1} \sup_{k\geq n} f_k,$$
$$\liminf_{n\to\infty} f_n = \sup_{n\geq 1} \inf_{k\geq n} f_k,$$

which are called the **limit supremum** and **limit infimum**, respectively.

Proposition 4.1.28 Let $(X, \mathcal{M})$ be a measurable space. Let $\{f_n\}$ be a sequence of real valued measurable functions defined on X. Then $\limsup_{n\to\infty}\{f_n\}$ and $\liminf_{n\to\infty}\{f_n\}$ are measurable.

[Proof] By Proposition 4.1.25, the proof is immediate. □

Definition 4.1.29 Let $(X, \mathcal{M}, \mu)$ be a measure space. Let $\{f_n\}$ be a sequence of real valued functions defined on X, let f a real valued function defined on X and let A a subset of X. Then we say that

(i) The sequence $\{f_n\}$ converges to f pointwise on A if for all $x \in A$,

$$\lim_{n\to\infty} f_n(x) = f(x).$$

(ii) The sequence $\{f_n\}$ converges to f pointwise a.e. on A if there exists a measurable subset B of A with $\mu(B) = 0$ such that for all $x \in A \setminus B$,

$$\lim_{n\to\infty} f_n(x) = f(x).$$

(iii) The sequence $\{f_n\}$ converges to f uniformly on A if for any $\epsilon > 0$, there exists $N \in \mathbb{N}$ such that

$$|f_n - f| < \epsilon \text{ on } A \text{ for all } n \geq N.$$

Proposition 4.1.30 Let $(X, \mathcal{M}, \mu)$ be a measure space. Let $\{f_n\}$ be a sequence of real valued measurable functions defined on X and f a real valued function defined on X. If one of the following conditions is satisfied:

(i) $\{f_n\}$ converges to f pointwise on all of X,

(ii) $(X, \mathcal{M}, \mu)$ is complete and $\{f_n\}$ converges to f pointwise a.e. on X,

then f is measurable.

[Proof] (i) Since, by assumption,

$$f = \limsup_{n\to\infty} f_n = \liminf_{n\to\infty} f_n,$$

the proof is immediate from Proposition 4.1.28.

(ii) Suppose that $(X, \mathcal{M}, \mu)$ is a complete measure space. Then there exists a measurable subset N of X with $\mu(N) = 0$ such that $\{f_n\}$ converges to f pointwise on all of $X_0 := X \setminus N$. Therefore, by (i), f is measurable on X_0 and so, since $(X, \mathcal{M}, \mu)$ is a complete measure space, by (i) in Proposition 4.1.20, f is measurable on all of X. □

Remark 4.1.31 A direct proof of (i) in Proposition 4.1.30 is the following. Suppose that $\{f_n\}$ converges to f pointwise on all of X. Then for any $c \in \mathbb{R}$, we have

$$\begin{aligned}\{x \in X \mid f(x) > c\} = \{x \in X \mid \liminf_{n\to\infty} f_n(x) > c\} &= \{x \in X \mid \sup_{n\geq 1} \inf_{k\geq n} f_k(x) > c\} \\ &= \bigcup_{n=1}^{\infty} \left\{x \in X \mid \inf_{k\geq n}\{f_k(x)\} > c\right\} \\ &= \bigcup_{n=1}^{\infty} \bigcap_{k\geq n} \{x \in X \mid f_k(x) > c\},\end{aligned}$$

which is measurable and so f is measurable.

Corollary 4.1.32 Let $\{f_n\}$ be a sequence of Lebesgue measurable functions defined on $\mathbb{R}$ and f a function defined on $\mathbb{R}$. If $\{f_n\}$ converges to f pointwise a.e. on $\mathbb{R}$, then f is Lebesgue measurable.

[Proof] Since the Lebesgue measure space $(\mathbb{R}, \mathfrak{M}_{\mathrm{L}}, m)$ is complete, the proof is immediate from Proposition 4.1.30. □

For a subset A of X, the **characteristic function** χ_A of A is defined by

$$\mathbf{1}_A(x) := \chi_A(x) := \begin{cases} 1, & x \in A, \\ 0, & x \in X \setminus A. \end{cases}$$

Remark 4.1.33 Let $(X, \mathcal{M})$ be a measurable space. For a given subset A of X, the characteristic function χ_A is measurable if and only if A is measurable. Therefore, for a nonmeasurable set A, χ_A is a nonmeasurable function.

Definition 4.1.34 Let $(X, \mathcal{M})$ be a measurable space. A real valued function ϕ defined on X is called a **simple function** if ϕ is measurable and $\phi(X)$ is a finite set.

Remark 4.1.35 Let $(X, \mathcal{M})$ be a measurable space. Let $\phi : X \to \mathbb{R}$ be a simple (measurable) function defined on X and let $\phi(X) = \{a_1, a_2, \cdots, a_n\}$. Put

$$X_i = \phi^{-1}(a_i), \qquad i = 1, 2, \cdots, n.$$

Then we have

$$\phi = \sum_{i=1}^{n} a_i \chi_{X_i}, \tag{4.1}$$

which is called the **canonical representation** of ϕ.

Remark 4.1.36 For a finite sequence $\{I_i\}_{i=1}^n$ of intervals and a finite sequence $\{a_i\}_{i=1}^n$ of real numbers,

$$\phi(x) = \sum_{i=1}^{n} a_i \chi_{I_i}$$

is called a **step function**. Since every interval is a (Borel) Lebesgue measurable set, every step function is Lebesgue measurable and so a simple function.

For a sequence $\{f_n\}_{n=1}^\infty$ of real valued functions defined on a set X, $\{f_n\}_{n=1}^\infty$ is called an increasing sequence if for any $n \in \mathbb{N}$ and $x \in X$,

$$f_n(x) \le f_{n+1}(x).$$

Theorem 4.1.37 **[Simple Approximation Theorem]** Let $(X, \mathcal{M})$ be a measurable space and f an extended real valued measurable function on X.

(i) If f is nonnegative and bounded, i.e., there exists $M \in \mathbb{N}$ (for convenience) such that $0 \le f < M$ (for convenience), then there exist increasing sequence $\{\phi_n\}$ and decreasing sequence $\{\psi_n\}$ of nonnegative simple functions defined on X such that

$$0 \le \phi_n \le f \le \psi_n \quad \text{and} \quad 0 \le \psi_n - \phi_n < \frac{1}{2^{n-1}} \quad \text{on } X$$

and $\{\phi_n\}$ converges to f pointwise on X. Moreover, $\{\phi_n\}$ converges to f uniformly on X.

(ii) There exists a sequence $\{\phi_n\}$ of simple functions defined on X such that

$$0 \le |\phi_1| \le |\phi_2| \le \cdots \le |f|$$

and $\{\phi_n\}$ converges to f pointwise on X. Moreover, $\{\phi_n\}$ converges to f uniformly on any set on which f is bounded.

(iii) If the measure space $(X, \mathcal{M}, \mu)$ is σ-finite, then in (ii) we can choose $\{\phi_n\}$ such that ϕ_n vanishes outside a set of finite measure.

[Proof] (i) For each $n \in \mathbb{N}$ and any $k = 0, 1, 2, \cdots, M2^n - 1$, put

$$E_{n,k} = f^{-1}\left(\left[k2^{-n}, (k+1)2^{-n}\right)\right),$$

which are measurable since f is measurable and every interval is a Borel set. For each $n \in \mathbb{N}$, define the simple functions ϕ_n and ψ_n by

$$\phi_n = \phi_{f,M,n} = \sum_{k=0}^{M2^n-1} \frac{k}{2^n}\chi_{E_{n,k}}, \qquad \psi_n = \psi_{f,M,n} = \sum_{k=0}^{M2^n-1} \frac{k+1}{2^n}\chi_{E_{n,k}}.$$

Then for any $x \in X$,

$$f(x) \in f(X) \subset [0, M) = \bigcup_{k=0}^{M2^n-1} \left[k2^{-n}, (k+1)2^{-n}\right)$$

and so there exists a unique $k \in \{0, 1, \cdots, M2^n - 1\}$ such that $f(x) \in [k2^{-n}, (k+1)2^{-n})$, i.e., $x \in f^{-1}([k2^{-n}, (k+1)2^{-n})) = E_{n,k}$, which implies that

$$\phi_n(x) = \frac{k}{2^n} \leq f(x) < \frac{k+1}{2^n} = \psi_n(x)$$

and $0 \leq \psi_n(x) - \phi_n(x) = 2^{-n} < 2^{-(n-1)}$.

(ii) By using the decomposition $f = f^+ - f^-$ of f, for each $n \in \mathbb{N}$, put

$$E_n^+ = \{x \in X \,|\, 0 \leq f^+(x) < 2^n\},$$
$$E_n^- = \{x \in X \,|\, 0 \leq f^-(x) < 2^n\},$$

and define the simple functions ϕ_n^+ and ϕ_n^- by

$$\phi_n^+ = \phi_{f^+\chi_{E_n^+}, 2^n, n} + 2^n\chi_{(E_n^+)^c},$$
$$\phi_n^- = \phi_{f^-\chi_{E_n^-}, 2^n, n} + 2^n\chi_{(E_n^-)^c},$$

and

$$\phi_n = \phi_n^+ - \phi_n^-.$$

Then for each $n \in \mathbb{N}$,

$$|\phi_n| = \phi_n^+ + \phi_n^- \leq \phi_{n+1}^+ + \phi_{n+1}^- = |\phi_{n+1}| \leq f^+ + f^- = |f|.$$

For any $x \in X$ with $|f(x)| < \infty$ there exists $M \in \mathbb{N}$ such that $|f(x)| < M$ and so for any $n \in \mathbb{N}$ with $2^n \geq M$,

$$|f(x) - \phi_n(x)| \leq \frac{2}{2^n} \leq \frac{1}{2^{n-1}},$$

which implies that $\lim_{n\to\infty} \phi_n(x) = f(x)$. Also, for $x \in X$ with $f(x) = \pm\infty$, for any $n \in \mathbb{N}$, $\phi_n(x) = \pm 2^n$, which implies that $\lim_{n\to\infty} \phi_n(x) = \lim_{n\to\infty} \pm 2^n = \pm\infty = f(x)$. Therefore, $\{\phi_n\}$ converges to f pointwise on X. The uniform convergence of $\{\phi_n\}$ to f on any set on which f is bounded is immediate from (i).

(iii) If X is σ-finite, then there exists an ascending sequence of A_n of measurable sets of finite measure such that $X = \cup_{n=1}^{\infty} A_n$. Therefore, by replacing ϕ_n by $\phi_n \cdot \chi_{A_n}$, we prove (iii). □

Theorem 4.1.38 **[Egoroff's Theorem]** Let $(X, \mathcal{M}, \mu)$ be a finite measure space and let $\{f_n\}_{n=1}^{\infty}$ a sequence of extended real valued measurable functions defined on X such that $\{f_n\}$ converges pointwise a.e. on X to a extended real valued measurable function f which is finite a.e. Then for each $\epsilon > 0$, there exists a measurable subset $E \in \mathcal{M}$ with $\mu(E) < \epsilon$ such that

$$f_n \longrightarrow f \text{ uniformly on } E^c.$$

[Proof] Let $\epsilon > 0$ be given. Without loss of generality, we may assume that $\{f_n\}$ converges pointwise everywhere on X to a measurable function f which is finite. For each $k \in \mathbb{N}$, define a descending sequence $\{E_{k,n}\}_{n=1}^{\infty}$ of measurable sets by

$$E_{k,n} = \bigcup_{m=n}^{\infty} \left\{ x \in X \;\middle|\; |f_m(x) - f(x)| \geq \frac{1}{k} \right\}.$$

Then since $\{f_n\}$ converges pointwise everywhere on X to a measurable function f, we have

$$\bigcap_{n=1}^{\infty} E_{k,n} = \emptyset,$$

which, by the continuity of the measure μ and $\mu(E_{k,1}) \leq \mu(X) < \infty$, implies that

$$\lim_{n \to \infty} \mu(E_{k,n}) = \mu\left(\bigcap_{n=1}^{\infty} E_{k,n} \right) = \mu(\emptyset) = 0.$$

Therefore, for each $k \in \mathbb{N}$, there exists $n_k \in \mathbb{N}$ such that for any $n \geq n_k$,

$$\mu(E_{k,n}) < \frac{\epsilon}{2^k}.$$

Put

$$E = \bigcup_{k=1}^{\infty} E_{k,n_k}.$$

Then by the countable subadditivity of μ, we have

$$\mu(E) \leq \sum_{k=1}^{\infty} \mu(E_{k,n_k}) < \sum_{k=1}^{\infty} \frac{\epsilon}{2^k} = \epsilon.$$

For the given $\epsilon > 0$, there exists a $k \in \mathbb{N}$ such that $1/k < \epsilon$ and for any $x \in E^c$,

$$x \in E_{k,n_k}^c = \bigcap_{n=n_k}^{\infty} \left\{ x \in X \;\middle|\; |f_n(x) - f(x)| < \frac{1}{k} \right\},$$

which means that for any $x \in E^c$ and $n \geq n_k$,

$$|f_n(x) - f(x)| < \frac{1}{k} < \epsilon.$$

Therefore, $\{f_n\}$ converges uniformly on E^c to f. □

Remark 4.1.39 In Theorem 4.1.38, the finiteness of the measure μ is necessary (see Exercise **4.11**).

Remark 4.1.40 In Theorem 4.1.38, the measurability of f is necessary. Therefore, if $(X, \mathcal{M}, \mu)$ is complete, then the measurability of f is automatic from Theorem 4.1.30. But if $(X, \mathcal{M}, \mu)$ is not complete, then the measurability of f should be assumed or it should be assumed that $\{f_n\}$ converges pointwise on X to f.

4.2 Littlewood's Three Principles

The Littlewood's three principles are roughly expressed as the following:

(1) Every Lebesgue measurable set is nearly a finite union of disjoint open intervals,

(2) Every Lebesgue measurable function is nearly continuous.

(3) Every pointwise convergent sequence of Lebesgue measurable functions is nearly uniformly convergent.

Remark 4.2.1 By Theorem 3.2.19, for each Lebesgue measurable set E with finite Lebesgue measure and each $\epsilon > 0$, there exists a finite disjoint family $\{I_k\}_{k=1}^{n_0}$ of open intervals such that for $O = \cup_{k=1}^{n_0} I_k$,

$$m(E \triangle O) < \epsilon,$$

where $E \triangle O = (E \setminus O) \cup (O \setminus E)$.

Remark 4.2.2 Let E be a Lebesgue measurable set with finite measure and let $\{f_n\}_{n=1}^{\infty}$ a sequence of Lebesgue measurable functions defined on E such that $\{f_n\}$ converges pointwise a.e. on E to a Lebesgue measurable on f which is finite a.e.. Then by the Egoroff's theorem (Theorem 4.1.38) for each $\epsilon > 0$, there exists a Lebesgue measurable subset D of E with $m(D) < \epsilon$ such that

$$f_n \longrightarrow f \text{ uniformly on } E \setminus D.$$

Theorem 4.2.3 **[Egoroff's Theorem]** Let E be a Lebesgue measurable set with finite measure and let $\{f_n\}_{n=1}^{\infty}$ a sequence of Lebesgue measurable functions defined on E such that $\{f_n\}$ converges pointwise a.e. on E to a Lebesgue measurable function f which is finite a.e on E. Then for each $\epsilon > 0$, there is a closed subset F of E such that

$$f_n \longrightarrow f \text{ uniformly on } F \text{ and } m(E \setminus F) < \epsilon.$$

[Proof] By the Egoroff's theorem (Theorem 4.1.38) for each $\epsilon > 0$, there exists a Lebesgue measurable subset D of E with $m(D) < \epsilon/2$ such that

$$f_n \longrightarrow f \text{ uniformly on } A := E \setminus D.$$

By (iii) in Theorem 3.2.18, there exists a closed set F contained in A such that

$$m(A \setminus F) < \frac{\epsilon}{2}$$

and so $\{f_n\}$ converges to f uniformly on F and

$$\begin{aligned} m(E \setminus F) &= m(E \setminus A) + m(A \setminus F) \\ &< \frac{\epsilon}{2} + \frac{\epsilon}{2} \\ &= \epsilon. \end{aligned}$$

□

Proposition 4.2.4 Let f be a simple function defined on a Lebesgue measurable set E. Then for each $\epsilon > 0$, there exist a continuous function g on $\mathbb{R}$ and a closed set F contained in E such that

$$f = g \text{ on } F \text{ and } m(E \setminus F) < \epsilon.$$

[Proof] Since f is a simple function defined on a Lebesgue measurable set E, let

$$f = \sum_{k=1}^{n} a_k \chi_{E_k}$$

be the canonical representation of f. Then for each $k = 1, 2, \cdots, n$, by (iii) in Theorem 3.2.18, there exists a closed set F_k contained in E_k such that $m(E_k \setminus F_k) < \epsilon/n$. Put

$$F = \bigcup_{k=1}^{n} F_k.$$

Then F is a closed set and

$$\begin{aligned} m(E \setminus F) &= m\left(\bigcup_{k=1}^{n} (E_k \setminus F_k)\right) \\ &= \sum_{k=1}^{n} m(E_k \setminus F_k) \\ &< \epsilon. \end{aligned}$$

Define the function g on F such that $g|_{F_k} = a_k$ for $k = 1, 2, \cdots, n$. Then g is continuous on the closed set F. In fact, for a point $x \in F$, there exists i such that $x \in F_i$ and so there exists an open interval I containing x such that $I \cap (\cup_{k\neq i} F_k) = \emptyset$. In fact, $x \in F_k \subset (\cup_{k\neq i} F_k)^c$ and $(\cup_{k\neq i} F_k)^c$ is open and so there exists an open interval I such that $x \in I \subset (\cup_{k\neq i} F_k)^c$. Hence $g|_{I\cap F} = g|_{I\cap F_i} = a_i$ and so g is continuous at x. Then the continuous function g on the closed set F can be extended to $\mathbb{R}$ as a continuous function (see, Tietze Extension Theorem). □

Theorem 4.2.5 [**Lusin's Theorem**] Let f be a real valued Lebesgue measurable function defined on a Lebesgue measurable set E. Then for each $\epsilon > 0$, there exist a continuous function g on $\mathbb{R}$ and a closed set F contained in E such that

$$f = g \text{ on } F \text{ and } m(E \setminus F) < \epsilon.$$

[Proof] Suppose that $m(E) < \infty$. By the simple approximation theorem, there is a sequence $\{\phi_n\}$ of simple functions such that $\{\phi_n\}$ converges pointwise on E to f. Then by Proposition 4.2.4, for each $n \in \mathbb{N}$, there exists a closed set F_n contained in E and a continuous g_n on $\mathbb{R}$ such that

$$\phi_n = g_n \text{ on } F_n \text{ and } m(E \setminus F_n) < \frac{\epsilon}{2^{n+1}}.$$

By the Egoroff's Theorem (Theorem 4.2.3) there exists a closed subset F_0 contained in E such that $m(E \setminus F_0) < \epsilon/2$ and $\{\phi_n\}$ converges uniformly on F_0 to f. Put $F = \cap_{n=0}^{\infty} F_n$ and then F is a closed set and we have

$$\begin{aligned} m(E \setminus F) &= m\left((E \setminus F_0) \cup \left(\bigcup_{n=1}^{\infty} (E \setminus F_n)\right)\right) \\ &< \frac{\epsilon}{2} + \sum_{n=1}^{\infty} \frac{\epsilon}{2^{n+1}} \\ &= \epsilon. \end{aligned}$$

Since $F \subset F_n$ and g_n is continuous on F_n for each n, $g_n = \phi_n$ is continuous on F. $g = f|_F$ is continuous on F since g is the uniform limit of continuous functions g_n on the closed set $F \subset F_0$. Then the continuous function g on the closed set F can be extended to $\mathbb{R}$ as a continuous function (see, Tietze Extension Theorem). □

Exercises

4.1 Let $(X, \mathcal{M})$ be a measurable space and Y a (nonempty) set and $f : X \to Y$ a function for a (nonempty) set Y. Prove that

$$\mathcal{F} = \{F \subset Y \mid f^{-1}(F) \in \mathcal{M}\}$$

is a σ-algebra.

4.2 Let $(X, \mathcal{M})$ and $(Y, \mathcal{N})$ be measurable spaces and $f : X \to Y$ a function. Let $E \in \mathcal{M}$ be a subset of X, i.e., E is a measurable subset of X. Prove that

(1) f is measurable on E if and only if $f|_E : E \to Y$ is $(\mathcal{M}_E, \mathcal{N})$-measurable, where $\mathcal{M}_E = \mathcal{M} \cap E$.

(2) f is measurable (on X) if and only if $f|_E$ is $(\mathcal{M}_E, \mathcal{N})$-measurable and $f|_{E^c}$ is $(\mathcal{M}_{E^c}, \mathcal{N})$-measurable.

4.3 Prove that every decreasing function defined on an interval is Borel measurable.

4.4 Let $(X, \mathcal{M})$ be a measurable space and $f : X \to \overline{\mathbb{R}}$ an extended real-valued function defined on X. Prove that f is measurable if and only if f is $(\mathcal{M}, \sigma(\mathcal{B}(\mathbb{R}) \cup \{\{-\infty\}, \{\infty\}\}))$-measurable.

4.5 Let $(X, \mathcal{M})$ be a measurable space and $f : X \to \overline{\mathbb{R}}$ an extended real-valued function defined on X. Prove that the following statements are equivalent:

(i) f is measurable,

(ii) for each $c \in \mathbb{R}$, $f^{-1}((c, \infty])$ is measurable,

(iii) for each $c \in \mathbb{R}$, $f^{-1}([c, \infty])$ is measurable,

(iv) for each $c \in \mathbb{R}$, $f^{-1}([-\infty, c))$ is measurable,

(v) for each $c \in \mathbb{R}$, $f^{-1}([-\infty, c])$ is measurable.

4.6 Prove that every continuous function from $\mathbb{R}$ into $\mathbb{R}$ is Lebesgue measurable.

4.7 Let $(X, \mathcal{M}, \mu)$ be a measure space and X_0 a measurable subset of X such that $\mu(X \setminus X_0) = 0$. Let $f : X \to \overline{\mathbb{R}}$ be a measurable function. Construct a function $g : X \to \overline{\mathbb{R}}$ such that $f = g$ a.e. on X, but g is not measurable.

4.8 Let $(X, \mathcal{M})$ be a measurable space. Let f be a real valued measurable function defined on X and g a Borel measurable function defined on $\mathbb{R}$. Prove that $g \circ f$ is measurable.

4.9 Let $(X, \mathcal{M})$ be a measurable space. Let f be a real valued measurable function defined on X and g a continuous function defined on $\mathbb{R}$. Prove that $g \circ f$ is measurable.

4.10 Let $\mathcal{E} = \{E_i\}_{i=1}^n$ be a finite disjoint family of nonempty subset of a set X such that $\cup_{i=1}^n E_i = X$ and let $\mathcal{M} = \sigma(\mathcal{E})$.

(i) Prove that if $E \in \mathcal{M}$, then there exists a subset $\{i_k\}_{k=1}^m$ of $\{1, 2, \cdots, n\}$ such that

$$E = \bigcup_{k=1}^{m} E_{i_k}.$$

(ii) Prove that if a function $f : X \to \mathbb{R}$ is measurable, then for each $i \in \{1, 2, \cdots, n\}$, $\#(f(E_i)) = 1$, i.e., f is a constant function on E_i.

(iii) Find all measurable function f from X into $\mathbb{R}$.

4.11 Construct a sequence $\{f_n\}_{n=1}^\infty$ of (real-valued) measurable functions on $\mathbb{R}$ and a function f on $\mathbb{R}$ such that $\{f_n\}$ converges pointwise on $\mathbb{R}$ to f, but for each $\epsilon > 0$, there is no measurable subset $E \in \mathfrak{M}_{\mathrm{L}}$ such that $\mu(E) < \epsilon$ and f_n converges uniformly on E^{c} to f. Therefore, in the Egoroff's theorem, finiteness of the measure is necessary.

Chapter 5

General Integration Theory

5.1 Integration over General Measure Spaces

Let $(X, \mathcal{M}, \mu)$ be a measure space. Put

$$M^+ = \text{ the set of all nonnegative measurable functions from } X \text{ into } [0, \infty].$$

Definition 5.1.1 [Integration]

(i) If $\phi \in M^+$ is a simple function with the canonical representation $\phi = \sum_{k=1}^{n} a_n \chi_{E_k}$, then the value:

$$\int_X \phi d\mu := \sum_{k=1}^{n} a_n \mu(E_k)$$

is called the **integral** of ϕ with respect to (the measure) μ, where the integral can be ∞ and, with the convention, $0 \cdot \infty = 0$. For a measurable subset E, the integral of ϕ over E with respect to μ is defined to be $\int_X \phi \cdot \chi_E d\mu$ and denoted by $\int_E \phi d\mu$.

(ii) If $f \in M^+$, then the value:

$$\int_X f d\mu = \sup \left\{ \int_X \phi d\mu \,\middle|\, 0 \leq \phi \leq f \text{ and } \phi \text{ is simple} \right\}$$

is called the **integral** of f with respect to μ. For a measurable subset E, the integral of f over E with respect to μ is defined to be $\int_X f \cdot \chi_E d\mu$ and denoted by $\int_E f d\mu$. If $\int_X f d\mu < \infty$, then f is said to be **integrable** over X with respect to μ.

(iii) If f is measurable and both of $\int_X f^+ d\mu$ and $\int_X f^- d\mu$ are finite, then f is said to be **integrable**. The space of all integrable functions is denoted by L^1, more precisely, by $L^1(X, \mathcal{M}, \mu)$. For an integrable function $f \in L^1$, the value:

$$\int_X f d\mu = \int_X f^+ d\mu - \int_X f^- d\mu$$

is called the **integral** of f with respect to μ. For a measurable subset E, if $f\mathbf{1}_E$ is integrable, then f is said to be **integrable** on E. The integral of f over E with respect to μ is defined to be $\int_X f \cdot \chi_E d\mu$ and denoted by $\int_E f d\mu$.

Lemma 5.1.2 Let $\phi \in M^+$. If there exist a finite disjoint family $\{E_i\}_{i=1}^n$ of measurable sets and a finite sequence $\{a_i\}_{i=1}^n$ of positive real numbers such that

$$\phi = \sum_{i=1}^n a_i \chi_{E_i},$$

then

$$\int_X \phi d\mu = \sum_{i=1}^n a_i \mu(E_i).$$

[Proof] Put

$$\phi(X) \setminus \{0\} = \{\lambda_j\}_{j=1}^m, \qquad A_j = \phi^{-1}(\{\lambda_j\}), \qquad j = 1, 2, \cdots, m,$$

where $\lambda_i \neq \lambda_j$ for $i \neq j$. Put $\{I_j\}_{j=1}^m$ is a partition of $\{1, 2, \cdots, n\}$ such that for any $i \in I_j$, $a_i = \lambda_j$. Then for each $j = 1, 2, \cdots, m$,

$$A_j = \bigcup_{i \in I_j} E_i,$$

and then we have the canonical representation of ϕ by

$$\phi = \sum_{j=1}^m \lambda_j \chi_{A_j}.$$

Therefore, we obtain that

$$\begin{aligned}\int_X \phi d\mu &= \sum_{j=1}^m \lambda_j \mu(A_j) = \sum_{j=1}^m \sum_{i \in I_j} \lambda_j \mu(E_i) = \sum_{j=1}^m \sum_{i \in I_j} a_i \mu(E_i) \\ &= \sum_{i=1}^n a_i \mu(E_i),\end{aligned}$$

which gives the proof. □

Remark 5.1.3 Let ϕ and ψ be nonnegative simple (measurable) functions defined on X and let

$$\phi = \sum_{i=1}^{l} a_i \chi_{E_i}, \qquad \psi = \sum_{j=1}^{m} b_i \chi_{F_i},$$

where $\cup_{i=1}^{l} E_i = \cup_{j=1}^{m} F_j = X$. For each $1 \leq i \leq l$ and $1 \leq j \leq m$, put

$$G_{ij} = E_i \bigcap F_j, \quad a_{ij} = a_i, \quad b_{ij} = b_j$$

Then we have

$$\phi = \sum_{i=1}^{l} a_i \chi_{E_i} = \sum_{i=1}^{l} a_i \left(\sum_{j=1}^{m} \chi_{G_{ij}} \right) = \sum_{i=1}^{l} \sum_{j=1}^{m} a_i \chi_{G_{ij}} = \sum_{i=1}^{l} \sum_{j=1}^{m} a_{ij} \chi_{G_{ij}}.$$

Similarly, we have

$$\psi = \sum_{i=1}^{l} \sum_{j=1}^{m} b_{ij} \chi_{G_{ij}}.$$

Proposition 5.1.4 Let ϕ and ψ be nonnegative simple functions defined on X. Then we have

(i) [**Linearity**] For any positive real numbers $\alpha, \beta > 0$,

$$\int_X (\alpha\phi + \beta\psi)\, d\mu = \alpha \int_X \phi d\mu + \beta \int_X \psi d\mu.$$

(ii) [**Monotonicity**] If $\phi \leq \psi$ on X, then

$$\int_X \phi d\mu \leq \int_X \psi d\mu.$$

[Proof] By Remark 5.1.3, we know that there exists a finite disjoint family $\{E_i\}_{i=1}^{n}$ of measurable sets and two finite sequences $\{a_i\}_{i=1}^{n}$ and $\{b_i\}_{i=1}^{n}$ of nonnegative real numbers such that

$$\phi = \sum_{i=1}^{n} a_i \chi_{E_i}, \qquad \psi = \sum_{i=1}^{n} b_i \chi_{E_i}. \tag{5.1}$$

(i) Then we have

$$\alpha\phi + \beta\psi = \sum_{i=1}^{n} (\alpha a_i + \beta b_i) \chi_{E_i}.$$

Therefore, by Lemma 5.1.2, we obtain that

$$\int_X (\alpha\phi + \beta\psi)\, d\mu = \sum_{i=1}^{n} (\alpha a_i + \beta b_i) \mu(E_i)$$

$$= \alpha \sum_{i=1}^{n} a_i \mu(E_i) + \beta \sum_{i=1}^{n} b_i \mu(E_i)$$
$$= \alpha \int_X \phi d\mu + \beta \int_X \psi d\mu,$$

where the second equality can be held as that both sides are infinite, which gives the proof.

(ii) Since $\phi \le \psi$ on X, $a_i \le b_i$ for all $i = 1, 2, \cdots, n$. Therefore, we have

$$\int_X \phi d\mu = \sum_{i=1}^{n} a_i \mu(E_i) \le \sum_{i=1}^{n} b_i \mu(E_i) = \int_X \psi d\mu,$$

which implies the proof. □

Remark 5.1.5 Let ϕ and ψ be nonnegative simple functions defined on X. If A and B are disjoint measurable sets, then we have that

$$\int_{A\cup B} \phi d\mu = \int_A \phi d\mu + \int_B \phi d\mu.$$

In fact, we may assume that ϕ and ψ are given as in (5.1) and then we have

$$\int_{A\cup B} \phi d\mu = \int_X \phi \cdot \chi_{A\cup B} d\mu = \int_X \phi \cdot \chi_A d\mu + \int_X \phi \cdot \chi_B d\mu = \int_A \phi d\mu + \int_B \phi d\mu.$$

In particular, if X_0 is a measurable set with $\mu(X \setminus X_0) = 0$, then we have

$$\int_X \phi d\mu = \int_{X_0} \phi d\mu + \int_{X\setminus X_0} \phi d\mu = \int_{X_0} \phi d\mu.$$

Furthermore, if $\phi \le \psi$ a.e. on A for a measurable set A, then there exists a measurable set $N \subset A$ with $\mu(N) = 0$ and so

$$\int_A \phi d\mu = \int_{A\setminus N} \phi d\mu \le \int_{A\setminus N} \psi d\mu = \int_A \psi d\mu.$$

Proposition 5.1.6 Let $(X, \mathcal{M}, \mu)$ be a measure space. Let f and g be nonnegative measurable functions defined on X.

(i) [**Homogeneity**] For any positive real number $\alpha > 0$,

$$\int_X \alpha f d\mu = \alpha \int_X f d\mu.$$

(ii) [**Monotonicity**] If $f \le g$ on X, then

$$\int_X f d\mu \le \int_X g d\mu.$$

[Proof] (i) By definition, we obtain that

$$\begin{aligned}
\int_X \alpha f d\mu &= \sup \left\{ \int_X \phi d\mu \,\middle|\, 0 \le \phi \le \alpha f \text{ and } \phi \text{ is simple} \right\} \\
&= \alpha \sup \left\{ \int_X \frac{1}{\alpha} \phi d\mu \,\middle|\, 0 \le \frac{1}{\alpha} \phi \le f \text{ and } \phi \text{ is simple} \right\} \\
&= \alpha \sup \left\{ \int_X \psi d\mu \,\middle|\, 0 \le \psi \le f \text{ and } \psi \text{ is simple} \right\} \\
&= \alpha \int_X f d\mu.
\end{aligned}$$

(ii) Since $0 \le f \le g$,

$$\{\phi \text{ is simple} \,|\, 0 \le \phi \le f\} \subset \{\phi \text{ is simple} \,|\, 0 \le \phi \le g\}.$$

Therefore, by definition, we obtain that

$$\begin{aligned}
\int_X f d\mu &= \sup \left\{ \int_X \phi d\mu \,\middle|\, 0 \le \phi \le f \text{ and } \phi \text{ is simple} \right\} \\
&\le \sup \left\{ \int_X \phi d\mu \,\middle|\, 0 \le \phi \le g \text{ and } \phi \text{ is simple} \right\} \\
&= \int_X g d\mu.
\end{aligned}$$

□

Theorem 5.1.7 **[Markov's Inequality]** Let $(X, \mathcal{M}, \mu)$ be a measure space. Let f be a nonnegative measurable functions on X. Then for any $\lambda > 0$, we have

$$\mu(\{x \in X \,|\, f(x) \ge \lambda\}) \le \frac{1}{\lambda} \int_X f d\mu. \tag{5.2}$$

[Proof] For any $\lambda > 0$, put

$$A_\lambda = \{x \in X \,|\, f(x) \ge \lambda\}$$

and $\varphi = \lambda \chi_{A_\lambda}$. Then φ is a simple function such that $\varphi \le f$. Therefore, by the definition of integral, we have

$$\begin{aligned}
\lambda \mu(A_\lambda) = \int_{A_\lambda} \varphi d\mu &\le \int_{A_\lambda} f d\mu \\
&\le \int_X f d\mu,
\end{aligned}$$

which implies (5.2). □

Proposition 5.1.8 Let $(X, \mathcal{M}, \mu)$ be a measure space. Let f be a nonnegative integrable function defined on X. Then f is finite a.e. on X and $\{x \in X \mid f(x) > 0\}$ is σ-finite.

[Proof] For any $n \in \mathbb{N}$, by applying Morkov's inequality (Theorem 5.1.7), we have

$$\mu(\{x \in X \mid f(x) = \infty\}) \leq \mu\left(\{x \in X \mid f(x) \geq n\}\right) \leq \frac{1}{n}\int_X f d\mu,$$

which, since $0 \leq \int_X f d\mu < \infty$, implies that $\mu(\{x \in X \mid f(x) = \infty\}) = 0$.

For each $n \in \mathbb{N}$, put

$$X_n = \left\{x \in X \,\middle|\, f(x) \geq \frac{1}{n}\right\}.$$

Then by the Morkov's inequality (Theorem 5.1.7), we have

$$\mu(X_n) \leq n \int_X f d\mu < \infty$$

and

$$\{x \in X \mid f(x) > 0\} = \bigcup_{n=1}^{\infty} X_n.$$

Therefore, $\{x \in X \mid f(x) > 0\}$ is σ-finite. □

Proposition 5.1.9 Let $(X, \mathcal{M}, \mu)$ be a measure space. Let f be a nonnegative integrable function defined on X. If $\int_X f d\mu = 0$, then $f = 0$ a.e. on X.

[Proof] Note that

$$\{x \in X \mid f(x) \neq 0\} = \bigcup_{n=1}^{\infty} \left\{x \in X \,\middle|\, f(x) \geq \frac{1}{n}\right\}.$$

Therefore, by the continuity of measure and the Markov's inequality, we see that

$$\begin{aligned}
\mu\left(\{x \in X \mid f(x) \neq 0\}\right) &= \mu\left(\bigcup_{n=1}^{\infty} \left\{x \in X \,\middle|\, f(x) \geq \frac{1}{n}\right\}\right) \\
&= \lim_{n\to\infty} \mu\left(\left\{x \in X \,\middle|\, f(x) \geq \frac{1}{n}\right\}\right) \\
&\leq \lim_{n\to\infty} n \int_X f \, d\mu \\
&= 0,
\end{aligned}$$

which gives the proof. □

Proposition 5.1.10 Let $(X, \mathcal{M}, \mu)$ be a measure space and let f a nonnegative measurable function. Let $\{\phi_n\}_{n=1}^{\infty}$ be an increasing sequence of simple functions such that $\{\phi_n\}_{n=1}^{\infty}$ converges to f pointwise on X. Then we have

$$\int_X f d\mu = \lim_{n\to\infty} \int_X \phi_n d\mu.$$

[Proof] Let $0 < \alpha < 1$ and φ be a nonnegative simple function such that $0 \le \varphi \le f$. For each $n \in \mathbb{N}$, put

$$X_n = \{x \in X \mid \phi_n(x) \ge \alpha\varphi(x)\}.$$

Then $\{X_n\}_{n=1}^{\infty}$ is an ascending sequence of measurable sets and $X = \cup_{n=1}^{\infty} X_n$. Suppose that $\varphi = \sum_{i=1}^{k} a_i \chi_{A_i}$ is the canonical representation. Then, by the continuity of the measure μ, we obtain that

$$\begin{aligned}
\lim_{n\to\infty} \int_X \alpha\varphi \cdot \chi_{X_n} d\mu &= \lim_{n\to\infty} \sum_{i=1}^{k} \alpha a_i \mu(A_i \cap X_n) = \sum_{i=1}^{k} \alpha a_i \left(\lim_{n\to\infty} \mu(A_i \cap X_n) \right) \\
&= \sum_{i=1}^{k} \alpha a_i \mu(A_i) \\
&= \alpha \int_X \varphi d\mu,
\end{aligned}$$

which, by (ii) (monotonicity) in Proposition 5.1.6, implies that

$$\alpha \int_X \varphi d\mu = \lim_{n\to\infty} \int_X \alpha\varphi \cdot \chi_{X_n} d\mu = \lim_{n\to\infty} \int_{X_n} \alpha\varphi d\mu \le \lim_{n\to\infty} \int_{X_n} \phi_n d\mu \le \lim_{n\to\infty} \int_X \phi_n d\mu$$

for arbitrary $\alpha \in (0, 1)$, which implies that

$$\int_X \varphi d\mu \le \lim_{n\to\infty} \int_X \phi_n d\mu,$$

which, by taking supremum on the left hand side, implies that

$$\int_X f d\mu \le \lim_{n\to\infty} \int_X \phi_n d\mu.$$

On the other hand, since $\{\phi_n\}_{n=1}^{\infty}$ is an increasing sequence of simple functions such that $\{\phi_n\}_{n=1}^{\infty}$ converges pointwise to f on X, it is obvious that

$$\lim_{n\to\infty} \int_X \phi_n d\mu \le \int_X f d\mu,$$

which gives the proof. □

Corollary 5.1.11 Let $(X, \mathcal{M}, \mu)$ be a measure space and let f a nonnegative measurable function. Then there exists an increasing sequence $\{\phi_n\}_{n=1}^{\infty}$ of simple functions such that $\{\phi_n\}_{n=1}^{\infty}$ converges pointwise to f on X and

$$\int_X f d\mu = \lim_{n\to\infty} \int_X \phi_n d\mu.$$

[Proof] The proof is immediate from the Simple Approximation Theorem (Theorem 4.1.37) and Proposition 5.1.10. □

Theorem 5.1.12 **[Linearity]** Let $(X, \mathcal{M}, \mu)$ be a measure space. Let f and g be nonegative measurable functions defined on X. Then for any positive real numbers $\alpha, \beta > 0$, it holds that

$$\int_X (\alpha f + \beta g)\, d\mu = \alpha \int_X f d\mu + \beta \int_X g d\mu.$$

[Proof] By Corollary 5.1.11, there exist increasing sequences $\{\phi_n\}_{n=1}^{\infty}$ and $\{\psi_n\}_{n=1}^{\infty}$ of simple functions such that $\{\phi_n\}_{n=1}^{\infty}$ and $\{\psi_n\}_{n=1}^{\infty}$ converge pointwise to f and g, respectively, on X and

$$\int_X f d\mu = \lim_{n\to\infty} \int_X \phi_n d\mu, \qquad \int_X g d\mu = \lim_{n\to\infty} \int_X \psi_n d\mu.$$

Then $\{\alpha\phi_n + \beta\psi_n\}_{n=1}^{\infty}$ is an increasing sequence of simple functions such that $\{\alpha\phi_n + \beta\psi_n\}_{n=1}^{\infty}$ converge pointwise to $\alpha f + \beta g$ on X. Therefore, by Propositions 5.1.10 and 5.1.4, we obtain that

$$\begin{aligned}\int_X (\alpha f + \beta g)\, d\mu &= \lim_{n\to\infty} \int_X (\alpha\phi_n + \beta\psi_n)\, d\mu = \alpha \lim_{n\to\infty} \int_X \phi_n d\mu + \beta \lim_{n\to\infty} \int_X \psi_n d\mu \\ &= \alpha \int_X f d\mu + \beta \int_X g d\mu,\end{aligned}$$

which gives the proof. □

Corollary 5.1.13 Let $(X, \mathcal{M}, \mu)$ be a measure space. Let f be a nonnegative measurable function defined on X. Suppose that A and B are disjoint measurable sets. Then we have

$$\int_{A\cup B} f d\mu = \int_A f d\mu + \int_B f d\mu.$$

[Proof] By Theorem 5.1.12, we obtain that

$$\int_{A\cup B} f\mu = \int_X f\cdot\chi_{A\cup B}\mu = \int_X f(\chi_A+\chi_B)\mu = \int_X f\chi_A\mu + \int_X f\cdot\chi_B\mu = \int_A f + \int_B f,$$

which gives the proof. □

Proposition 5.1.14 [**Monotonicity**] Let $(X, \mathcal{M}, \mu)$ be a measure space. Let f and g be nonegative measurable functions defined on X. If $f \leq g$ a.e. on X, then

$$\int_X f d\mu \leq \int_X g d\mu.$$

[Proof] Since $f \leq g$ a.e. on X, there exists a measurable set $N \in \mathcal{M}$ with $\mu(N) = 0$ such that $f \leq g$ on $X \setminus N$, therefore, by (ii) in Proposition 5.1.6, we have

$$\begin{aligned}\int_X f d\mu &= \int_{X\setminus N} f d\mu + \int_N f d\mu = \int_{X\setminus N} f d\mu \leq \int_{X\setminus N} g d\mu = \int_{X\setminus N} g d\mu + \int_N g d\mu \\ &\leq \int_X g d\mu,\end{aligned}$$

which gives the proof. □

Theorem 5.1.15 [**Monotone Convergence Theorem**] Let $(X, \mathcal{M}, \mu)$ be a measure space. Let $\{f_n\}$ be an increasing sequence of nonnegative measurable functions on X. If $\{f_n\}$ converges to f pointwise on X, then

$$\lim_{n\to\infty} \int_X f_n d\mu = \int_X f d\mu.$$

[Proof] For any $n \in \mathbb{N}$, $f_n \leq f$ and so $\int_X f_n d\mu \leq \int_X f d\mu$ and $\lim_{n\to\infty} \int_X f_n d\mu \leq \int_X f d\mu$. For each $n \in \mathbb{N}$, there exists an increasing sequence $\left\{\phi_k^{(n)}\right\}_{k=1}^{\infty}$ of nonnegative simple functions $\phi_k^{(n)}$ such that $\lim_{k\to\infty} \phi_k^{(n)}(x) = f_n(x)$ for any $x \in X$. For each $n \in \mathbb{N}$, put

$$\psi_n = \max\left\{\phi_n^{(k)} \mid 1 \leq k \leq n\right\}.$$

Then $\{\psi_n\}_{n=1}^{\infty}$ is an increasing sequence of simple functions. In fact, it holds that

$$\begin{aligned}\psi_n &= \max\{\phi_n^{(k)} \,|\, 1 \leq k \leq n\} \leq \max\{\phi_{n+1}^{(k)} \,|\, 1 \leq k \leq n\} \\ &\leq \max\{\phi_{n+1}^{(k)} \,|\, 1 \leq k \leq n+1\} \\ &= \psi_{n+1}.\end{aligned}$$

Put

$$\psi(x) = \lim_{n\to\infty} \psi_n(x), \qquad x \in X.$$

Then ψ is a nonnegative measurable function. Moreover, for each $n \in \mathbb{N}$ and $1 \le k \le n$, $\phi_n^{(k)} \le f_k$, and so $\psi_n \le f_n \le f$, which implies that $\psi \le f$. On the other hand, for any $1 \le k \le n$, we have that $\phi_n^{(k)} \le \psi_n \le \psi$ and so, by taking limit in n, we see that $f_k \le \psi$ and so $f \le \psi$. Therefore, $\psi = f$, and so by Proposition 5.1.10, we see that

$$\int_X f d\mu = \int_X \psi d\mu = \lim_{n\to\infty} \int_X \psi_n d\mu.$$

Also, since $\psi_n \le f_n \le f$,

$$\lim_{n\to\infty} \int_X \psi_n d\mu \le \lim_{n\to\infty} \int_X f_n d\mu \le \int_X f d\mu.$$

Hence we prove that

$$\int_X f d\mu \le \lim_{n\to\infty} \int_X f_n d\mu \le \int_X f d\mu,$$

which gives the proof. □

Example 5.1.16 Let $X = [0, \infty)$ and let $\mu = m$ the Lebesgue measure. Let $f_n = \chi_{(n,n+1)}$ for each $n \in \mathbb{N}$. Then $\{f_n\}$ converges pointwise to $f := 0$ on X and

$$\int_X f d\mu = 0 < 1 = \lim_{n\to\infty} \int_X f_n d\mu.$$

Example 5.1.17 Let $(\mathbb{N}, \mathcal{P}(\mathbb{N}), c)$ be the measure space with the counting measure c. Let $f_n = \chi_{\{n\}}$ for each $n \in \mathbb{N}$. Then $\{f_n\}$ converges pointwise to $f := 0$ on $\mathbb{N}$ and

$$\int_{\mathbb{N}} f dc = 0 < 1 = \lim_{n\to\infty} \int_{\mathbb{N}} f_n dc.$$

Theorem 5.1.18 **[Fatou's Lemma]** Let $(X, \mathcal{M}, \mu)$ be a measure space. Let $\{f_n\}$ be a sequence of nonnegative measurable functions on X. Then it holds that

$$\int_X (\liminf_{n\to\infty} f_n) d\mu \le \liminf_{n\to\infty} \int_X f_n d\mu.$$

[Proof] For each $n \in \mathbb{N}$ and for all $j \ge n$, $g_n := \inf_{k\ge n} f_k \le f_j$. Therefore, for each $n \in \mathbb{N}$ and for all $j \ge n$, by (ii) (monotonicity) in Proposition 5.1.6, we have

$$\int_X \inf_{k\ge n} f_k d\mu \le \int_X f_j d\mu,$$

which implies that

$$\int_X \inf_{k\ge n} f_k d\mu \le \inf_{j\ge n} \int_X f_j d\mu.$$

Therefore, by the Monotone Convergence Theorem (Theorem 5.1.15), we have

$$\int_X \liminf_{n\to\infty} f_k d\mu = \lim_{n\to\infty} \int_X \inf_{k\geq n} f_k d\mu \leq \liminf_{n\to\infty} \int_X f_n d\mu$$

which gives the proof. □

Corollary 5.1.19 **[Fatou's Lemma]** Let $(X, \mathcal{M}, \mu)$ be a measure space. Let $\{f_n\}$ be a sequence of nonnegative measurable functions on X such that $\{f_n\}$ converges pointwise to a function f a.e. on X. Suppose that f is measurable and nonnegative. Then it holds that

$$\int_X f d\mu \leq \liminf_{n\to\infty} \int_X f_n d\mu.$$

[Proof] There exists a measurable set N with $\mu(N) = 0$ such that $\{f_n\}$ converges pointwise to the measurable function f on $X \setminus N$. Then by applying Corollary 5.1.13, Proposition 5.1.9 and the Fatou's Lemma (Theorem 5.1.18), we have

$$\int_X f d\mu = \int_{X\setminus N} f d\mu = \int_{X\setminus N} (\liminf_{n\to\infty} f_n) d\mu = \int_X (\liminf_{n\to\infty} f_n) d\mu \leq \liminf_{n\to\infty} \int_X f_n d\mu,$$

which gives the proof. □

Remark 5.1.20 In Corollary 5.1.19, if the sequence $\{f_n\}$ converges pointwise to a function f on X or $(X, \mathcal{M}, \mu)$ is a complete measure space, then, by Proposition 4.1.30, f is measurable.

Theorem 5.1.21 **[Beppo Levi's Lemma]** Let $(X, \mathcal{M}, \mu)$ be a measure space. Let $\{f_n\}$ be an increasing sequence of nonnegative measurable functions on X. If the sequence $\{\int_X f_n d\mu\}$ is bounded, then there exists a measurable function f on X which is finite a.e. on X such that $\{f_n\}$ converges to f pointwise on X and

$$\lim_{n\to\infty} \int_X f_n d\mu = \int_X f d\mu < \infty.$$

[Proof] Define the measurable function f on X by

$$f(x) = \lim_{n\to\infty} f_n(x) \quad \text{for all } x \in X.$$

In this case, if $\{f_n(x)\}$ is bounded, then $f(x)$ is finite and if $\{f_n(x)\}$ is unbounded, then $f(x) = \infty$. Then $\{f_n\}$ converges to f pointwise on X to f on X and by the Monotone Convergence Theorem (Theorem 5.1.15) we have

$$\lim_{n\to\infty} \int_X f_n d\mu = \int_X f d\mu,$$

and, since $\{\int_X f_n d\mu\}$ is a bounded increasing sequence, the limit $\int_X f d\mu$ is finite. Therefore, by Proposition 5.1.8, f is finite a.e. on X. □

For a function defined on X, we define

$$f^+(x) = \max\{f(x), 0\}, \qquad f^-(x) = \max\{-f(x), 0\}, \qquad |f| = f^+ + f^-.$$

Then we note that

$$f = f^+ - f^-.$$

Let $(X, \mathcal{M})$ be a measurable space and let f a measurable function defined on X. Then by Proposition 4.1.27, f^+, f^- and $|f|$ are measurable.

Proposition 5.1.22 Let $(X, \mathcal{M}, \mu)$ be a measure space. Let f be a measurable function defined on X such that for an integrable function g defined on X, $|f| \leq g$ a.e. on X. Then f is integrable over X with respect to μ and

$$\left|\int_X f d\mu\right| \leq \int_X |f| d\mu \leq \int_X g d\mu.$$

[Proof] By Proposition 5.1.14, f is integrable over X with respect to μ. Note that, since f is finite a.e.,

$$\begin{aligned}
\left|\int_X f d\mu\right| &= \left|\int_X f^+ d\mu - \int_X f^- d\mu\right| \\
&\leq \int_X f^+ d\mu + \int_X f^- d\mu \\
&= \int_X |f| d\mu \\
&\leq \int_X g d\mu,
\end{aligned}$$

which implies the proof. □

Remark 5.1.23 Let $(X, \mathcal{M}, \mu)$ be a measure space. Let f and g be integrable functions on X. Then by Proposition 5.1.8, f and g are finite a.e. on X. Then there exists a measurable subset $N \subset X$ with $\mu(N) = 0$ such that f and g are finite on $X_0 := X \setminus N$. Therefore,$f + g$ is well-defined on X_0 (see Remark 4.1.24). Moreover, by Propositions 4.1.20 and 4.1.23, $f + g$ is measurable with respect to the measurable space $(X_0, \mathcal{M}_{X_0})$. Therefore, we can extend $f+g$ to X as a measurable function on X and then, by Corollary 5.1.13, we have

$$\int_X (f+g) d\mu = \int_{X_0} (f+g) d\mu,$$

of which the integral does not defend on the extension of $f + g$ on N. Therefore, for two integrable functions f and g defined on X, $f+g$ is considered as any measurable extension which agrees with $f + g$ on $X_0 = X \setminus N$ for some measurable set with $\mu(N) = 0$.

Theorem 5.1.24 [**Linearity and Monotonicity**] Let $(X, \mathcal{M}, \mu)$ be a measure space. Let f and g be integrable functions over X. Then we have

(i) [**Linearity**] For any $\alpha, \beta \in \mathbb{R}$, $\alpha f + \beta g$ is integrable over X and

$$\int_X (\alpha f + \beta g)\, d\mu = \alpha \int_X f d\mu + \beta \int_X g d\mu.$$

(ii) [**Monotonicity**] If $f \le g$ a.e. on X, then

$$\int_X f d\mu \le \int_X g d\mu.$$

[Proof] (i) For any $\alpha \in \mathbb{R}$, we obtain that

$$(\alpha f)^+ = \begin{cases} 0, \ \alpha = 0 \\ \alpha f^+, \ \alpha > 0 \\ -\alpha f^-, \ \alpha < 0 \end{cases}, \qquad (\alpha f)^- = \begin{cases} 0, \ \alpha = 0 \\ \alpha f^-, \ \alpha > 0 \\ -\alpha f^+, \ \alpha < 0 \end{cases}.$$

Therefore, by the homogeneity (Proposition 5.1.6), we have

$$\int_X \alpha f d\mu = \int_X (\alpha f)^+ d\mu - \int_X (\alpha f)^- d\mu = \alpha \int_X f d\mu.$$

On the other hand, since f and g are integrable over X and $|f+g| \le |f| + |g|$, by the linearity (Theorem 5.1.12), $f+g$ is integrable over X. Also, by Proposition 5.1.8, f and g are finite a.e. on X, and so by Corollary 5.1.13 we may assume that f and g are finite on X. Then we have

$$(f+g)^+ - (f+g)^- = f + g = (f^+ - f^-) + (g^+ - g^-),$$

which implies that

$$(f+g)^+ + f^- + g^- = (f+g)^- + f^+ + g^+,$$

and so by the linearity (Theorem 5.1.12)), we have

$$\int_X (f+g)^+ d\mu + \int_X f^- d\mu + \int_X g^- d\mu = \int_X (f+g)^- d\mu + \int_X f^+ d\mu + \int_X g^+ d\mu,$$

which proves that

$$\int_X (f+g) d\mu = \int_X f d\mu + \int_X g d\mu.$$

(ii) Put $h = g - f$ on the set of all $x \in X$ at which f and g are finite. Then we have

$$\int_X g d\mu - \int_X f d\mu = \int_X (g-f) d\mu = \int_X h d\mu \ge 0,$$

which implies the proof. □

Corollary 5.1.25 Let $(X, \mathcal{M}, \mu)$ be a measure space. Let f be an integrable function defined over X. Suppose A and B are disjoint measurable subsets of X. Then we have

$$\int_{A\cup B} f d\mu = \int_A f d\mu + \int_B f d\mu.$$

[Proof] For any measurable subset $E \subset X$, since f is integrable and $f \cdot \chi_E$ is integrable over X. Then we have

$$\begin{aligned}\int_{A\cup B} f d\mu &= \int_X f \cdot \chi_{A\cup B} d\mu = \int_X f \cdot (\chi_A + \chi_B) d\mu = \int_X f \cdot \chi_A d\mu + \int_X f \cdot \chi_B d\mu \\ &= \int_A f d\mu + \int_B f d\mu,\end{aligned}$$

which gives the proof. □

Theorem 5.1.26 **[Lebesgue Dominated Convergence Theorem]** Let $(X, \mathcal{M}, \mu)$ be a measure space. Let $\{f_n\}$ be a sequence of measurable functions defined on X. Suppose that there exists a (nonnegative) integrable function g on X such that $|f_n| \leq g$ a.e. on X for any $n \in \mathbb{N}$. If $\{f_n\}$ converges to a function f pointwise a.e. on X and f is measurable, then f is integrable over X and

$$\lim_{n\to\infty} \int_X f_n d\mu = \int_X f d\mu.$$

[Proof] For almost all $x \in X$, we have

$$|f(x)| = \lim_{n\to\infty} |f_n(x)| \leq g(x).$$

Therefore, by Proposition 5.1.14, f_n and f are integrable over X. Then since $\{f_n + g\}$ is a sequence of nonnegative measurable functions such that $\{f_n + g\}$ converges to $f + g$ pointwise a.e. on X, by Fadou's Lemma (Corollary 5.1.19) and Theorem 5.1.24, we have

$$\int_X f d\mu + \int_X g d\mu = \int_X (f+g) d\mu \leq \liminf_{n\to\infty} \int_X (f_n + g) d\mu = \liminf_{n\to\infty} \int_X f_n d\mu + \int_X g d\mu,$$

which implies that

$$\int_X f d\mu \leq \liminf_{n\to\infty} \int_X f_n d\mu.$$

Similarly, we have

$$\int_X g d\mu - \int_X f d\mu = \int_X (g - f) d\mu \leq \liminf_{n\to\infty} \int_X (g - f_n) d\mu = \int_X g d\mu - \limsup_{n\to\infty} \int_X f_n d\mu,$$

which implies that

$$\limsup_{n\to\infty} \int_X f_n d\mu \le \int_X f d\mu.$$

Finally, we have

$$\int_X f d\mu \le \liminf_{n\to\infty} \int_X f_n d\mu \le \limsup_{n\to\infty} \int_X f_n d\mu \le \int_X f d\mu,$$

which implies the proof. □

Remark 5.1.27 (i) Let $\{a_n\}$ and $\{b_n\}$ be sequences of real numbers such that

$$-\infty < \lim_{n\to\infty} \inf_{k\ge n} a_k, \quad \lim_{n\to\infty} \sup_{j\ge n} b_j < \infty,$$

then we have

$$\begin{aligned}\liminf_{n\to\infty}(a_n+b_n) &= \sup_{n\ge 1}\inf_{k\ge n}(a_k+b_k) = \lim_{n\to\infty}\inf_{k\ge n}(a_k+b_k) \le \lim_{n\to\infty}\inf_{k\ge n}(a_k + \sup_{j\ge n} b_j)\\ &= \lim_{n\to\infty}\left(\inf_{k\ge n} a_k + \sup_{j\ge n} b_j\right)\\ &= \lim_{n\to\infty}\inf_{k\ge n} a_k + \lim_{n\to\infty}\sup_{k\ge n} b_k.\end{aligned}$$

(ii) Let $(X, \mathcal{M}, \mu)$ be a measure space. Let $\{f_n\}$ be a sequence of measurable functions defined on X and let g be a nonnegative integrable function over X. Suppose that $|f_n| \le g$ a.e. on X for all $n \in \mathbb{N}$. Then we have

$$\int_X \liminf f_n d\mu \le \liminf \int_X f_n d\mu \le \limsup \int_X f_n d\mu \le \int_X \limsup f_n d\mu.$$

In fact, we may assume that $|f_n| \le g$ on X for all $n \in \mathbb{N}$ and then by Fadou's Lemma (Theorem 5.1.18), we obtain that

$$\begin{aligned}\int_X g d\mu + \int_X \liminf_{n\to\infty} f_n d\mu &= \int_X \liminf_{n\to\infty}(g+f_n)d\mu \le \liminf_{n\to\infty}\int_X (g+f_n)d\mu\\ &= \int_X g d\mu + \liminf_{n\to\infty}\int_X f_n d\mu,\\ \int_X g d\mu - \int_X \limsup_{n\to\infty} f_n d\mu &= \int_X \liminf_{n\to\infty}(g-f_n)d\mu \le \liminf_{n\to\infty}\int_X (g-f_n)d\mu\\ &= \int_X g d\mu - \limsup_{n\to\infty}\int_X f_n d\mu,\end{aligned}$$

which imply the proof of Theorem 5.1.26.

Example 5.1.28 Let $(\mathbb{R}, \mathfrak{M}_{\mathrm{L}}, m)$ be the Lebesgue measure space and let $h : \mathbb{R} \to \mathbb{R}$ be a function defined by

$$h(x) = \frac{1}{\sqrt{2\pi}} e^{-\frac{x^2}{2}}, \qquad x \in \mathbb{R}.$$

For each $n \in \mathbb{N}$, put $f_n(x) = h(x-n)$. Then f_n is given by

$$f_n(x) = \frac{1}{\sqrt{2\pi}} e^{-\frac{(x-n)^2}{2}}, \qquad x \in \mathbb{R},$$

and so f_n is continuous and Lebesgue measurable. On the other hand, $\{f_n\}_{n=1}^{\infty}$ converges to $f = 0$ pointwise on $\mathbb{R}$ and $\int_{\mathbb{R}} f_n dm = 1$, but $\int_{\mathbb{R}} f dm = 0$. Therefore,

$$\lim_{n\to\infty} \int_{\mathbb{R}} f_n dm = 1 \neq 0 = \int_{\mathbb{R}} f dm.$$

If $g : \mathbb{R} \to \mathbb{R}$ is a Lebesgue measurable function such that $|f_n| = f_n \leq g$ a.e on $\mathbb{R}$ for all $n \in \mathbb{N}$, then we can easily see that $f \leq g$ a.e. on $(-\infty, 0]$ and $g \geq e^{-1/8}/\sqrt{2\pi}$ a.e. on $[0, \infty)$, and then g is not integrable.

Theorem 5.1.29 [General Lebesgue Dominated Convergence Theorem] Let $(X, \mathcal{M}, \mu)$ be a measure space. Let $\{f_n\}$ be a sequence of measurable functions defined on X which converges to f pointwise a.e. on X. Suppose that f is measurable and there exists a sequence $\{g_n\}$ of nonnegative measurable functions g_n on X which converges to a measurable function g pointwise a.e. on X such that $|f_n| \leq g_n$ a.e. on X for any $n \in \mathbb{N}$. If

$$\lim_{n\to\infty} \int_X g_n d\mu = \int_X g d\mu < \infty,$$

then

$$\lim_{n\to\infty} \int_X f_n d\mu = \int_X f d\mu.$$

[Proof] By assumption, g is nonnegative a.e. on X. We may assume that g is nonnegative on X. For almost all $x \in X$, we have

$$|f(x)| = \lim_{n\to\infty} |f_n(x)| \leq \lim_{n\to\infty} g_n(x) = g(x).$$

Therefore, by Theorem 5.1.24, f is integrable over X. Then since $\{f_n + g_n\}$ is a sequence of nonnegative measurable functions such that $\{f_n + g_n\}$ converges to $f + g$ pointwise a.e. on X, by Fadou's Lemma (Corollary 5.1.19), Theorem 5.1.24 and Remark 5.1.27, we have

$$\int_X f d\mu + \int_X g d\mu = \int_X (f+g) d\mu \leq \liminf_{n\to\infty} \int_X (f_n + g_n) d\mu = \liminf_{n\to\infty} \int_X f_n d\mu + \int_X g d\mu,$$

which implies that

$$\int_X f d\mu \leq \liminf_{n\to\infty} \int_X f_n d\mu.$$

Similarly, we have

$$\int_X g d\mu - \int_X f d\mu = \int_X (g-f) d\mu \leq \liminf_{n\to\infty} \int_X (g_n - f_n) d\mu = \int_X g d\mu - \limsup_{n\to\infty} \int_X f_n d\mu,$$

which implies that

$$\limsup_{n\to\infty}\int_X f_n d\mu \le \int_X f d\mu.$$

Finally, we have

$$\int_X f d\mu \le \liminf_{n\to\infty}\int_X f_n d\mu \le \limsup_{n\to\infty}\int_X f_n d\mu \le \int_X f d\mu,$$

which implies the proof. □

Theorem 5.1.30 Let $(X, \mathcal{M}, \mu)$ be a measure space. Let f be an integrable function over X and let $\{X_n\}$ a disjoint countable family of measurable subsets of X. Then we have

$$\int_{\cup_{i=1}^{\infty} X_i} f d\mu = \sum_{i=1}^{\infty}\int_{X_i} f d\mu.$$

[Proof] By the Lebesgue Dominated Convergence Theorem (Theorem 5.1.26), we obtain that

$$\int_{\cup_{i=1}^{\infty} X_i} f d\mu = \int_X f\cdot\chi_{\cup_{i=1}^{\infty} X_i} d\mu = \int_X \lim_{n\to\infty} f\cdot\chi_{\cup_{i=1}^{n} X_i} d\mu = \lim_{n\to\infty}\sum_i^n \int_X f\cdot\chi_{X_i} d\mu,$$

which gives the proof. □

Theorem 5.1.31 Let $(X, \mathcal{M}, \mu)$ be a measure space. Let f be an integrable function over X.

(i) If $\{X_n\}_{n=1}^{\infty}$ is an increasing sequence of measurable subsets of X, then

$$\int_{\cup_{n=1}^{\infty} X_n} f d\mu = \lim_{n\to\infty}\int_{X_n} f d\mu.$$

(ii) If $\{X_n\}_{n=1}^{\infty}$ is a decreasing sequence of measurable subsets of X, then

$$\int_{\cap_{n=1}^{\infty} X_n} f d\mu = \lim_{n\to\infty}\int_{X_n} f d\mu.$$

[Proof] (i) By the Lebesgue Dominated Convergence Theorem (Theorem 5.1.26), we have

$$\int_{\cup_{n=1}^{\infty} X_n} f d\mu = \int_X \lim_{n\to\infty} f\cdot\chi_{X_n} d\mu = \lim_{n\to\infty}\int_X f\cdot\chi_{X_n} d\mu = \lim_{n\to\infty}\int_{X_n} f d\mu.$$

(2) By the Lebesgue Dominated Convergence Theorem (Theorem 5.1.26), we have

$$\int_{\bigcap_{n=1}^{\infty} X_n} f d\mu = \int_X \lim_{n\to\infty} f \cdot \chi_{X_n} d\mu = \lim_{n\to\infty} \int_X f \cdot \chi_{X_n} d\mu = \lim_{n\to\infty} \int_{X_n} f d\mu.$$

□

5.2 Lebesgue Integration

Let $(\mathbb{R}, \mathfrak{M}_{\mathrm{L}}, m)$ be the Lebesgue measure space, $f : \mathbb{R} \to \overline{\mathbb{R}}$ a (Lebesgue) measurable function and E a Lebesgue measurable set, i.e., $E \in \mathfrak{M}_{\mathrm{L}}$. If the integral:

$$\int_E f dm = \int_{\mathbb{R}} f^+ \chi_E dm - \int_{\mathbb{R}} f^- \chi_E dm \tag{5.3}$$

of f over E with respect to m is well-defined, i.e., the right hand side of (5.3) has a meaning, then it is called the **Lebesgue integral** of f over E with respect to the Lebesgue measure m, or simply **Lebesgue integral**. If $\int_E |f| dm < \infty$, then f is said to be **Lebesgue integrable** over E.

Let f be a bounded real valued function defined on a measurable set E of finite measure. We define the **lower Lebesgue integral** of f over E by

$$\underline{\int_E} f \, dm = \sup \left\{ \int_E \phi \, dm \,\middle|\, \phi \text{ a simple function and } \phi \leq f \right\}$$

and, similarly, we define the **upper Lebesgue integral** of f over E by

$$\overline{\int_E} f \, dm = \inf \left\{ \int_E \phi \, dm \,\middle|\, \phi \text{ a simple function and } \phi \geq f \right\}.$$

Proposition 5.2.1 Let $E \in \mathfrak{M}_{\mathrm{L}}$ with $m(E) < \infty$ and $f : E \to \mathbb{R}$ a bounded measurable function on E. Then the lower and upper Lebesgue integrals of f are equal, i.e.,

$$\underline{\int_E} f \, dm = \overline{\int_E} f \, dm.$$

[Proof] For any $n \in \mathbb{N}$, by the simple approximation theorem (Theorem 4.1.37), there are simple functions ϕ_n and ψ_n defined on E such that

$$\phi_n \leq f \leq \psi_n \text{ and } 0 \leq \psi_n - \phi_n < \frac{1}{2^{n-1}} \text{ on } E.$$

Therefore, by the monotonicity of the integral ((ii) in Theorem 5.1.24), we obtain that

$$0 \leq \inf \left\{ \int_E \psi \, dm \,\middle|\, \psi \text{ simple}, \psi \geq f \right\} - \sup \left\{ \int_E \phi \, dm \,\middle|\, \phi \text{ simple}, \phi \leq f \right\}$$

$$\begin{aligned}&\leq \int_E \psi_n \, dm - \int_E \phi_n \, dm = \int_E (\psi_n - \phi_n) \, dm \leq \int_E \frac{1}{2^{n-1}} \, dm \\ &= \frac{1}{2^{n-1}} m(E),\end{aligned}$$

which converges to 0 as $n \to \infty$. Therefore, we have

$$\inf \left\{ \int_E \psi \, dm \, \middle| \, \psi \text{ simple}, \psi \geq f \right\} = \sup \left\{ \int_E \phi \, dm \, \middle| \, \phi \text{ simple}, \phi \leq f \right\},$$

which is the assertion. □

Proposition 5.2.2 Let $E \in \mathfrak{M}_{\mathrm{L}}$ with $m(E) < \infty$ and $f, g : E \to \mathbb{R}$ bounded measurable functions on E. Then we have

$$\begin{aligned}\underline{\int}_E f \, dm + \underline{\int}_E g \, dm &= \underline{\int}_E (f+g) \, dm \\ &= \overline{\int}_E (f+g) \, dm \\ &= \overline{\int}_E f \, dm + \overline{\int}_E g \, dm.\end{aligned}$$

[Proof] For any simple functions ϕ and ψ with $f \geq \phi$ and $g \geq \psi$, $\phi + \psi$ is simple and $f + g \geq \phi + \psi$, which implies that

$$\underline{\int}_E (f+g) \, dm \geq \int_E (\phi + \psi) \, dm = \int_E \phi \, dm + \int_E \psi \, dm,$$

and so by taking supremum for the right hand side of the previous equation, we have

$$\underline{\int}_E (f+g) \, dm \geq \underline{\int}_E f \, dm + \underline{\int}_E g \, dm.$$

Similarly, we see that

$$\overline{\int}_E (f+g) \, dm \leq \overline{\int}_E f \, dm + \overline{\int}_E g \, dm.$$

Therefore, by applying Proposition 5.2.1, we prove the assertion. □

Proposition 5.2.3 Let $E \in \mathfrak{M}_{\mathrm{L}}$ with $m(E) < \infty$ and $f : E \to \mathbb{R}$ a bounded measurable function on E. Then we have

$$\int_E f \, dm = \underline{\int}_E f \, dm = \overline{\int}_E f \, dm.$$

[Proof] The proof of the first equality is similar to the proof of the second equality. We now prove the second equality. In fact, by applying Proposition 5.2.2, we obtain that

$$\begin{aligned}\int_E f\,dm &= \int_E f^+\,dm - \int_E f^-\,dm \\ &= \underline{\int}_E f^+\,dm - \underline{\int}_E f^-\,dm \\ &= \overline{\int}_E f^+\,dm + \overline{\int}_E (-f^-)\,dm \\ &= \overline{\int}_E f\,dm,\end{aligned}$$

which gives the proof. □

5.3 Riemann Integral as Lebsgue Integral

Let f be a real valued bounded function defined on a bounded closed interval $[a, b]$. Let $P = \{x_0, x_1, \cdots, x_n\}$ be a partition of $[a, b]$, i.e.,

$$P : a = x_0 < x_1 < \cdots < x_n = b.$$

Define the **lower Riemann sum** $L(f, P)$ and **upper Riemann sum** $U(f, P)$ for f with respect to the partition P by

$$\begin{aligned}L(f, P) &= \sum_{i=1}^{n} m_i(x_i - x_{i-1}), \\ U(f, P) &= \sum_{i=1}^{n} M_i(x_i - x_{i-1}),\end{aligned}$$

where

$$m_i = \inf\{f(x) \mid x_{i-1} < x < x_i\}, \qquad M_i = \sup\{f(x) \mid x_{i-1} < x < x_i\}$$

Then the **lower Riemann integral** $(R)\underline{\int}_a^b f dx$ and **upper Riemann integral** $(R)\overline{\int}_a^b f dx$ of f over $[a, b]$ are defined by

$$\begin{aligned}(R)\underline{\int}_a^b f\,dx &= \sup\left\{L(f, P) \mid P \text{ a partition of } [a, b]\right\}, \\ (R)\overline{\int}_a^b f\,dx &= \inf\left\{U(f, P) \mid P \text{ a partition of } [a, b]\right\}.\end{aligned}$$

If the lower Riemann integral and upper Riemann integral are equal, then we say that f is **Riemann integrable** over $[a, b]$ and the common value of the lower and upper Riemann integrals is called the **Riemann integral** of f over $[a, b]$ and denoted by

$$(R)\int_a^b f dx.$$

A real valued function ϕ defined on $[a,b]$ is called a **step function** if there exist a partition $\{I_i\}_{i=1}^n$ of $[a,b]$, i.e., $I_i \neq \emptyset$ for all $i = 1, 2, \cdots, n$, $I_i \cap I_j = \emptyset$ for $i \neq j$ and $\cup_{i=1}^n I_i = [a,b]$, and real numbers $a_1, a_2, \cdots, a_n$ such that

$$\phi(x) = a_i, \qquad x \in I_i, \qquad 1 \le i \le n.$$

For the step function ϕ and the partition P of $[a,b]$, we have

$$L(\phi, P) = \sum_{i=1}^n a_i(x_i - x_{i-1}) = U(\phi, P),$$

where $x_{i-1} = \inf\{x \mid x \in I_i\}$ and $x_i = \sup\{x \mid x \in I_i\}$. Therefore, ϕ is Riemann integrable and

$$(R)\int_a^b \phi\, dx = \sum_{i=1}^n a_i(x_i - x_{i-1})$$

and so we can reformulate the **lower Riemann integral** $(R)\underline{\int}_a^b f dx$ and **upper Riemann integral** $(R)\overline{\int}_a^b f dx$ of f over $[a,b]$ by

$$(R)\underline{\int}_a^b f\, dx = \sup\left\{ (R)\int_a^b \phi\, dx \,\middle|\, \phi \text{ a step function and } \phi \le f \text{ on } [a,b] \right\},$$
$$(R)\overline{\int}_a^b f\, dx = \inf\left\{ (R)\int_a^b \phi\, dx \,\middle|\, \phi \text{ a step function and } \phi \ge f \text{ on } [a,b] \right\}.$$

Remark 5.3.1 Let $[0,1] \cap \mathbb{Q} = \{x_1, x_2, \cdots, x_n, \cdots\}$. For each $n \in \mathbb{N}$, define

$$f_n(x) = \begin{cases} 1, & x \in \{x_1, x_2, \cdots, x_n\}, \\ 0, & x \in [0,1] \setminus \{x_1, x_2, \cdots, x_n\} \end{cases}$$

and

$$f(x) = \begin{cases} 1, & x \in [0,1] \cap \mathbb{Q}, \\ 0, & x \in [0,1] \cap \mathbb{Q}^c. \end{cases}$$

Then for each $n \in \mathbb{N}$, f_n is Riemann integrable and $\{f_n\}$ converges to f pointwise on $[0,1]$. But f is not Riemann integrable.

Theorem 5.3.2 Let f be a real valued bounded function defined on a bounded closed interval $[a,b]$. If f is Riemann integrable over $[a,b]$, then f is Lebesgue integrable over $[a,b]$ and it holds that

$$(R)\int_a^b f dm = \int_a^b f dm. \tag{5.4}$$

[Proof] Suppose that f is Riemann integral. Then there exist an increasing sequence $\{\phi_n\}_{n=1}^{\infty}$ of step functions and a decreasing sequence $\{\psi_n\}_{n=1}^{\infty}$ of step functions such that $\phi_n \le f$ and $f \le \psi_n$ for all $n \in \mathbb{N}$, and

$$\lim_{n\to\infty} \int_a^b \phi_n \, dx = \lim_{n\to\infty} \int_a^b \psi_n \, dx = (R) \int_a^b f \, dm.$$

Put

$$g = \lim_{n\to\infty} \phi_n, \qquad h = \lim_{n\to\infty} \psi_n,$$

and then $g \le f \le h$ and by the dominated convergence theorem, we see that

$$\int_{[a,b]} g dm = \int_{[a,b]} h dm = (R) \int_{[a,b]} f dm.$$

Therefore, $h - g \ge 0$ and by the linearity of the integral

$$\int_{[a,b]} (h-g) \, dm = 0,$$

which implies that $h - g = 0$ a.e. and so $f = h = g$ a.e. and so by Proposition 4.1.20, f is measurable since $(\mathbb{R}, \mathfrak{M}_{\mathrm{L}}, m)$ is a complete measure space. Therefore, f is Lebesgue integrable. Since f is Riemann integral, we have

$$\begin{aligned} &\sup\left\{ (R)\int_a^b \phi dx \,\middle|\, \phi \text{ a step function }, \phi \le f \right\} \\ &= \inf\left\{ (R)\int_a^b \phi dx \,\middle|\, \phi \text{ a step function }, \phi \ge f \right\}. \end{aligned}$$

Note that all step functions are simple functions. Therefore, we obtain that

$$\begin{aligned} (R)\underline{\int_a^b} f dm &= \sup\left\{ (R)\int_a^b \phi dx \,\middle|\, \phi \text{ a step function }, \phi \le f \right\} \\ &\le \sup\left\{ (R)\int_a^b \phi dx \,\middle|\, \phi \text{ a simple function }, \phi \le f \right\} \\ &= \underline{\int_a^b} f dm = \int_a^b f dm = \overline{\int_a^b} f dm \\ &= \inf\left\{ \int_a^b \phi dx \,\middle|\, \phi \text{ a simple function }, \phi \ge f \right\} \\ &\le (R)\overline{\int_a^b} f dm, \end{aligned}$$

which implies (5.4). □

5.4 Lebesgue's Theorem for Riemann Integrability

Let $f : [a,b] \to \mathbb{R}$ be a function. For each $\delta > 0$ and $x \in [a,b]$, put

$$M(f,x,\delta) = \sup\{f(y) \,|\, y \in (x-\delta, x+\delta) \cap [a,b]\},$$

$$m(f, x, \delta) = \inf\{f(y) \,|\, y \in (x - \delta, x + \delta) \cap [a, b]\},$$

and

$$v(f, x) = \inf\{M(f, x, \delta) - m(f, x, \delta) \,|\, \delta > 0\}.$$

Then $M(f, x, \delta)$ and $m(f, x, \delta)$ are decreasing and increasing, respectively, in δ.

Remark 5.4.1 Let $A \subset \mathbb{R}$. If for any $\epsilon > 0$, there exists a countable family $\{(a_n, b_n)\}_{n=1}^{\infty}$ of open intervals such that

(i) $A \subset \bigcup_{n=1}^{\infty}(a_n, b_n)$,

(ii) $\sum_{n=1}^{\infty}(b_n - a_n) < \epsilon$,

then by Definition 3.1.2, $m^*(A) = 0$ and so by Proposition 3.2.3, A is (Lebesgue) measurable and then, by the definition of Lebesgue measure, we have $\mathfrak{m}(A) = 0$.

Lemma 5.4.2 A function $f : [a, b] \to \mathbb{R}$ is discontinuous at $x_0 \in [a, b]$ if and only if $v(f, x_0) > 0$.

[Proof] $\Rightarrow$) Suppose that f is discontinuous at x_0. Then there exists $\epsilon_0 > 0$ such that for any $\delta > 0$, there exists $x_\delta \in [a, b]$ with $|x_0 - x_\delta| < \delta$ and $|f(x_\delta) - f(x_0)| \geq \epsilon_0$. Therefore,

$$M(f, x_0, \delta) - m(f, x_0, \delta) \geq \epsilon_0,$$

which implies that

$$v(f, x_0) > 0.$$

$\Leftarrow$) Suppose that f is continuous at x_0. Then for any $\epsilon > 0$, there exists $\delta > 0$ such that for any $y \in [a, b]$ with $|x_0 - y| < \delta$, $|f(y) - f(x_0)| < \epsilon$. Therefore, for any $\epsilon > 0$, there exists $\delta > 0$ such that

$$M(f, x_0, \delta) - m(f, x_0, \delta) \leq 2\epsilon,$$

and so

$$\begin{aligned} v(f, x_0) &= \inf\{M(f, x_0, \delta) - m(f, x_0, \delta) \,|\, \delta > 0\} \\ &= 0. \end{aligned}$$

Therefore, if $v(f, x_0) > 0$, i.e., $v(f, x_0) \neq 0$, then f is discontinuous at x_0. $\square$

Theorem 5.4.3 **[Lebesgue's Theorem]** Let a function $f : [a,b] \to \mathbb{R}$ be bounded on the closed and bounded interval $[a,b]$. Then f is Riemann integrable over $[a,b]$ if and only if the set

$$D = \{x \in [a,b] \mid f \text{ is discontinuous at } x\}$$

is with measure zero, i.e., $m(D) = 0$.

[Proof] $\Rightarrow$) Suppose that f is Riemann integrable. For each $k \in \mathbb{N}$, put

$$D_k = \left\{ x \in [a,b] \;\middle|\; v(f,x) \geq \frac{1}{k} \right\}. \tag{5.5}$$

Then by Lemma 5.4.2, we have

$$D = \bigcup_{k=1}^{\infty} D_k. \tag{5.6}$$

Now, we prove that $m(D_k) = 0$ for any $k \in \mathbb{N}$. Let $k \in \mathbb{N}$ be given. Since f is Riemann integrable, for any $\epsilon > 0$, there exists a partition $P = \{x_0, x_1, \cdots, x_n\}$ of $[a,b]$ such that

$$U(f,P) - L(f,P) < \frac{\epsilon}{k}.$$

Then the number of elements of $D_k \cap P$ is less than equal to $n+1$ and $m(D_k \cap P) = 0$. Put

$$D_k' = \bigcup_{j=1}^{n} (x_{j-1}, x_j) \cap D_k.$$

and take $x \in D_k'$ for $D_k' \neq \emptyset$, then by (5.5), we have

$$\frac{1}{k} \leq v(f,x) \leq M(f,I_j) - m(f,I_j).$$

Therefore, we have

$$\begin{aligned} \sum_{\substack{j\in\{1,2,\cdots,n\};\\ D_k\cap(x_{j-1},x_j)\neq\emptyset}} \frac{1}{k}\Delta_j &\leq \sum_{\substack{j\in\{1,2,\cdots,n\};\\ D_k\cap(x_{j-1},x_j)\neq\emptyset}} \left(M(f,I_j) - m(f,I_j)\right)\Delta_j \\ &\leq U(f,P) - L(f,P) \\ &< \frac{\epsilon}{k}, \end{aligned}$$

which implies that

$$\sum_{\substack{j\in\{1,2,\cdots,n\};\\ D_k\cap(x_{j-1},x_j)\neq\emptyset}} \Delta_j < \epsilon.$$

Hence $m(D_m') = 0$ and so

$$m(D_k) = m\left(D_k' \cup (D_k \cap P)\right) = m\left(D_k'\right) + m\left(D_k \cap P\right) = 0.$$

Note that a countable union of measure zero sets is a set with measure zero. Thus by (5.6), $m(D) = 0$.

$\Leftarrow$) Since f is bounded, there exists $M > 0$ such that

$$|f(x)| \le M, \qquad x \in [a, b].$$

Suppose that $m(D) = 0$. There exists a countable family $\{(a_n, b_n)\}_{n=1}^{\infty}$ such that

(i) $D \subset \cup_{n=1}^{\infty}(a_n, b_n)$,

(ii) $\sum_{n=1}^{\infty}(b_n - a_n) < \frac{\epsilon}{4M}$.

Note that for any $x \in [a, b]$ at which f is continuous, there exists an open interval (α_x, β_x) centered by x such that for any $y \in (\alpha_x, \beta_x)$

$$|f(x) - f(y)| \le \frac{\epsilon}{4(b-a)}.$$

Then the family

$$\mathcal{F} = \{(a_n, b_n) \,|\, n \in \mathbb{N}\} \bigcup \{(\alpha_x, \beta_x) \,|\, x \in [a, b] \setminus D\}$$

of open intervals covers the compact set $[a, b]$. Therefore, there exists a finite subfamily $\mathcal{G}$ of $\mathcal{F}$ such that $\mathcal{G}$ covers $[a, b]$. The end points of open intervals in $\mathcal{G}$ belonging to $[a, b]$ and the end points a, b of $[a, b]$ becomes a partition of $[a, b]$ which is denoted by

$$\begin{aligned} P &= \{x_0, x_1, \cdots, x_n\} \\ &= (\{\text{the end points of open intervals in } \mathcal{G}\} \cap [a, b]) \cup \{a, b\}. \end{aligned}$$

Then the set of all subindexes k such that the interval (x_{k-1}, x_k) belonging to (α_x, β_x) in the subfamily $\mathcal{G}$ is denoted by J, and put

$$I = \{1, 2, \cdots, n\} \setminus J.$$

Then we have

$$\begin{aligned} U(f, P) - L(f, P) &= \sum_{k \in I} (M(f, I_k) - m(f, I_k))\, \Delta_k + \sum_{k \in J} (M(f, I_k) - m(f, I_k))\, \Delta_k \\ &\le 2M \frac{\epsilon}{4M} + (b-a)\frac{\epsilon}{2(b-a)} \\ &< \epsilon. \end{aligned}$$

Therefore, f is Riemann integrable. □

Exercises

5.1 Construct a function $f : [0,1] \to \mathbb{R}$ which is Lebesgue integrable, but not Riemann integrable.

5.2 Construct a nonnegative measurable function $f : \mathbb{R} \to \mathbb{R}$ such that $\int_{\mathbb{R}} f(x)dm(x) < \infty$, but $\lim_{|x|\to\infty} f(x) \neq 0$.

5.3 For each $n \in \mathbb{N}$, put

$$f_n(x) = \frac{1}{\sqrt{2\pi}} e^{-\frac{(x-n)^2}{2}}, \qquad x \in \mathbb{R}.$$

Let $g : \mathbb{R} \to \mathbb{R}$ be a Lebesgue measurable function such that $f_n \leq g$ a.e on $\mathbb{R}$ for all $n \in \mathbb{N}$. Prove that g is not integrable.

5.4 Prove that if f is a Lebesgue integrable function on $[a, b]$, then

$$\lim_{\epsilon \downarrow 0} \int_{a+\epsilon}^{b-\epsilon} f\, dm = \int_a^b f\, dm,$$

where ϵ is varying over $(0, b-a)$.

5.5 Let $\{g_n\}_{n=1}^{\infty}$ be a sequence of nonnegative measurable functions on a measure space $(X, \mathcal{M}, \mu)$. Prove that

$$\int_X \left(\sum_{n=1}^{\infty} g_n \right) d\mu = \sum_{n=1}^{\infty} \int_X g_n\, d\mu.$$

5.6 Let $\{f_n\}_{n=1}^{\infty}$ be a sequence of integrable functions on a complete measure space $(X, \mathcal{M}, \mu)$ such that

$$\sum_{n=1}^{\infty} \int_X |f_n| d\mu < \infty.$$

Prove that

(1) the series $f(x) := \sum_{n=1}^{\infty} f_n$ converges pointwise a.e. on X,

(2) f is integrable and

$$\int_X f d\mu = \sum_{n=1}^{\infty} \int_X f_n d\mu.$$

5.7 Let $(X, \mathcal{M}, \mu)$ be a measure space and $g : [a, b] \times X \to \mathbb{R}$ a function. Assume that

(i) for each $x \in X$, the function

$$g(\cdot, x) : [a, b] \ni t \longmapsto g(t, x) \in \mathbb{R}$$

is continuous in $t \in [a, b]$;

(ii) there exists an integrable function $h \geq 0$ such that

$$|g(t,x)| \leq h(x), \qquad (t,x) \in [a,b] \times X.$$

Prove that

(1) the function G defined by

$$G(t) = \int_X g(t,x)\, d\mu(x), \qquad t \in [a,b]$$

is continuous in $t \in [a,b]$,

(2) and prove that the following equality:

$$\int_a^b \left(\int_X g(t,x)\, d\mu(x) \right) dt = \int_X \left(\int_a^b g(t,x)\, dt \right) d\mu(x)$$

holds.

5.8 Let $(X, \mathcal{M}, \mu)$ be a measure space and $g : [a,b] \times X \to \mathbb{R}$ a function. Assume that

(i) there exists an element $t_0 \in [a,b]$ such that the function

$$g(t_0, \cdot) : X \ni x \longmapsto g(t_0, x) \in \mathbb{R}$$

is integrable with respect to μ;

(ii) for each $x \in X$, the function

$$g(\cdot, x) : [a,b] \ni t \longmapsto g(t,x) \in \mathbb{R}$$

is differentiable in $t \in [a,b]$;

(iii) there exists an integrable function $h \geq 0$ on X such that

$$\left| \frac{\partial g(t,x)}{\partial t} \right| \leq h(x), \qquad (t,x) \in [a,b] \times X.$$

Prove that for each $t \in [a,b]$, $g(t,\cdot)$ is integrable with respect to μ, the function G defined by

$$G : [a,b] \ni t \longmapsto G(t) = \int_X g(t,x)\, d\mu(x) \in \mathbb{R}$$

is differentiable and

$$\frac{d}{dt} \int_X g(t,x) d\mu = \int_X \frac{\partial g(t,x)}{\partial t}\, d\mu.$$

5.9 Prove that the following equality:

$$\int_0^\infty x^n e^{-x} dx = n!$$

holds.

5.10 Let f be an integrable function on a measure space $(X, \mathcal{M}, \mu)$. Define a set function μ_f on $\mathcal{M}$ by

$$\nu_f(E) = \int_E f d\mu, \qquad E \in \mathcal{M}.$$

Prove that

(1) ν_f is a signed measure,

(2) Find a Hahn decomposition of X for ν_f,

(3) Find the Jordan decomposition of ν_f,

(4) Compute the total variation of ν_f.

5.11 Let $[a, b]$ be a bounded interval and let $f : [a, b] \to \mathbb{R}$ be a bounded measurable function. Prove that

$$\lim_{n\to\infty} \int_a^b f(x)\sin(nx)dx = 0.$$

5.12 Let $f : [a, b] \to \mathbb{R}$ be a bounded measurable function and $F : \mathbb{R} \to \mathbb{R}$ be a function defined by

$$F(t) = \int_a^b f(x)\sin(tx)dx, \qquad t \in \mathbb{R}.$$

Prove that F is continuous.

5.13 Let $(X, \mathcal{M}, \mu)$ be a measure space and $f : X \to \overline{\mathbb{R}}$ an integrable function. Prove that $\int_E f d\mu = 0$ for all $E \in \mathcal{M}$, then $f = 0$ a.e. on X.

Chapter 6

Differentiation and Integration

For a real-valued function f defined on the closed bounded interval $[a, b]$ and $\delta > 0$, we extend f to $[a, b+\delta]$ by

$$f(x) = \begin{cases} f(x), \ a \le x \le b, \\ f(b), \ b < x \le b + \delta. \end{cases}$$

Then for any $x \in [a, b]$ and $0 < h \le \delta$, put

$$D_h f(x) = \frac{f(x+h) - f(x)}{h}, \qquad M_h f(x) = \frac{1}{h} \int_x^{x+h} f(t)\, dt \tag{6.1}$$

if the integral is well-defined.

Remark 6.0.1 Let f ba a continuous function on $[a, b]$. Then by the change of variable formula of integration, we obtain that

$$\begin{aligned} \int_a^b D_h f\, dm &= \frac{1}{h} \left(\int_a^b f(x+h)\, dm(x) - \int_a^b f(x)\, dm(x) \right) \\ &= \frac{1}{h} \left(\int_{a+h}^{b+h} f(x)\, dm(x) - \int_a^b f(x)\, dm(x) \right) \\ &= \frac{1}{h} \int_b^{b+h} f(x)\, dm(x) - \frac{1}{h} \int_a^{a+h} f(x)\, dm(x) \\ &= M_h f(b) - M_h f(a). \end{aligned}$$

On the other hand, by the mean value theorem of integration, we have

$$\lim_{h \to 0+} (M_h f(b) - M_h f(a)) = f(b) - f(a).$$

Therefore, to obtain the fundamental theorem of integral and differential calculus:

$$\int_a^b f'\, dm = f(b) - f(a), \tag{6.2}$$

we have to justify that when does the limit:

$$\lim_{h \to 0+} D_h f(x) = \lim_{h \to 0+} \frac{f(x+h) - f(x)}{h}$$

exists?, and when does the equality:

$$\lim_{h\to 0+}\int_a^b D_h f\, dm = \int_a^b \left(\lim_{h\to 0+} D_h f\right) dm$$

holds?

6.1 Uniform Integrability and Tightness

Proposition 6.1.1 Let $(X, \mathcal{M}, \mu)$ be a measure space. Let f be a measurable function on X.

(i) If f is nonnegative and integrable over X, then for any $\epsilon > 0$, there exists a $\delta > 0$ such that for any $A \in \mathcal{M}$ with $m(A) < \delta$,

$$\int_A f d\mu < \epsilon.$$

(ii) If f is integrable over X, then for any $\epsilon > 0$, there exists a $\delta > 0$ such that for any $A \in \mathcal{M}$ with $m(A) < \delta$,

$$\int_A |f| d\mu < \epsilon.$$

(iii) If f is integrable over X, then for each $\epsilon > 0$, there exists a measurable subset X_0 of X with $\mu(X_0) < \infty$ such that

$$\int_{X\setminus X_0} |f| d\mu < \epsilon.$$

[Proof] (i) For any given $\epsilon > 0$, by the definition of integral, there exists a simple function ψ such that $0 \le \psi \le f$ and

$$\int_X f d\mu - \int_X \psi d\mu < \frac{\epsilon}{2}.$$

Then for any measurable subset A of X, we have

$$\int_A f d\mu - \int_A \psi d\mu \le \int_X (f - \psi)\, d\mu < \frac{\epsilon}{2},$$

and so for $M = \sup\{\psi(x) \,|\, x \in X\}$, we have

$$\int_A f d\mu \le \int_A \psi d\mu + \frac{\epsilon}{2} \le M \times \mu(A) + \frac{\epsilon}{2}.$$

Therefore, if we take $\delta = \epsilon/(2M+1)$, then for any measurable subset A of X with $m(A) < \delta$, $\int_A f d\mu < \epsilon$.

(ii) Suppose f is integrable over X. Then $f^{\pm}$ are integrable over X and by (i) for any given $\epsilon > 0$, there exist $\delta_1 > 0$ and $\delta_2 > 0$ such that for any measurable subsets A_1, A_2 of X with $m(A_1) < \delta_1$ and $m(A_2) < \delta_2$

$$\int_{A_1} f^+ d\mu < \frac{\epsilon}{2}, \qquad \int_{A_2} f^- d\mu < \frac{\epsilon}{2}.$$

Therefore, by taking $\delta = \min\{\delta_1, \delta_2\}$, for any measurable subset A of X with $m(A) < \delta$

$$\int_A |f| d\mu = \int_A f^+ d\mu + \int_A f^- d\mu < \frac{\epsilon}{2} + \frac{\epsilon}{2} = \epsilon.$$

(iii) Since f is integrable over X. Then $f^{\pm}$ are integrable over X. Therefore, by the definition of integral, for each $\epsilon > 0$, there exist simple functions ϕ and ψ such that $0 \le \phi \le f^+$ and $0 \le \psi \le f^-$, and

$$\int_X (f^+ - \phi) d\mu < \frac{\epsilon}{2}, \qquad \int_X (f^- - \psi) d\mu < \frac{\epsilon}{2}.$$

Put

$$X_0 = \{x \in X \mid \phi(x) > 0 \text{ or } \psi(x) > 0\}.$$

Then it is obvious that $\mu(X_0) < \infty$ and $\phi = \psi = 0$ on $X \setminus X_0$ and so we obtain that

$$\begin{aligned}\int_{X \setminus X_0} |f| d\mu &= \int_{X \setminus X_0} f^+ d\mu + \int_{X \setminus X_0} f^- d\mu \\ &= \int_{X \setminus X_0} (f^+ - \phi)\, d\mu + \int_{X \setminus X_0} (f^- - \psi)\, d\mu \\ &\le \int_X (f^+ - \phi)\, d\mu + \int_X (f^- - \psi)\, d\mu \\ &< \frac{\epsilon}{2} + \frac{\epsilon}{2} = \epsilon,\end{aligned}$$

which gives the proof. □

Definition 6.1.2 Let $(X, \mathcal{M}, \mu)$ be a measure space. A family $\mathcal{F}$ of integrable functions on X is said to be **uniformly integrable** over X if for any $\epsilon > 0$ there exists a $\delta > 0$ such that for any $f \in \mathcal{F}$ and any measurable subset A of X with $\mu(A) < \delta$,

$$\int_A |f|\, d\mu < \epsilon.$$

Remark 6.1.3 Let $(X, \mathcal{M}, \mu)$ be a measure space. Let g be nonnegative integrable over X and let

$$\mathcal{F} = \{f : \text{measurable} \mid |f| \le g\}.$$

By (ii) in Proposition 6.1.1, for any given $\epsilon > 0$, there exists a $\delta > 0$ such that for any measurable subset A of X with $\mu(A) < \delta$,

$$\int_A g\, d\mu < \epsilon,$$

which implies that

$$\int_A |f|\, d\mu \leq \int_A g\, d\mu < \epsilon.$$

Therefore, $\mathcal{F}$ is uniform integrable.

Remark 6.1.4 Let $(X, \mathcal{M}, \mu)$ be a measure space. Let $\mathcal{F}$ and $\mathcal{G}$ be two families of measurable functions on X which are uniformly integrable over X.

(1) For each $\alpha \in \mathbb{R}$, $\alpha\mathcal{F} = \{\alpha f \,|\, f \in \mathcal{F}\}$ is uniformly integrable over X.

(2) $\mathcal{F} + \mathcal{G} = \{f + g \,|\, f \in \mathcal{F},\ \ g \in \mathcal{G}\}$ is uniformly integrable over X.

(3) $\mathcal{F} \cup \mathcal{G} = \{h \,|\, h \in \mathcal{F} \text{ or } h \in \mathcal{G}\}$ is uniformly integrable.

Proposition 6.1.5 Let $(X, \mathcal{M}, \mu)$ be a measure space. Let $\{f_k\}_{k=1}^n$ be a finite sequence of integrable functions over X. Then $\{f_k\}_{k=1}^n$ is uniformly integrable over X.

[Proof] Let $\epsilon > 0$ be given. For each $1 \leq k \leq n$, by (ii) in Proposition 6.1.1, there exists a $\delta_k > 0$ such that for any measurable subset A of X with $m(A) < \delta_k$,

$$\int_A |f_k|\, d\mu < \epsilon.$$

Put $\delta = \min\{\delta_1, \cdots, \delta_k\}$. Then for any measurable subset A of X with $m(A) < \delta$ and $1 \leq k \leq n$,

$$\int_A |f_k|\, d\mu < \epsilon.$$

Therefore, $\{f_k\}_{k=1}^n$ is uniform integrable. □

Definition 6.1.6 Let $(X, \mathcal{M}, \mu)$ be a measure space. Let $\mathcal{F}$ be a family of measurable functions on X. Then $\mathcal{F}$ is said to be **tight** over X if for any $\epsilon > 0$ there exists a measurable subset X_0 of X of finite measure, i.e., $m(X_0) < \infty$ such that for any $f \in \mathcal{F}$,

$$\int_{X \setminus X_0} |f|\, d\mu < \epsilon.$$

Remark 6.1.7 Let $(X, \mathcal{M}, \mu)$ be a measure space. Let $\{f_k\}_{k=1}^n$ be a finite sequence of integrable functions over X. Then $\{f_k\}_{k=1}^n$ is tight over X. In fact, for a given $\epsilon > 0$ and for each $1 \le k \le n$, by (iii) in Proposition 6.1.1, there exists a measurable subset $X_k \subset X$ of finite measure such that

$$\int_{X \setminus X_k} |f_k|\, d\mu < \epsilon.$$

Put $X_0 = \cup_{k=1}^n X_k$. Then $m(X_0) \le \sum_{k=1}^n m(X_k) < \infty$ and any $1 \le k \le n$, we have

$$\int_{X \setminus X_0} |f_k|\, d\mu \le \int_{X \setminus X_k} |f_k|\, d\mu < \epsilon.$$

Therefore, $\{f_k\}_{k=1}^n$ is tight.

Remark 6.1.8 Let $(X, \mathcal{M}, \mu)$ be a measure space. Let $\mathcal{F}$ and $\mathcal{G}$ be two families of measurable functions on X which are tight over X.

(1) For each $\alpha \in \mathbb{R}$, $\alpha\mathcal{F} = \{\alpha f \,|\, f \in \mathcal{F}\}$ is tight over X.

(2) $\mathcal{F} + \mathcal{G} = \{f + g \,|\, f \in \mathcal{F},\ \ g \in \mathcal{G}\}$ is tight over X.

(3) $\mathcal{F} \cup \mathcal{G} = \{h \,|\, h \in \mathcal{F} \text{ or } h \in \mathcal{G}\}$ is tight .

Theorem 6.1.9 **[Vitali Convergence Theorem]** Let $(X, \mathcal{M}, \mu)$ be a measure space. Let $\{f_n\}_{n=1}^\infty$ be a sequence of integrable functions over X such that $\{f_n\}_{n=1}^\infty$ is uniformly integrable and tight over X. Suppose that $\{f_n\}$ converges pointwise a.e. on X to an integrable function f over X. Then

$$\lim_{n \to \infty} \int_X f_n d\mu = \int_X f d\mu.$$

[Proof] Let $\epsilon > 0$ be given. Then by the tightness of $\{f_n\}_{n=1}^\infty \cup \{f\}$, there exists a measurable subset $X_0 \subset X$ of finite measure such that for any $n \in \mathbb{N}$

$$\int_{X \setminus X_0} |f_n| < \frac{\epsilon}{6}, \qquad \int_{X \setminus X_0} |f| < \frac{\epsilon}{6}.$$

By the uniform integrability of $\{f_n\}_{n=1}^\infty \cup \{f\}$, there exists a $\delta > 0$ such that for any $g \in \{f_n\}_{n=1}^\infty \cup \{f\}$ and any measurable subset $A \subset X_0$ with $\mu(A) < \delta$,

$$\int_A |g| d\mu < \frac{\epsilon}{6}.$$

Since f is integrable, f is finite a.e. on X. By Egoroff's Theorem (Theorem 4.1.38), there exists a measurable subset X_1 of X_0 such that $\mu(X_0 \setminus X_1) < \delta$ and $\{f_n\}_{n=1}^\infty$ converges uniformly to f on X_1. Therefore, there exists a $N \in \mathbb{N}$ such that for any $n \ge N$,

$$\sup\{|f_n(x) - f(x)| \,|\, x \in X_1\} < \frac{\epsilon}{3\mu(X_1) + 1}.$$

Hence for any $n \geq N$, we obtain that

$$\begin{aligned}\left|\int_X f_n d\mu - \int_X f d\mu\right| &\leq \int_{X_1} |f_n - f| d\mu + \int_{X\setminus X_0} (|f_n| + |f|)\, d\mu + \int_{X_0\setminus X_1} (|f_n| + |f|)\, d\mu \\ &< \frac{\epsilon\mu(X_1)}{3\mu(X_1)+1} + \frac{2\epsilon}{3} \\ &< \epsilon,\end{aligned}$$

which gives the proof. □

Example 6.1.10 In Theorem 6.1.9, the integrability of f is necessary. Let X be a nonempty set with a proper nonempty subset E and $\mathcal{M} = \{X, \emptyset, E, E^c\}$. Define a measure μ on $\mathcal{M}$ by

$$\mu(\emptyset) = 0, \quad \mu(E) = 1, \quad \mu(E^c) = 1.$$

For each $n \in \mathbb{N}$, put $f_n = n\chi_E - n\chi_{E^c}$. Then we can easily see that $\{f_n\}_{n=1}^{\infty}$ is uniformly integrable and tight over X, and $\{f_n\}_{n=1}^{\infty}$ converges pointwise to $f = \infty\chi_E - \infty\chi_{E^c}$ on X. But f is not integrable over X, and $\lim_{n\to\infty} \int_X f_n d\mu = 0$.

Corollary 6.1.11 Let $(X, \mathcal{M}, \mu)$ be a measure space. Let $\{f_n\}_{n=1}^{\infty}$ be a sequence of nonnegative integrable functions over X. Suppose $\{f_n\}$ converges to $f = 0$ pointwise on X. Then

$$\lim_{n\to\infty} \int_X f_n d\mu = 0$$

if and only if $\{f_n\}$ is uniformly integrable and tight over X.

[Proof] $\Leftarrow$) It is immediate from the Vitali's convergence theorem (Theorem 6.1.9).

$\Rightarrow$) For any given $\epsilon > 0$, there exists $N \in \mathbb{N}$ such that for any $n > N$,

$$\int_X f_n\, d\mu < \epsilon.$$

By (i) in Proposition 6.1.1, for each $1 \leq k \leq N$, there exists $\delta_k > 0$ such that for any $A \in \mathcal{M}$ with $\mu(A) < \delta_k$, $\int_A f_k d\mu < \epsilon$. Put $\delta = \min\{\delta_i \,|\, i = 1, 2, \cdots, N\}$. Then for any $k \in \mathbb{N}$ and $A \in \mathcal{M}$ with $\mu(A) < \delta$,

$$\int_A f_k d\mu < \epsilon,$$

i.e., $\{f_n\}$ is uniform integrable. Also, for each $1 \leq k \leq N$, there exists $X_k \in \mathcal{M}$ with $\mu(X_k) < \infty$ such that for any $k \in \mathbb{N}$, $\int_{X\setminus X_k} f_k d\mu < \epsilon$. Put $X_0 = \cup_{k=1}^{N} X_k$. Then $\mu(X_0) < \infty$ and for any $k \in \mathbb{N}$,

$$\int_{X\setminus X_0} f_k d\mu \leq \int_{X\setminus X_k} f_k d\mu < \epsilon,$$

where for $k > N$, $X_k = \emptyset$. Therefore, $\{f_n\}_{n=1}^{\infty}$ is tight. □

6.2 Radon-Nikodym Theorem

Let $(X, \mathcal{M}, \mu)$ be a measure space and f a nonnegative (integrable) function over X. Then the set function ν on $\mathcal{M}$ is well-defined by

$$\nu(E) = \int_E f d\mu, \qquad E \in \mathcal{M}, \tag{6.3}$$

and then by applying the monotone convergence theorem (Theorem 5.1.15), we can see that ν is a measure satisfying that

$$\text{if } E \in \mathcal{M} \text{ with } \mu(E) = 0, \text{ then } \nu(E) = 0. \tag{6.4}$$

Remark 6.2.1 If f is nonegative integrable function on a measure space $(X, \mathcal{M}, \mu)$, then the measure ν defined as in (6.3) is finite.

Definition 6.2.2 Let μ and ν be measures on a measurable space $(X, \mathcal{M})$. Then we say that ν is **absolutely continuous** with respect to μ if (6.4) holds, in this case, we write $\nu \ll \mu$.

Theorem 6.2.3 Let μ and ν be measures on a measurable space $(X, \mathcal{M})$. Suppose that ν is a finite measure. Then ν is absolutely continuous with respect to μ if and only if for any $\epsilon > 0$, there exists a $\delta > 0$ such that for any $E \in \mathcal{M}$ with $\mu(E) < \delta$,

$$\nu(E) < \epsilon.$$

[Proof] $\Rightarrow$) Suppose not, i.e., there exists a fixed $\epsilon_0 > 0$ such that for any $\delta > 0$, there exists $E_\delta \in \mathcal{M}$ with $\mu(E_\delta) < \delta$ and $\nu(E_\delta) \geq \epsilon_0$. Therefore, for any $n \in \mathbb{N}$, there exists $E_n \in \mathcal{M}$ such that $\mu(E_n) < \frac{1}{2^n}$ and $\nu(E_n) \geq \epsilon_0$. For each $n \in \mathbb{N}$, put

$$A_n = \bigcup_{k \geq n} E_k, \qquad A = \bigcap_{n=1}^{\infty} A_n, \qquad \text{i.e., } A = \limsup_{n\to\infty} E_n.$$

Then $\{A_n\}_{n=1}^{\infty}$ is a descending sequence of measurable sets and by the countable subaddititivity, we see that

$$\mu(A_n) \leq \sum_{k=n}^{\infty} \mu(E_k) < 2^{-(n-1)}$$

and so $\mu(A_1) < 1$ and, by the continuity of measure,

$$\mu(A) = \mu\left(\bigcap_{n=1}^{\infty} A_n\right) = \lim_{n\to\infty} \mu(A_n) = 0.$$

Since ν is absolutely continuous with respect to μ, $\nu(A) = 0$. On the other hand, since ν is finite, by the continuity of measure again,

$$\nu(A) = \nu\left(\bigcap_{n=1}^{\infty} A_n\right) = \lim_{n\to\infty} \nu(A_n) \geq \lim_{n\to\infty} \nu(E_n) \geq \epsilon_0 > 0,$$

which is a contradiction.

$\Leftarrow$) Let $E \in \mathcal{M}$ with $\mu(E) = 0$ and let $\epsilon > 0$ be given. By assumption, there exists $\delta > 0$ such that for any $A \in \mathcal{M}$ with $\mu(A) < \delta$, $\nu(A) < \epsilon$. Therefore, since $\mu(E) = 0 < \delta$, $\nu(E) < \epsilon$. Since $\epsilon > 0$ is arbitrary, $\nu(E) = 0$. Hence ν is absolutely continuous with respect to μ. □

Remark 6.2.4 Let $(X, \mathcal{M}, \mu)$ be a measure space and f a non-negative integrable function over X. Then the finite measure ν on the measurable space $(X, \mathcal{M})$ defined by

$$\nu(E) = \int_E f\,d\mu, \qquad E \in \mathcal{M}$$

is absolutely continuous with respect to μ. Therefore, by Theorem 6.2.3, for any ϵ, there exists $\delta > 0$ such that for any $A \in \mathcal{M}$ with $\mu(A) < \delta$,

$$\int_A f\,d\mu = \nu(A) < \epsilon,$$

which gives another proof of (i) in Proposition 6.1.1.

Lemma 6.2.5 Let μ and ν be finite measures defined on a measurable space $(X, \mathcal{M})$. Suppose that $\nu \neq 0$ and ν is absolutely continuous with respect to μ. Then there exists a nonnegative measurable function $f : X \to [0, \infty]$ such that $\int_X f d\mu > 0$ and

$$\int_E f d\mu \leq \nu(E), \qquad E \in \mathcal{M}$$

[Proof] For each $n \in \mathbb{N}$, consider the finite signed measure $\nu - n^{-1}\mu$ and let $\{P_n, N_n\}$ be the Hahn decomposition of X for $\nu - n^{-1}\mu$. Then for some $n_0 \in \mathbb{N}$, $\mu(P_{n_0}) > 0$. If $\mu(P_n) = 0$ for any $n \in \mathbb{N}$, then for any $E \in \mathcal{M}$,

$$\mu(E) = \mu(P_n \cap E) + \mu(N_n \cap E) = \mu(N_n \cap E)$$

and

$$(\nu - n^{-1}\mu)(N_n \cap E) \leq 0,$$

which implies that

$$\nu(E) = \nu(P_n \cap E) + \nu(N_n \cap E) = \nu(N_n \cap E) \leq n^{-1}\mu(N_n \cap E) = n^{-1}\mu(E),$$

which implies $\nu = 0$ on $\mathcal{M}$ and it is a contradiction. Therefore, there exists a $n_0 \in \mathbb{N}$ such that $\mu(P_{n_0}) > 0$ and the for $f = n_0^{-1}\chi_{P_{n_0}}$, we have

$$\int_X f d\mu = n_0^{-1}\mu(P_{n_0}) > 0$$

and for any $E \in \mathcal{M}$,

$$\int_E f d\mu = n_0^{-1}\mu(P_{n_0} \cap E) \leq \nu(P_{n_0} \cap E) \leq \nu(E).$$

Therefore, we complete the proof. □

Theorem 6.2.6 **[Radon-Nikodym Theorem]** Let μ and ν be σ-finite measures defined on a measurable space $(X, \mathcal{M})$. Suppose that ν is absolutely continuous with respect to μ. Then there exists a unique (up to sets of μ-measure zero, i.e., μ-a.e. on X) nonnegative measurable function f such that

$$\nu(E) = \int_E f \, d\mu \tag{6.5}$$

for all $E \in \mathcal{M}$. In this case, f is called the **Radon-Nikodym derivative** of ν with respect to μ and it is denoted by $\frac{d\nu}{d\mu}$.

[Proof] We prove only the case that μ and ν are finite measures. If $\nu = 0$ on $\mathcal{M}$, then it is obvious that (6.5) holds with $f = 0$ on X. Suppose that $\nu \neq 0$. Put

$$\mathcal{G} = \left\{ g : X \to [0, \infty] \text{ is measurable} \,\middle|\, \int_X g d\mu > 0 \text{ and } \nu(E) \geq \int_E g d\mu \text{ for all } E \in \mathcal{M} \right\}.$$

Then by Lemma 6.2.5, $\mathcal{G}$ is nonempty. If $g, h \in \mathcal{G}$, then $\max\{g, h\} \in \mathcal{G}$. In fact, we have

$$\int_X \max\{g, h\} d\mu \geq \int_X g d\mu > 0$$

and for any $E \subset \mathcal{M}$,

$$\begin{aligned}\int_E \max\{g, h\} d\mu &= \int_{E_1} \max\{g, h\} d\mu + \int_{E_2} \max\{g, h\} d\mu \leq \int_{E_1} g d\mu + \int_{E_2} h d\mu \\ &\leq \nu(E_1) + \nu(E_2) = \nu(E),\end{aligned}$$

where $E_1 = \{x \in E \,|\, h(x) \leq g(x)\}$ and $E_2 = \{x \in E \,|\, h(x) > g(x)\}$. Put

$$S = \sup_{g \in \mathcal{G}} \int_X g d\mu \leq \nu(X) < \infty.$$

Then there exists a sequence $\{g_n\}_{n=1}^{\infty} \subset \mathcal{G}$ such that $S = \lim_{n\to\infty} \int_X g_n d\mu$. In this case, we may assume that $\{g_n\}_{n=1}^{\infty}$ is an (pointwisely) increasing sequence. In fact, by replacing each g_n by $\max\{g_1, \cdots, g_n\}$, we can take an increasing sequence $\{g_n\}_{n=1}^{\infty} \subset \mathcal{G}$ such that $S = \lim_{n\to\infty} \int_X g_n d\mu$. Put

$$f(x) = \lim_{n\to\infty} g_n(x) \geq 0.$$

Then $\int_X f d\mu \geq \int_X g_1 d\mu > 0$ and by the monotone convergence theorem, for any $E \in \mathcal{M}$,

$$\int_E f d\mu = \lim_{n\to\infty} \int_E g_n d\mu \leq \nu(E),$$

which implies that $f \in \mathcal{G}$ and f is finite a.e. on X. Define a set function $\nu_0 : \mathcal{M} \to [0, \infty]$

$$\nu_0(E) = \nu(E) - \int_E f d\mu$$

for all $E \in \mathcal{M}$. Then, since $\nu(X) < \infty$, ν_0 is a measure which is absolutely continuous with respect to μ. We want to show that $\nu_0 = 0$ and then f is the desired function. Suppose ν_0 is not zero. Then by Lemma 6.2.5, there exists a nonnegative measurable function h such that

$$\int_X h d\mu > 0 \quad \text{and} \quad \int_E h d\mu \leq \nu_0(E) = \nu(E) - \int_E f d\mu$$

for all $E \in \mathcal{M}$. Therefore, $h + f \in \mathcal{G}$ and

$$\sup_{g\in\mathcal{G}} \int_X g d\mu = S \geq \int_X (h + f) d\mu > \int_X f d\mu = S = \sup_{g\in\mathcal{G}} \int_X g d\mu,$$

which is a contradiction. Therefore, $\nu_0 = 0$ on $\mathcal{M}$, i.e., for all $E \in \mathcal{M}$,

$$\nu(E) = \int_E f d\mu.$$

To show uniqueness, suppose that there exists another nonnegative measurable function f_1 such that

$$\nu(E) = \int_E f_1 d\mu$$

for all $E \in \mathcal{M}$. Then for all $E \in \mathcal{M}$, $\int_E (f - f_1) d\mu = 0$. Therefore, $f = f_1$ μ-a.e. on X. □

Remark 6.2.7 In Theorem 6.2.6, if ν is finite, then the Radon-Nikodym derivative f is integrable.

Remark 6.2.8 The assumption that the measures μ and ν are σ-finite is crucial in Theorem 6.2.6.

Example 6.2.9 Let $(\mathbb{R}, \mathfrak{M}, m)$ be the Lebesgue measure space. Let μ be the counting measure on $\mathfrak{M}$. Then μ is not σ-finite, however m is absolutely continuous with respect to the counting measure μ. Suppose that there exists a nonnegative measurable function f on $\mathbb{R}$ such that

$$m(E) = \int_E f d\mu$$

for any $E \in \mathfrak{M}$. Then for any $x \in \mathbb{R}$, we have

$$0 = m(\{x\}) = \int_{\{x\}} f d\mu = f(x)\mu(\{x\}) = f(x),$$

which implies that $f = 0$, and then for any $a < b$,

$$b - a = m([a, b]) = \int_{[a,b]} f d\mu = 0,$$

which is a contradiction.

Let ν be a signed measure and μ a measure on a measurable space $(X, \mathcal{M})$. Then by the Jordan Decomposition (Theorem 2.4.11), there exist two (positive) measures ν^+ and ν^- such that

$$\nu = \nu^+ - \nu^-,$$

and then $|\nu| = \nu^+ + \nu^-$. The signed measure ν is said to be **absolutely continuous** with respect to the measure μ if $|\nu| \ll \mu$.

Theorem 6.2.10 Let μ be a σ-finite measure on a measurable space $(X, \mathcal{M})$ and ν a σ-finite signed measure on measurable space $(X, \mathcal{M})$. Suppose that ν is absolutely continuous with respect to the measure μ. Then there exists a unique (up to sets of μ-measure zero) measurable function f such that

$$\nu(E) = \int_E f d\mu$$

for all $E \in \mathcal{M}$. Furthermore, if ν is a finite signed measure, then f is integrable.

[Proof] The proof is left as an exercise for the reader (see Exercise **6.4**). □

Theorem 6.2.11 [**Lebesgue Decomposition Theorem**] Let μ and ν be σ-finite measures on a measurable space $(X, \mathcal{M})$. There exist measures ν_0 and ν_1 on $\mathcal{M}$ such that ν_0 is singular with respect to μ, ν_1 is absolutely continuous with respect to μ, and

$$\nu = \nu_0 + \nu_1.$$

The measures ν_0 and ν_1 are unique.

[Proof] Let λ be a measure defined on the measurable space $(X, \mathcal{M})$ by $\lambda = \mu + \nu$. Since μ and ν are σ-finite measures, λ is a σ-finite measure. The measure μ is absolutely continuous with respect to λ. Therefore, by the Radon-Nikodym Theorem (Theorem 6.2.6), there exists a nonnegative measurable function f such that

$$\mu(E) = \int_E f d\lambda = \int_E f d\mu + \int_E f d\nu$$

for all $E \in \mathcal{M}$. Put

$$X_+ = \{x \in \mid f(x) > 0\}, \qquad X_0 = \{x \in X \mid f(x) = 0\},$$

and define the measures ν_0 and ν_1 on $\mathcal{M}$ by

$$\nu_0(E) = \nu(E \cap X_0), \qquad \nu_1(E) = \nu(E \cap X_+)$$

for all $E \in \mathcal{M}$. Then

$$\nu = \nu_0 + \nu_1.$$

Also, ν_0 and μ are mutually singular. In fact, since $f = 0$ on X_0, $\mu(X_0) = \int_{X_0} f d\lambda = 0$ and $\nu_0(X_+) = \nu(X_+ \cap X_0) = \nu(\emptyset) = 0$. We now show that $\nu_1 \ll \mu$. Suppose that $\mu(E) = 0$. Then since $f = 0$ on X_0, it holds that

$$\begin{aligned}\int_{E\cap X_+} f d\nu &= \int_{E\cap X_0} f d\nu + \int_{E\cap X_+} f d\nu = \int_E f d\nu = \int_E f d\mu + \int_E f d\nu \\ &= \mu(E) \\ &= 0,\end{aligned}$$

which implies that $\int_{E\cap X_+} f d\nu = 0$. On the other hand, since $f > 0$ on $E\cap X_+$, $\nu(E\cap X_+) = 0$, i.e., $\nu_1(E) = 0$. Therefore, ν_1 is absolutely continuous with respect to μ. □

6.3 Functions of Bounded Variation

Let $[a, b]$ be a closed bounded interval and let $f; [a, b] \to \mathbb{R}$ be a function. Let $P : a = x_0 < x_1 < \cdots < x_n = b$ be a partition of $[a, b]$. The **variation** $V(f, P)$ of f with respect to P is defined by

$$V(f, P) = \sum_{k=1}^{n} |f(x_k) - f(x_{k-1})|,$$

and then the **total variation** $TV(f)$ of f on $[a, b]$ is defined by

$$TV(f) = \sup\{V(f, P) | P \text{ a partition of } [a, b]\}.$$

Definition 6.3.1 Let $[a, b]$ be a closed bounded interval and let $f; [a, b] \to \mathbb{R}$ be a function. Then f is said to be of **bounded variation** if

$$TV(f) < \infty.$$

Example 6.3.2 Let $[a, b]$ be a closed bounded interval and let $f; [a, b] \to \mathbb{R}$ be an increasing function. Then f is of bounded variation on $[a, b]$. In fact, for any partition $P : a = x_0 < x_1 < \cdots < x_n = b$ of $[a, b]$, it holds that

$$V(f, P) = \sum_{k=1}^{n} |f(x_k) - f(x_{k-1})| = \sum_{k=1}^{n} (f(x_k) - f(x_{k-1}))$$

$$= f(b) - f(a),$$

which implies that $TV(f) = f(b) - f(a) < \infty$.

Example 6.3.3 Consider the function f defined on $[0, 1]$ by

$$f(x) = \begin{cases} x\cos(\pi/x), & 0 < x \le 1, \\ 0, & x = 0. \end{cases}$$

and the partition

$$P_n = \left\{0, \frac{1}{2n}, \frac{1}{2n-1}, \cdots, \frac{1}{3}, \frac{1}{2}, 1\right\}$$

of $[0, 1]$ for each $n \in \mathbb{N}$. Then we have

$$\begin{aligned} V(f, P_n) &= \left|\frac{1}{2n}\cos(2n\pi) - 0\right| + \left|\frac{1}{2n-1}\cos((2n-1)\pi) - \frac{1}{2n}\cos(2n\pi)\right| \\ &+ \left|\frac{1}{2n-2}\cos((2n-2)\pi) - \frac{1}{2n-1}\cos((2n-1)\pi)\right| \\ &+ \left|\frac{1}{2n-3}\cos((2n-3)\pi) - \frac{1}{2n-2}\cos((2n-2)\pi)\right| + \cdots \\ &+ \left|\cos\pi - \frac{1}{2}\cos(2\pi)\right| \\ &= 2\left(\frac{1}{2n} + \frac{1}{2n-1} + \cdots + \frac{1}{2}\right) + 1, \end{aligned}$$

which diverges as $n \to \infty$ and so f is not of bounded variation on $[0, 1]$.

Example 6.3.4 Let $[a, b]$ be a closed bounded interval and let $f; [a, b] \to \mathbb{R}$ be a differentiable function. If f' is bounded on (a, b), then f is of bounded variation on $[a, b]$. In fact, for any partition $P : a = x_0 < x_1 < \cdots < x_n = b$, by applying mean value theorem in calculus, it holds that

$$\begin{aligned} V(f, P) &= \sum_{k=1}^{n} |f(x_k) - f(x_{k-1})| \\ &= \sum_{k=1}^{n} |f'(x_k')|\,(x_k - x_{k-1}), \quad x_k' \in (x_{k-1}, x_k) \\ &\le M \sum_{k=1}^{n} (x_k - x_{k-1}), \quad M = \sup_{a<x<b} \{|f'(x)|\} \\ &= M(b - a), \end{aligned}$$

which implies that f is of bounded variation on $[a, b]$.

Remark 6.3.5 Let $[a, b]$ be a closed bounded interval and let $f; [a, b] \to \mathbb{R}$ be a function. Let $a < c < b$ and let P' is the refinement of a partition P of $[a, b]$ by adjoining c to P.

Then by the triangle inequality, we have $V(f,P) \leq V(f,P')$. In fact, if $x_{k-1} < c < x_k$, then we have

$$|f(x_k) - f(x_{k-1})| \leq |f(x_k) - f(c)| + |f(c) - f(x_{k-1})|.$$

(i) For any partition $P : a = x_0 < x_1 < \cdots < x_n = b$ of $[a,b]$, we construct the partitions P_1 and P_2 of $[a,c]$ and $[c,b]$, respectively, by

(1) if $c \in \{x_0, \cdots, x_n\}$, i.e., there exists $1 \leq l \leq n$ such that $x_l = c$, then

$$P_1 : a = x_0 < \cdots < x_l = c, \qquad P_2 : c = x_l < \cdots < x_n = b.$$

(2) if $c \notin \{x_0, \cdots, x_n\}$, then there exists $0 \leq l \leq n-1$ such that $x_l < c < x_{l+1}$ and so we can construct P_1 and P_2 as

$$P_1 : a = x_0 < \cdots < x_l < c, \qquad P_2 : c < x_{l+1} < \cdots < x_n = b.$$

Then we have

$$V(f_{[a,b]}, P) \leq V(f_{[a,c]}, P_1) + V(f_{[a,c]}, P_2),$$

where $f_{[u,v]}$ is the restriction of f to $[u,v]$ for a subinterval $[u,v]$ of $[a,b]$. Therefore, by taking supremum on the right hand side of the above inequality, we have

$$V(f_{[a,b]}, P) \leq TV(f_{[a,c]}) + TV(f_{[c,b]})$$

and then by taking supremum on the left hand side of the above inequality, we have

$$TV(f_{[a,b]}) \leq TV(f_{[a,c]}) + TV(f_{[c,b]})$$

(ii) for any partitions $P_1 : a = x_0 < \cdots < x_l = c$ and $P_2 : c = y_0 < \cdots < \cdots < y_m = b$ of $[a,c]$ and $[c,b]$, respectively, we have a partition

$$P : a = x_0 < \cdots < x_l = c = y_0 < \cdots < \cdots < y_m = b$$

of $[a,b]$ and then we have

$$V(f_{[a,b]}, P) = V(f_{[a,c]}, P_1) + V(f_{[c,b]}, P_2),$$

which implies that

$$TV(f_{[a,b]}) \geq V(f_{[a,c]}, P_1) + V(f_{[c,b]}, P_2)$$

and

$$TV(f_{[a,b]}) \geq TV(f_{[a,c]}) + TV(f_{[c,b]}).$$

Finally, we have

$$TV(f_{[a,b]}) = TV(f_{[a,c]}) + TV(f_{[c,b]}). \tag{6.6}$$

Therefore, if f is of bounded variation on $[a,b]$, then for any $a \leq u < v \leq b$, we have

$$TV(f_{[a,v]}) - TV(f_{[a,u]}) = TV(f_{[u,v]}). \tag{6.7}$$

Moreover, we have

$$f(u) - f(v) \leq |f(v) - f(u)| \leq TV(f_{[u,v]}) = TV(f_{[a,v]}) - TV(f_{[a,u]})$$

and so

$$f(u) + TV(f_{[a,u]}) \leq f(v) + TV(f_{[a,v]}).$$

Theorem 6.3.6 [**Jordan's Theorem**] A function f is of bounded variation on a bounded closed interval $[a,b]$ if and only if f is the difference of two increasing functions on $[a,b]$.

[Proof] $\Rightarrow$) By Remark 6.3.5, the functions

$$[a,b] \ni x \longmapsto f(x) + TV(f_{[a,x]}), \qquad [a,b] \ni x \longmapsto TV(f_{[a,x]})$$

are increasing on $[a,b]$ and then we have the following explicit decomposition of f as the difference of two increasing functions on $[a,b]$:

$$f(x) = \big[f(x) + TV(f_{[a,x]})\big] - TV(f_{[a,x]}) \tag{6.8}$$

for all $a \le x \le b$.

$\Leftarrow$) If $f = g - h$ with two increasing functions g and h on $[a,b]$, then for any partition $P : a = x_0 < \cdots < x_n = b$ of $[a,b]$, we have

$$\begin{aligned}\sum_{k=1}^{n} |f(x_k) - f(x_{k-1})| &\le \sum_{k=1}^{n} [g(x_k) - g(x_{k-1})] + \sum_{k=1}^{n} [h(x_k) - h(x_{k-1})] \\ &= g(b) - g(a) + h(b) - h(a),\end{aligned}$$

which implies that

$$TV(f) \le g(b) + h(b) - g(a) - h(a).$$

□

An expression of function of bounded variation as the difference of two increasing functions is called a **Jordan Decomposition** of the function.

6.4 Differentiability of Functions of Bounded Variation

Theorem 6.4.1 Let f be a monotone function on the open interval (a,b). Then the set

$$D_{a,b} = \{x \in (a,b) \mid f \text{ is discontinuous at } x\}$$

is at most countable.

[Proof] If the statement of this theorem is true for the increasing function f, then the case of decreasing function f is obvious by considering the increasing function $-f$. Therefore, we now prove the case of increasing function f.

(i) Suppose that (a, b) is bounded and $f(a+), f(b-)$ are finite. For each $x \in (a, b)$, put

$$f(x-) = \lim_{y \to x-} f(y) = \sup\{f(x) \mid a < y < x\},$$
$$f(x+) = \lim_{y \to x+} f(y) = \inf\{f(x) \mid x < y < b\}.$$

Then since f is increasing, $f(x-) \le f(x+)$ for $x \in (a, b)$. Therefore, f is discontinuous at x if and only if $f(x-) < f(x+)$. For each $n \in \mathbb{N}$, put

$$D_{a,b}^n = \left\{ x \in D_{a,b} \,\middle|\, f(x+) - f(x-) > \frac{1}{n} \right\}.$$

Then since $[f(a+), f(b-)]$ is a bounded interval, $D_{a,b}^n$ is a finite set and

$$D_{a,b} = \bigcup_{n=1}^{\infty} D_{a,b}^n$$

is a countable union of finite sets, and so $D_{a,b}$ is countable.

(ii) If (a, b) is unbounded or $\lim_{x \to a+} f(x)$ and $\lim_{x \to b-} f(x)$ are not finite. Then we take an increasing (ascending) sequence $\{(a_n, b_n)\}$ of bounded intervals such that

$$(a, b) = \bigcup_{n=1}^{\infty} (a_n, b_n)$$

and f is increasing on (a_n, b_n). Then by applying (i), the set of discontinuity points of f in (a_n, b_n) is countable, and so the set of discontinuity points of f in (a, b) is countable. □

Proposition 6.4.2 Let D be a countable subset of the open interval (a, b). Then there is an increasing function f on (a, b) such that

$$\{x \in (a, b) \mid f \text{ is continuous at } x\} = (a, b) \setminus D.$$

[Proof] (i) If D is a finite set, the proof is obvious.

(ii) Suppose that D is countably infinite. Let $D = \{d_1, d_2, \cdots, d_n, \cdots\}$ be the enumeration of D. For each given $x \in (a, b)$, put

$$L_x = \{n \mid d_n \le x\}.$$

Let $0 < r < 1$ be a given constant. Define a function f on (a, b) by

$$f(x) = \sum_{i \in L_x} r^i$$

where if $L_x = \emptyset$, then the series is 0, and for other case, the series is well-defined. Then for any $a < u < v < b$, we have

$$L_u \subseteq L_v$$

and

$$f(v) - f(u) = \sum_{i \in L_v} r^i - \sum_{i \in L_u} r^i = \begin{cases} \sum_{i \in L_v \setminus L_u} r^i, & \text{if } L_v \neq L_u \\ 0, & \text{if } L_v = L_u \end{cases}$$
$$\geq 0,$$

which implies that f is increasing.

For each $x_0 \in D$, there exists $n \in \mathbb{N}$ such that $x_0 = d_n \in D$, and then for all $x < x_0$

$$f(x_0) - f(x) = \sum_{i \in L_{x_0} \setminus L_x} r^i$$
$$\geq r^n$$

which implies that f is discontinuous at x_0.

Let $x_0 \notin D$ and let $\epsilon > 0$ be given. Then there exists $n \in \mathbb{N}$ such that $r^{n+1}/(1-r) < \epsilon$ and there exists an open interval I containing x_0 such that $q_k \notin I$ for $k = 1, 2, \cdots, n$. Therefore, for any $x \in I$,

$$|f(x) - f(x_0)| < \sum_{i=n+1}^{\infty} r^i = \frac{r^{n+1}}{1-r} < \epsilon,$$

which implies that f is continuous at x_0. □

A closed bounded interval $[c, d]$ is said to be **nondegenerate** if $c < d$.

Definition 6.4.3 Let E be a subset of $\mathbb{R}$. Then a family $\mathcal{F}$ of nondegenerate closed intervals is said to **cover the set E in the sense of Vitali** (simply, called a **Vitali covering** of E) if for any $x \in E$ and $\epsilon > 0$, there exists an interval $I \in \mathcal{F}$ such that $x \in I$ and $\ell(I) < \epsilon$.

Remark 6.4.4 Let E be a subset of $\mathbb{R}$ and $\mathcal{F}$ be a family of nondegenerate closed intervals. If $\mathcal{F}$ is a Vitali covering of E, then for each $\lambda > 0$,

$$\mathcal{F}_\lambda = \{I \in \mathcal{F} \,|\, \ell(I) < \lambda\}$$

is also a Vitali covering of E.

Theorem 6.4.5 **[Viltali Covering Theorem]** Let E be a subset of $\mathbb{R}$ with finite Lebesgue outer measure and $\mathcal{F}$ a family of nondegenerate closed intervals which is a Vitali covering of E. Then for each $\epsilon > 0$, there exists a finite disjoint subfamily $\{I_k\}_{k=1}^n$ of $\mathcal{F}$ such that

$$m^*\left(E \setminus \bigcup_{k=1}^n I_k\right) < \epsilon.$$

[Proof] Let $\epsilon > 0$ be given. Since $m^*(E) < \infty$, there exists an open set containing E such that $m^*(O) = m(O) < \infty$. Since $\mathcal{F}$ is a Vitali covering of E,

$$\mathcal{F}_O = \{I \in \mathcal{F} \mid I \subset O\}$$

is also a Vitali covering of E. Now, we choose a sequence $\{I_k\}_{k=1}^{\infty}$ of disjoint intervals in $\mathcal{F}_O$ by induction as follows:

(1) let I_1 be any interval in $\mathcal{F}_O$;

(2) suppose $\{I_1, I_2, \cdots, I_n\}$ is already chosen as finite disjoint intervals in $\mathcal{F}_O$;

(3) if $E \not\subseteq \cup_{k=1}^n I_k$, then we choose $I_{n+1} \in \mathcal{F}_O$ such that

  (i) $I_{n+1} \cap (\cup_{k=1}^n I_k) = \emptyset$. In fact, if $x \in E \setminus \cup_{k=1}^n I_k$, then since $\cup_{k=1}^n I_k$ is a closed set and $\mathcal{F}_O$ is a Vitali covering of E, there exists a closed interval $I_{n+1} \in \mathcal{F}_O$ such that $x \in I_{n+1}$ and $I_{n+1} \cap (\cup_{k=1}^n I_k) = \emptyset$.

  (ii) $\ell(I_{n+1}) > \frac{1}{2} s_n = \frac{1}{2} \sup\{\ell(I) \mid I \in \mathcal{F}_O \text{ and } I \cap (\cup_{k=1}^n I_k) = \emptyset\}$, where $s_n \leq m(O) < \infty$ since $\ell(I) \leq m(O)$ for all $I \in \mathcal{F}_O$.

For the sequence $\{I_k\}_{k=1}^{\infty}$, we see that

$$\sum_{k=1}^{\infty} \ell(I_k) = m\left(\bigcup_{k=1}^{\infty} I_k\right) \leq m(O) < \infty,$$

and so there exists $N \in \mathbb{N}$ such that

$$\sum_{k=N+1}^{\infty} \ell(I_k) < \frac{\epsilon}{5}.$$

For each $k \in \mathbb{N}$, let J_k be the interval with the same midpoint as I_k and $\ell(J_k) = 5\ell(I_k)$. Then we claim that for any $n \in \mathbb{N}$,

$$F_n := E \setminus \left(\bigcup_{k=1}^{n} I_k\right) \subset \bigcup_{k=n+1}^{\infty} J_k. \tag{6.9}$$

Let $n \in \mathbb{N}$ be fixed and $x \in F_n$. Then there exists an interval $I \in \mathcal{F}_O$ such that $x \in I$ and $I \cap (\cup_{k=1}^n I_k) = \emptyset$, and so $\ell(I) \leq s_n < 2\ell(I_{n+1})$. If $I \cap I_k = \emptyset$ for any $k \in \mathbb{N}$, then $\ell(I) \leq s_k < 2\ell(I_{k+1})$ and $\lim_{k \to \infty} \ell(I_k) = 0$ $(\sum_{k=1}^{\infty} \ell(I_k) \leq m(O) < \infty)$, which implies that $\ell(I) = 0$ which contradicts that I is nondegenerate, i.e., $\ell(I) > 0$. Therefore, there exists $k \in \mathbb{N}$ with $k > n$ such that $I \cap I_k \neq \emptyset$. Put

$$m = \min\{k \in \mathbb{N} \mid I \cap I_k \neq \emptyset\}.$$

Then it holds that $m > n$ and $\ell(I) \leq s_{m-1} \leq 2\ell(I_m)$. Since $I \cap I_m \neq \emptyset$, $I \subset J_m$. Therefore, (6.9) holds, and so we obtain that

$$m^*\left(E \setminus \bigcup_{k=1}^{N} I_k\right) \leq m\left(\bigcup_{k=N+1}^{\infty} J_k\right) \leq \sum_{k=N+1}^{\infty} \ell(J_k)$$

$$= 5 \sum_{k=N+1}^{\infty} \ell(I_k)$$
$$< \epsilon,$$

which gives the proof. □

Definition 6.4.6 Let f be a real-valued function defined on an open interval (a,b) and $x \in (a,b)$. Then the **upper derivative** $\overline{D}f(x)$ and the **lower derivative** $\underline{D}f(x)$ of f at x are defined by

$$\overline{D}f(x) = \lim_{h\to 0+} \left\{ \sup_{0<|t|\le h} \frac{f(x+t)-f(x)}{t} \right\},$$
$$\underline{D}f(x) = \lim_{h\to 0+} \left\{ \inf_{0<|t|\le h} \frac{f(x+t)-f(x)}{t} \right\}.$$

If $\underline{D}f(x)$, $\overline{D}f(x)$ are finite and $\underline{D}f(x) = \overline{D}f(x)$, then we say that f is **differentiable** at x and we write

$$f'(x) = \underline{D}f(x) = \overline{D}f(x).$$

For any function $f : (a,b) \to \mathbb{R}$ and $x \in (a,b)$, it is obvious that

$$\underline{D}f(x) \le \overline{D}f(x).$$

Let $[a,b]$ be a bounded closed interval and let $f : [a,b] \to \mathbb{R}$ be an increasing function. For each $\alpha > 0$, put

$$D_\alpha = \left\{ x \in (a,b) \,|\, \overline{D}f(x) \ge \alpha \right\}. \tag{6.10}$$

Lemma 6.4.7 Let $[a,b]$ be a bounded closed interval and let $f : [a,b] \to \mathbb{R}$ be an increasing function. Let $\alpha > 0$. Then for any $0 < \alpha' < \alpha$, the family

$$\mathcal{F}_{\alpha'} = \left\{ [c,d] : \text{ nondegenerate} \,\middle|\, [c,d] \subset [a,b], \ \frac{f(d)-f(c)}{d-c} \ge \alpha' \right\} \tag{6.11}$$

is a Vitali covering of D_α given as in (6.10).

[Proof] Let $x \in \mathcal{D}_\alpha$ and $\epsilon > 0$ be given. Then by the definition of D_α, it holds that

$$\overline{D}f(x) = \lim_{h\to 0+} \left\{ \sup_{0<|t|\le h} \frac{f(x+t)-f(x)}{t} \right\} \ge \alpha.$$

Therefore, for any sufficiently small $h > 0$ with $h < \epsilon$,

$$\sup_{0<|t|\le h} \frac{f(x+t)-f(x)}{t} \ge \alpha'', \qquad 0 < \alpha' < \alpha'' < \alpha,$$

and so there exists $t \in \mathbb{R}$ with $0 < |t| < h$,

$$\frac{f(x+t)-f(x)}{t} \ge \alpha'.$$

Therefore, $\mathcal{F}$ is a Vitali covering of $\mathcal{D}_\alpha$. □

Remark 6.4.8 Let $[a, b]$ be a bounded closed interval and let $f : [a, b] \to \mathbb{R}$ be a continuous function. If f is differentiable on (a, b) with $f' \ge \alpha > 0$ on (a, b), then by the mean value theorem in calculus, it holds that

$$\alpha \le \frac{f(b)-f(a)}{b-a} \quad \Leftrightarrow \quad b-a \le \frac{1}{\alpha}\left(f(b)-f(a)\right).$$

The following lemma gives a generalization of Remark 6.4.8.

Lemma 6.4.9 Let $[a, b]$ be a bounded closed interval and let $f : [a, b] \to \mathbb{R}$ be an increasing function. Then for each $\alpha > 0$,

$$m^*\left(D_\alpha\right) \le \frac{1}{\alpha}\left(f(b)-f(a)\right), \tag{6.12}$$

where D_α is given as in (6.10).

[Proof] Let $\alpha > 0$ be given. For any $0 < \alpha' < \alpha$, by Lemma 6.4.7, $F_{\alpha'}$ given as in (6.11) is a Vitali covering of $\mathcal{D}_\alpha$. Therefore, by the Viltali Covering Theorem (Theorem 6.4.5), for any $\epsilon > 0$, there exists a finite disjoint intervals $\{[c_k, d_k]\}_{k=1}^n \subset \mathcal{F}_{\alpha'}$ such that

$$m^*\left(\mathcal{D}_\alpha \setminus \bigcup_{k=1}^n [c_k, d_k]\right) < \epsilon.$$

On the other hand,

$$\mathcal{D}_\alpha \subset \left(\bigcup_{k=1}^n [c_k, d_k]\right) \bigcup \left(\mathcal{D}_\alpha \setminus \bigcup_{k=1}^n [c_k, d_k]\right).$$

Therefore, it holds that

$$m^*(\mathcal{D}_\alpha) \le \sum_{k=1}^n (d_k - c_k) + \epsilon \le \frac{1}{\alpha'}\sum_{k=1}^n (f(d_k) - f(c_k)) + \epsilon \le \frac{1}{\alpha'}(f(b)-f(a)) + \epsilon,$$

which proves (6.12). □

Corollary 6.4.10 Let $[a, b]$ be a bounded closed interval and let $f : [a, b] \to \mathbb{R}$ be an increasing function. Then

$$m^* \left(\left\{ x \in (a, b) \,|\, \overline{D}f(x) = \infty \right\} \right) = 0. \tag{6.13}$$

[Proof] For each $n \in \mathbb{N}$,

$$\left\{ x \in (a, b) \,|\, \overline{D}f(x) = \infty \right\} \subset \mathcal{D}_n = \left\{ x \in (a, b) \,|\, \overline{D}f(x) \geq n \right\},$$

and so from (6.12) we obtain that

$$m^* \left(\left\{ x \in (a, b) \,|\, \overline{D}f(x) = \infty \right\} \right) \leq m^*(\mathcal{D}_n) \leq \frac{1}{n}(f(b) - f(a)),$$

and then by taking limit as $n \to \infty$, we have (6.13). □

Let $f : (a, b) \to \mathbb{R}$ be an increasing function on the bounded interval (a, b). For fixed $(u, v) \in \mathbb{Q}_+ \times \mathbb{Q}_+$ with $u < v$, where $\mathbb{Q}_+ = \mathbb{Q} \cap (0, \infty)$, put

$$\mathcal{D}_{u,v} = \{x \in (a, b) \,|\, \underline{D}f(x) < u < v < \overline{D}f(x)\}. \tag{6.14}$$

Since $m^*(\mathcal{D}_{u,v}) \leq \ell((a, b)) = b - a$, there exists an open set O such that $\mathcal{D}_{u,v} \subset O$ and

$$m(O) < m^*(\mathcal{D}_{u,v}) + \epsilon \tag{6.15}$$

for a given $\epsilon > 0$.

Lemma 6.4.11 Let $f : (a, b) \to \mathbb{R}$ be an increasing function on the bounded interval (a, b) and let $(u, v) \in \mathbb{Q}_+ \times \mathbb{Q}_+$ with $u < v$ be fixed. Then the family

$$\mathcal{F}_O = \{[c, d] : \text{ nondegenerate } | \, [c, d] \subset O, \;\; f(d) - f(c) < u(d - c)\} \tag{6.16}$$

is a Vitali covering of $D_{u,v}$ given as in (6.14), where O is an open subset of $\mathbb{R}$ containing $D_{u,v}$ satisfying (6.15).

[Proof] For any $x \in \mathcal{D}_{u,v}$, it holds that

$$u > \underline{D}f(x) = \lim_{h \to 0+} \left\{ \inf_{0<|t|\leq h} \frac{f(x + t) - f(x)}{t} \right\}.$$

Therefore, for any $\epsilon > 0$, there exists a sufficiently small $h_0 > 0$ with $h_0 < \epsilon$, such that for any $0 < h < h_0 < \epsilon$,

$$\inf_{0<|t|\leq h} \frac{f(x + t) - f(x)}{t} < u,$$

which implies that there exists $t \in \mathbb{R}$ with $0 < |t| < h < h_0 < \epsilon$ such that

$$\frac{f(x + t) - f(x)}{t} < u.$$

Therefore, $\mathcal{F}_O$ is a Vitali covering of $\mathcal{D}_{u,v}$. □

Theorem 6.4.12 [**Lebesgue's Theorem**] Let $f : (a,b) \to \mathbb{R}$ be a monotone function. Then f is differentiable almost everywhere on (a,b).

[Proof] We prove the case of nondecreasing function f. Also, we assume that (a,b) is bounded. For the general case, we can take an ascending sequence of bounded open intervals of which the union is the given unbounded interval. Note that f is differentiable at $x \in (a,b)$ if $\underline{D}f(x) \geq \overline{D}f(x)$. Therefore, it is enough to prove that $m^*(\mathcal{D}) = 0$ for

$$\mathcal{D} = \{x \in (a,b) \,|\, \underline{D}f(x) < \overline{D}f(x)\}.$$

Note that

$$\mathcal{D} = \bigcup_{(u,v)\in\mathbb{Q}_+\times\mathbb{Q}_+;u<v} \mathcal{D}_{u,v},$$

where $D_{u,v}$ is given as in (6.14). Therefore, it is enough to prove that $m^*(\mathcal{D}_{u,v}) = 0$ for any $(u,v) \in \mathbb{Q}_+ \times \mathbb{Q}_+$ with $u < v$.

Let $(u,v) \in \mathbb{Q}_+ \times \mathbb{Q}_+$ with $u < v$ be fixed. Since $m^*(\mathcal{D}_{u,v}) \leq \ell((a,b)) = b - a$, there exists an open set O such that $\mathcal{D}_{u,v} \subset O$ and

$$m(O) < m^*(\mathcal{D}_{u,v}) + \epsilon.$$

Then by Lemma 6.4.11, $\mathcal{F}_O$, given as in (6.16), is a Vitali covering of $\mathcal{D}_{u,v}$. Hence by the Viltali Covering Theorem (Theorem 6.4.5), for any given $\epsilon > 0$, there exists a finite disjoint intervals $\{[c_k, d_k]\}_{k=1}^n \subset \mathcal{F}_O$ such that

$$m^*\left(\mathcal{D}_{u,v} \setminus \bigcup_{k=1}^n [c_k, d_k]\right) < \epsilon.$$

Therefore, by applying Lemma 6.4.9,

$$m^*\left(\mathcal{D}_{u,v} \cap [c_k, d_k]\right) \leq \frac{1}{v}(f(d_k) - f(c_k)),$$

and so we obtain that

$$\begin{aligned}
m^*(\mathcal{D}_{u,v}) &= m^*\left(\mathcal{D}_{u,v} \cap \left(\bigcup_{k=1}^n [c_k, d_k]\right)\right) + m^*\left(\mathcal{D}_{u,v} \setminus \bigcup_{k=1}^n [c_k, d_k]\right) \\
&< \sum_{k=1}^n m^*\left(\mathcal{D}_{u,v} \cap [c_k, d_k]\right) + \epsilon \\
&\leq \frac{1}{v}\left(\sum_{k=1}^n (f(d_k) - f(c_k))\right) + \epsilon \\
&\leq \frac{u}{v}\left(\sum_{k=1}^n (d_k - c_k)\right) + \epsilon \\
&\leq \frac{u}{v} m(O) + \epsilon
\end{aligned}$$

$$\leq \frac{u}{v} m^*(\mathcal{D}_{u,v}) + \frac{u}{v}\epsilon + \epsilon,$$

which implies that

$$m^*(\mathcal{D}_{u,v}) \leq \frac{u}{v} m^*(\mathcal{D}_{u,v}).$$

Therefore, since $u/v < 1$ and $0 \leq m^*(\mathcal{D}_{u,v}) < \infty$, $m^*(\mathcal{D}_{u,v}) = 0$. □

Corollary 6.4.13 Let $[a, b]$ be a bounded closed interval and $f : [a, b] \to \mathbb{R}$ be an increasing function. Then f' is integrable over $[a, b]$ and

$$\int_a^b f' \, dm \leq f(b) - f(a).$$

[Proof] Since f is an increasing function on $[a, b]$, by the Lebesgue's Theorem (Theorem 6.4.12), f is differentiable almost everywhere on $[a, b]$. Therefore, $\{D_{1/n}f\}$ is a sequence of nonnegative measurable functions which converges to f' pointwise almost everywhere on $[a, b]$, where $D_h f$ is given as in (6.1). By the Fatou's Lemma, we have

$$\int_a^b f' \, dm \leq \liminf_{n\to\infty} \int_a^b D_{1/n} f \, dm.$$

On the other hand, by the change of variable formula, we obtain that

$$\begin{aligned}\int_a^b D_{1/n} f \, dm &= \frac{1}{1/n} \int_b^{b+1/n} f \, dm - \frac{1}{1/n} \int_a^{a+1/n} f \, dm \\ &= f(b) - \frac{1}{1/n} \int_a^{a+1/n} f \, dm \\ &\leq f(b) - f(a),\end{aligned}$$

where $f(x) = f(b)$ for $b \leq x \leq b + 1/n$, and so it holds that

$$\int_a^b f' \, dm \leq \liminf_{n\to\infty} \int_a^b D_{1/n} f \, dm \leq f(b) - f(a),$$

which gives the proof. □

Theorem 6.4.14 Let $[a, b]$ be a bounded closed interval and $f : [a, b] \to \mathbb{R}$ be of bounded variation. Then f is differentiable almost everywhere on $[a, b]$ and f' is integrable on $[a, b]$.

[Proof] By Jordan's Theorem 6.3.6, f is expressed by the difference of two increasing functions g and h, i.e., $f = g - h$. Then by Lebesgue's Theorem (Theorem 6.4.12), g and h are differentiable almost everywhere on $[a, b]$ and so f is differentiable almost everywhere on $[a, b]$, and then by Corollary 6.4.13, f' is integrable over $[a, b]$. □

6.5 Differentiation of Integral

Let f be an integrable function on $[a, b]$. Define the function F on $[a, b]$ as the indefinite integral of f by

$$F(x) = \int_a^x f\, dm, \qquad x \in [a, b].$$

In this section, we prove that the derivative of the indefinite integral of an integrable function equals the integrand almost everywhere.

Proposition 6.5.1 Let f be an integrable function on a bounded closed interval $[a, b]$. Then the function F defined by

$$F(x) = \int_a^x f\, dm, \qquad x \in [a, b]$$

is continuous and of bounded variation on $[a, b]$.

[Proof] The continuity follows from (ii) in Proposition 6.1.1. We now prove that F is of bounded variation. Let $P : a = x_0 < x_1 < \cdots < x_n = b$ be a partition of $[a, b]$. Then it holds that

$$\begin{aligned}\sum_{k=1}^{n} |F(x_k) - F(x_{k-1})| &= \sum_{k=1}^{n} \left| \int_{x_{k-1}}^{x_k} f\, dm \right| \\ &\le \sum_{k=1}^{n} \int_{x_{k-1}}^{x_k} |f|\, dm \\ &= \int_a^b |f|\, dm,\end{aligned}$$

which implies that

$$TV(F) \le \int_a^b |f|\, dm < \infty,$$

and so F is of bounded variation. □

Lemma 6.5.2 Let f be an integrable function on a bounded closed interval $[a, b]$. Then $f = 0$ almost everywhere on $[a, b]$ if and only if

$$\int_{x_1}^{x_2} f\, dm = 0$$

for all $(x_1, x_2) \subset [a, b]$.

[Proof] $\Rightarrow$) It is clear.

$\Leftarrow$) Put

$$\mathcal{F} = \{A \subset [a,b] \mid A \text{ is measurable and } \int_A f\,dm = 0\}.$$

Then for any open subset $O \subset [a,b]$, O is the disjoint union of a countable collection $\{(a_n, b_n)\}$ of open intervals (see Theorem 1.4.6), and so by the Lebesgue Dominated Convergence Theorem (Theorem 5.1.26), we have

$$\int_O f\,dm = \sum \int_{a_n}^{b_n} f\,dm = 0,$$

which implies that $O \in \mathcal{F}$. Let $G \subset [a,b]$ be a G_δ set, i.e., there exists a sequence $\{O_n\}_{n=1}^{\infty}$ of open sets such that $G = \cap_{n=1}^{\infty} O_n$. Put $U_n = \cap_{k=1}^{n} O_k$ and then $\{U_n\}$ is a descending sequence of open sets and $G = \cap_{n=1}^{\infty} U_n$ and so again by the Lebesgue Dominated Convergence Theorem (Theorem 5.1.26), we obtain that

$$\int_G f\,dm = \int_{\cap_{n=1} U_n} f\,dm = \int_a^b \chi_{\cap_{n=1} U_n} f\,dm = \lim_{n\to\infty} \int_{U_n} f\,dm = 0,$$

which implies that $G \in \mathcal{F}$. Let $E \subset [a,b]$ be a measurable subset. Then by Theorem 3.2.18, there exists a G_δ set G such that $m(G \setminus E) = 0$ and so we have

$$\int_E f\,dm = \int_E f\,dm + \int_{G \setminus E} f\,dm = \int_G f\,dm = 0,$$

which implies that $E \in \mathcal{F}$. Put

$$F^+ = \{x \in [a,b] \mid f(x) \geq 0\}, \qquad F^- = \{x \in [a,b] \mid f(x) \leq 0\},$$

and then

$$\int_a^b f^+\,dm = \int_{F^+} f\,dm = 0, \qquad \int_a^b f^-\,dm = \int_{F^-} (-f)\,dm = -\int_{F^-} f\,dm = 0,$$

which, by Proposition 5.1.9, implies that $f^+ = 0$ and $f^- = 0$ almost everywhere on $[a,b]$ and so $f = 0$ almost everywhere on $[a,b]$. □

Lemma 6.5.3 Let f be a bounded and measurable function on a bounded closed interval $[a,b]$ and F be a function on $[a,b]$ defined by

$$F(x) = \int_a^x f\,dm, \qquad x \in [a,b].$$

Then $F' = f$ almost everywhere on $[a,b]$.

[Proof] By Proposition 6.5.1, F is of bounded variation over $[a, b]$, and so, by Theorem 6.4.14, F' exists almost everywhere on $[a, b]$. Since f is bounded, suppose that $|f| \leq K$. Then by setting

$$f_n(x) = \frac{F(x + 1/n) - F(x)}{1/n},$$

we have

$$f_n(x) = \frac{1}{1/n} \int_x^{x+1/n} f \, dm,$$

and so

$$|f_n| \leq K.$$

Since

$$f_n(x) \to F'(x) \quad \text{for almost all } x \in [a, b],$$

by the Lebesgue Dominated Convergence Theorem (Theorem 5.1.26), for any $(x_1, x_2) \subset [a, b]$, it holds that

$$\begin{aligned} \int_{x_1}^{x_2} F' \, dm &= \lim_{n \to \infty} \int_{x_1}^{x_2} f_n \, dm \\ &= \lim_{n \to \infty} \frac{1}{1/n} \int_{x_1}^{x_2} (F(x + 1/n) - F(x)) dx \\ &= \lim_{n \to \infty} \left[\frac{1}{1/n} \int_{x_2}^{x_2+1/n} F \, dm - \frac{1}{1/n} \int_{x_1}^{x_1+h} F \, dm \right] \\ &= F(x_2) - F(x_1) = \int_{x_1}^{x_2} f \, dm, \end{aligned}$$

since F is continuous. Hence for any $(x_1, x_2) \subset [a, b]$,

$$\int_{x_1}^{x_2} (F' - f) \, dm = 0,$$

and so by Lemma 6.5.2, $F' = f$ almost everywhere on $[a, b]$. □

Theorem 6.5.4 Let f be an integrable function on a bounded closed interval $[a, b]$ and F be a function on $[a, b]$ defined by

$$F(x) = \int_a^x f \, dm, \qquad x \in [a, b].$$

Then $F' = f$ almost everywhere on $[a, b]$.

[Proof] Suppose that $f \geq 0$. For each $n \in \mathbb{N}$, let

$$f_n(x) = \begin{cases} f(x), & 0 \leq f(x) \leq n, \\ n, & f(x) > n. \end{cases}$$

Then $f - f_n \geq 0$, and so

$$G_n(x) := \int_a^x (f - f_n)\, dm$$

is an increasing function of x and, by Lebesgue's Theorem (Theorem 6.4.12), G_n is differentiable almost everywhere on $[a, b]$, and its derivative is nonnegative. Since f_n is bounded, by Lemma 6.5.3, it holds that

$$\frac{d}{dx}\left(\int_a^x f_n\, dm\right) = f_n(x) \quad \text{almost all } x \in [a, b],$$

and so

$$F'(x) = \frac{d}{dx}G_n(x) + \frac{d}{dx}\left(\int_a^x f_n\, dm\right) \geq f_n(x) \quad \text{almost all } x \in [a, b].$$

Since n is arbitrary,

$$F'(x) \geq f(x) \quad \text{almost all } x \in [a, b].$$

Consequently, for any $(x_1, x_2) \subset [a, b]$, we have

$$\int_{x_1}^{x_2} F'\, dm \geq \int_{x_1}^{x_2} f\, dm = F(x_2) - F(x_1).$$

On the other hand, since F is increasing, by Corollary 6.4.13, it holds that

$$\int_{x_1}^{x_2} F'\, dm \leq F(x_2) - F(x_1).$$

Thus for any $(x_1, x_2) \subset [a, b]$, we have

$$\int_{x_1}^{x_2} F'\, dm = F(x_2) - F(x_1) = \int_{x_1}^{x_2} f\, dm$$

and

$$\int_{x_1}^{x_2} (F' - f)\, dm = 0.$$

and so by Lemma 6.5.2, $F' = f$ almost everywhere on $[a, b]$. For general integrable function f, we have $f = f^+ - f^-$ and

$$F(x) = \int_a^x f\, dm = \int_a^x f^+\, dm - \int_a^x f^-\, dm$$

and so F is differentiable almost everywhere on $[a, b]$ and

$$F' = f^+ - f^- = f$$

almost everywhere on $[a, b]$. □

6.6 Absolutely Continuous Functions

Definition 6.6.1 Let f be a real-valued function on a bounded closed interval $[a, b]$. Then f is said to be **absolutely continuous** on $[a, b]$ if for any $\epsilon > 0$, there exists a $\delta > 0$ such that for any finite disjoint collection $\{(x_i, y_i)\}_{i=1}^n$ of subintervals of $[a, b]$ with

$$\sum_{i=1}^n |y_i - x_i| < \delta,$$

it holds that

$$\sum_{i=1}^n |f(y_i) - f(x_i)| < \epsilon. \tag{6.17}$$

Example 6.6.2 (1) Every absolutely continuous function defined on the bounded closed interval $[a, b]$ is uniform continuous on $[a, b]$.

(2) The function defined by the indefinite integral of an integrable function is absolutely continuous, i.e., for an integrable function f on $[a, b]$, the function F defined by

$$F(x) = \int_a^x f\, dm, \qquad x \in [a, b]$$

is absolutely continuous. In fact, for any given $\epsilon > 0$, by (ii) in Proposition 6.1.1, there exists a $\delta > 0$ such that for any $A \subset E$ with $m(A) < \delta$,

$$\int_A |f|\, dm < \epsilon.$$

Therefore, for any finite disjoint collection $\{(x_i, y_i)\}_{i=1}^n$ of subintervals of $[a, b]$ with

$$\sum_{i=1}^n |y_i - x_i| < \delta,$$

it holds that

$$\sum_{i=1}^n |F(y_i) - F(x_i)| \le \sum_{i=1}^n \int_{x_i}^{y_i} |f|\, dm = \int_{\cup_{i=1}^n (x_i, y_i)} |f|\, dm < \epsilon,$$

and so F is absolutely continuous on $[a, b]$.

(3) Every finite linear combination of absolutely continuous functions on the bounded closed interval $[a, b]$ is absolutely continuous on $[a, b]$ (see Exercise **6.10**).

Proposition 6.6.3 Let f be a real-valued function defined on a bounded closed interval $[a, b]$. If f is Lipschitz on $[a, b]$, then f is absolutely continuous on $[a, b]$.

[Proof] Since f is Lipschitz on $[a, b]$, there exists a constant $c > 0$ such that

$$|f(y) - f(x)| \leq c|y - x|$$

for any $x, y \in [a, b]$. Therefore, for given $\epsilon > 0$, by taking $\delta = \epsilon/(c+1)$, for any finite disjoint collection $\{(x_i, y_i)\}_{i=1}^n$ of subintervals of $[a, b]$ with

$$\sum_{i=1}^{n} |y_i - x_i| < \delta,$$

it holds that

$$\sum_{i=1}^{n} |f(y_i) - f(x_i)| \leq c \sum_{i=1}^{n} |y_i - x_i| < c\delta < \epsilon,$$

and so f is absolutely continuous on $[a, b]$.

□

Proposition 6.6.4 Let $[a, b]$ be a bounded closed interval and $f : [a, b] \to \mathbb{R}$ be an absolutely continuous function. Then f is of bounded variation on $[a, b]$.

[Proof] Since f is absolutely continuous, there exists a $\delta > 0$ (for the given $\epsilon = 1$) such that (6.17) holds. Let $\{(a_k, b_k)\}_{k=1}^N$ be a finite disjoint family of subintervals of $[a, b]$ such that $|b_k - a_k| < \delta$ and

$$[a, b] = \bigcup_{k=1}^{N} [a_k, b_k].$$

On the other hand, by applying (6.6), we see that

$$TV(f_{[a,b]}) = \sum_{k=1}^{N} TV(f_{[a_k, b_k]})$$

and, since $|b_k - a_k| < \delta$, $TV(f_{[a_k,b_k]}) \leq 1 (= \epsilon)$ for any $k = 1, 2, \cdots, N$. Therefore, we have $TV(f_{[a,b]}) \leq N < \infty$ and so f is of bounded variation on $[a, b]$. □

Proposition 6.6.5 Let $[a, b]$ be a bounded closed interval and $f : [a, b] \to \mathbb{R}$ be an absolutely continuous function. Then the function $[a, b] \ni x \mapsto TV(f_{[a,x]})$ is absolutely continuous on $[a, b]$.

[Proof] Let $\epsilon > 0$ be given. Then since f is absolutely continuous, there exists a $\delta > 0$ such that for any finite disjoint collection $\{(x_i, y_i)\}_{i=1}^{\ell}$ of subintervals of $[a, b]$ with

$$\sum_{i=1}^{\ell} |y_i - x_i| < \delta,$$

it holds that

$$\sum_{i=1}^{\ell} |f(y_i) - f(x_i)| < \frac{\epsilon}{2}. \tag{6.18}$$

Let $\{(a_k, b_k)\}_{k=1}^{n}$ be a finite disjoint sequence of subintervals of $[a, b]$ such that $\sum_{k=1}^{n}(b_k - a_k) < \delta$ and let P_k be a partition of $[a_k, b_k]$ $(k = 1, 2, \cdots, n)$. Then we have

$$\sum_{k=1}^{n} V(f_{[a_k,b_k]}, P_k) < \frac{\epsilon}{2}$$

and then, by taking supremum and (6.7), we have

$$\begin{aligned}\sum_{k=1}^{n} |TV(f_{[a,b_k]}) - TV(f_{[a,a_k]})| &= \sum_{k=1}^{n} \big(TV(f_{[a,b_k]}) - TV(f_{[a,a_k]})\big) \\ &= \sum_{k=1}^{n} TV(f_{[a_k,b_k]}) \\ &\leq \frac{\epsilon}{2} \\ &< \epsilon,\end{aligned}$$

which implies that the function $[a, b] \ni x \mapsto TV(f_{[a,x]})$ is absolutely continuous on $[a, b]$. □

Theorem 6.6.6 Let $[a, b]$ be a bounded closed interval and $f : [a, b] \to \mathbb{R}$ be a function. Then f is absolutely continuous if and only if f is the difference of two increasing absolutely continuous functions on $[a, b]$.

[Proof] $\Leftarrow$) It is clear.

$\Rightarrow$) By (6.8), it holds that

$$f(x) = \big(f(x) + TV(f_{[a,x]})\big) - TV(f_{[a,x]}), \qquad x \in [a, b]. \tag{6.19}$$

On the other hand, by Proposition 6.6.5, the function $[a, b] \ni x \mapsto TV(f_{[a,x]})$ is absolutely continuous on $[a, b]$, and a sum of two absolutely continuous functions on $[a, b]$ is absolutely continuous. Therefore, from (6.19), f is absolutely continuous. □

As a corollary of Theorem 6.6.6, we have the following theorem.

Theorem 6.6.7 Let $[a,b]$ be a bounded closed interval and $f:[a,b]\to\mathbb{R}$ be an absolutely continuous function. Then f is differentiable almost everywhere on $[a,b]$.

[Proof] The proof is immediate from the Lebesgue's Theorem (Theorem 6.4.12) and Theorem 6.6.6. □

Proposition 6.6.8 Let $[a,b]$ be a bounded closed interval and $f:[a,b]\to\mathbb{R}$ be an absolutely continuous function. If $f'=0$ almost everywhere on $[a,b]$, then f is a constant function.

[Proof] Let $c\in(a,b)$ and E be a measurable subset of $[a,c]$ such that $m(E)=c-a$ and $f'=0$ on E. Let $\epsilon>0$ and $\lambda>0$ be given arbitrarily. Put

$$\mathcal{F}=\{[d,e]:\ \text{nondegenerate}\,|\,[d,e]\subset[a,c]\ \text{and}\ |f(e)-f(d)|<\lambda(e-d)\}.$$

Then for any $x\in E$, there exists $h>0$ such that $[x,x+h]\subset[a,c]$ and

$$\left|\frac{f(x+h)-f(x)}{h}\right|<\lambda$$

and so $\mathcal{F}$ is a Vitali covering of E. Since f is absolutely continuous on $[a,c]$, there exists $\delta>0$ such that for any finite disjoint family $\{[a_k,c_k]\}_{k=1}^{n}$ of subintervals of $[a,c]$ with

$$\sum_{k=1}^{n}|c_k-a_k|<\delta,$$

it holds that

$$\sum_{k=1}^{n}|f(c_k)-f(a_k)|<\epsilon.$$

Since $\mathcal{F}$ is a Vitali covering of E, by Viltali Covering Theorem (Theorem 6.4.5), there exists a finite disjoint family $\{[d_k,e_k]\}_{k=1}^{\ell}$ of intervals in $\mathcal{F}$ such that

$$m^*\left(E\setminus\bigcup_{k=1}^{\ell}[d_k,e_k]\right)<\delta.$$

Suppose that $d_k\le d_{k+1}$ for $k=1,2,\cdots,\ell-1$, and put $d_0=e_0=a$ and $d_{\ell+1}=c$, then it holds that

$$d_0=e_0=a\le d_1<e_1<d_2<\cdots<d_\ell<e_\ell\le c=d_{\ell+1}$$

and

$$
\begin{aligned}
|f(c)-f(a)| &= \left|\sum_{k=0}^{\ell}(f(d_{k+1})-f(d_k))\right| \\
&= \left|\sum_{k=0}^{\ell}(f(d_{k+1})-f(e_k))+\sum_{k=0}^{\ell}(f(e_k)-f(d_k))\right| \\
&\le \sum_{k=0}^{\ell}|f(d_{k+1})-f(e_k)|+\sum_{k=0}^{\ell}|f(e_k)-f(d_k)| \\
&< \epsilon+\lambda\sum_{k=0}^{\ell}|d_k-e_k| \\
&< \epsilon+\lambda(c-a),
\end{aligned}
$$

which implies that $f(a)=f(c)$ and so f is constant. □

Theorem 6.6.9 Let $[a,b]$ be a bounded closed interval and $F:[a,b]\to\mathbb{R}$ be a function. Then F is absolutely continuous on $[a,b]$ if and only if F is differentiable almost everywhere on $[a,b]$ with its integrable derivative and for each $x\in[a,b]$, $F(x)$ is the indefinite integral of its derivative over $[a,x]$, i.e.,

$$F(x)=\int_a^x F'\,dm+F(a), \tag{6.20}$$

[Proof] $\Leftarrow$) The proof is immediate from (2) in Example 6.6.2.

$\Rightarrow$) Suppose that F is absolutely continuous on $[a,b]$. Then F is of bounded variation and so by the Jordan's Theorem (Theorem 6.3.6), there exist two increasing functions F_1 and F_2 such that

$$F(x)=F_1(x)-F_2(x).$$

Then by Lebesgue Theorem (Theorem 6.4.12), F is differentiable almost everywhere on $[a,b]$ and

$$|F'(x)|\le F_1'(x)+F_2'(x),$$

and so by Corollary 6.4.13, F' is integrable. For each $x\in[a,b]$, put

$$G(x)=\int_a^x F'\,dm.$$

Then G is absolutely continuous on $[a,b]$ and so the function $f:=F-G$ is absolutely continuous on $[a,b]$, and by Theorem 6.5.4, $f'=F'-G'=0$ almost everywhere on $[a,b]$. Therefore, by Proposition 6.6.8, f is constant, i.e., there exists $C\in\mathbb{R}$ such that

$$F(x)=G(x)+C=\int_a^x F'\,dm+C,$$

which implies that $F(a)=C$ and so it holds that (6.20). □

Corollary 6.6.10 Let f be a monotone function defined on a bounded closed interval $[a, b]$. Then f is absolutely continuous on $[a, b]$ if and only if

$$\int_a^b f'\,dm = f(b) - f(a). \tag{6.21}$$

[Proof] $\Rightarrow$) It is obvious from (6.20).

$\Leftarrow$) Suppose that f is increasing and (6.21) holds. Then for each $x \in [a, b]$, we have

$$\begin{aligned} 0 &= \int_a^b f'\,dm - (f(b) - f(a)) \\ &= \left(\int_a^x f'\,dm - (f(x) - f(a))\right) + \left(\int_x^b f'\,dm - (f(b) - f(x))\right), \end{aligned}$$

and by Corollary 6.4.13, we have

$$\int_a^x f'\,dm - (f(x) - f(a)) \leq 0, \qquad \int_x^b f'\,dm - (f(b) - f(x)) \leq 0.$$

Then since sum of non-positive numbers is zero,

$$f(x) = \int_a^x f'\,dm + f(a),$$

i.e., f is the indefinite integral of its derivative and so by Theorem 6.6.9, f is absolutely continuous on $[a, b]$. □

Definition 6.6.11 Let $[a, b]$ be a bounded closed interval and let f be of bounded variation on $[a, b]$. Then f is said to be **singular** if $f' = 0$ almost everywhere on $[a, b]$.

Let f be of bounded variation on a bounded closed interval $[a, b]$. Then by Jordan's Theorem (Theorem 6.3.6),

$$f = f_1 - f_2$$

for two increasing functions f_1 and f_2 and so by Lebesgue's Theorem (Theorem 6.4.12), f is differentiable almost everywhere on $[a, b]$. Then by Corollary 6.4.13, f' is integrable over $[a, b]$. Put

$$g(x) = \int_a^x f'\,dm, \qquad h(x) = f(x) - \int_a^x f'\,dm, \qquad x \in [a, b].$$

Then g is absolutely continuous and h is singular, and

$$f = g + h,$$

which is referred to a **Lebesgue decomposition** of f as a sum of an absolutely continuous function g and a singular function h.

6.7 Riemann-Stieltjes Integral

We first study the Riemann-Stirltjes integral for increasing functions, and then extend it to the general case.

Let $f : I = [a, b] \to \mathbb{R}$ be a bounded function and let $\mathcal{P} : a = t_0 < t_1 < \cdots < t_n = b$ be a partition of I. For each $k = 1, 2, \cdots, n$, put $I_k = [t_{k-1}, t_k]$, and

$$m_k = \inf_{t \in I_k} f(t), \qquad M_k = \sup_{t \in I_k} f(t).$$

Let $\alpha : I \to \mathbb{R}$ be a increasing function. Then the lower and upper (Riemann-Stieltjes) sums of f corresponding to $\mathcal{P}$ with respect to α is defined by

$$L(f, \alpha, \mathcal{P}) = \sum_{k=1}^{n} m_k \Delta\alpha_k, \qquad U(f, \alpha, \mathcal{P}) = \sum_{k=1}^{n} M_k \Delta\alpha_k,$$

respectively, where $\Delta\alpha_k = \alpha(t_k) - \alpha(t_{k-1})$. It is easy to see that if $\mathcal{P}'$ ia a refinement of $\mathcal{P}$, then

$$L(f, \alpha, \mathcal{P}) \leq L(f, \alpha, \mathcal{P}'), \qquad U(f, \alpha, \mathcal{P}) \geq U(f, \alpha, \mathcal{P}').$$

Moreover, for any partitions $\mathcal{P}$ and $\mathcal{P}'$ of I, we obtain that

$$L(f, \alpha, \mathcal{P}) \leq L(f, \alpha, \mathcal{P} \cup \mathcal{P}') \leq U(f, \alpha, \mathcal{P} \cup \mathcal{P}') \leq U(f, \alpha, \mathcal{P}').$$

Therefore, as same as the lower and upper Riemann integrals, it is natural to define the lower and upper Riemann-Stieltjes integrals of f with respect to α by

$$L(f, \alpha) = \sup_{\mathcal{P}}\{L(f, \alpha, \mathcal{P})\}, \qquad U(f, \alpha) = \sup_{\mathcal{P}}\{U(f, \alpha, \mathcal{P})\},$$

respectively, where the supremum and infimum are taken over all finite partitions $\mathcal{P}$ of I. Then it is obvious that

$$L(f, \alpha) \leq U(f, \alpha).$$

Definition 6.7.1 If $L(f, \alpha) = U(f, \alpha)$, then we say that f is **Riemann-Stieltjes integrable** with respect to α over I and we write $f \in \mathcal{RS}(\alpha)$ $(= \mathcal{RS}(\alpha, I))$. In this case, the common value is called the **Riemann-Stieltjes integral** with respect to α over I and denoted by

$$\int_a^b f d\alpha = \int_a^b f(x) d\alpha(x) = L(f, \alpha) = U(f, \alpha).$$

Let $f : I \to \mathbb{R}$ be a bounded function and $\alpha : I \to \mathbb{R}$ be an increasing function. Then by the definition, it is obvious that $f \in \mathcal{RS}(\alpha, I)$ if and only if for any $\epsilon > 0$, there exists a partition $\mathcal{P}$ of I such that

$$U(f, \alpha, \mathcal{P}) - L(f, \alpha, \mathcal{P}) < \epsilon. \tag{6.22}$$

Proposition 6.7.2 Let $f : I \to \mathbb{R}$ be a bounded function and $\alpha : I \to \mathbb{R}$ be an increasing function. If $f \in \mathcal{RS}(\alpha, I)$, then $|f| \in \mathcal{RS}(\alpha, I)$.

[Proof] Let $\epsilon > 0$ be given. There exists a partition $\mathcal{P} : a = x_0 < x_1 < \cdots < x_n = b$ of I such that (6.22) holds. On the other hand, there exists $x_k', x_k'' \in [x_{k-1}, x_k]$ such that

$$\begin{aligned} &\sup_{x_{k-1} \le x \le x_k} \{|f(x)|\} - \inf_{x_{k-1} \le x \le x_k} \{|f(x)|\} \\ &\le |f(x_k')| - |f(x_k'')| + \epsilon \le |f(x_k') - f(x_k'')| + \epsilon \\ &\le \sup_{x_{k-1} \le x \le x_k} \{f(x)\} - \inf_{x_{k-1} \le x \le x_k} \{f(x)\}, \end{aligned}$$

which implies that

$$U(|f|, \alpha, \mathcal{P}) - L(|f|, \alpha, \mathcal{P}) < \epsilon.$$

Therefore, $|f|$ is Riemann-Stieltjes integrable. □

Proposition 6.7.3 Suppose that $f \in \mathcal{RS}(\alpha_1, I) \cap \mathcal{RS}(\alpha_2, I)$ and λ is a nonnegative real number. Then $f \in \mathcal{RS}(\alpha_1 + \lambda\alpha_2, I)$ and it holds that

$$\int_I f d(\alpha_1 + \lambda\alpha_2) = \int_I f d\alpha_1 + \lambda \int_I f d\alpha_2.$$

Also, if $f_1, f_2 \in \mathcal{RS}(\alpha, I)$, then for any real number λ, $f_1 + \lambda f_2 \in \mathcal{RS}(\alpha, I)$ and

$$\int_I (f_1 + \lambda f_2) d\alpha = \int_I f_1 d\alpha + \lambda \int_I f_2 d\alpha.$$

By applying Proposition 6.7.3, the following corollary is immediate.

Corollary 6.7.4 Let $I = [a, b]$ and $f \in \mathcal{RS}(\alpha, I)$. Then for any subinterval J of I, $f \in \mathcal{RS}(\alpha, J)$. Moreover, if $a < c < b$, then it holds that

$$\int_a^b f d\alpha = \int_a^c f d\alpha + \int_c^b f d\alpha.$$

Proposition 6.7.5 Let $f : I \to \mathbb{R}$ be a continuous function and $\alpha : I \to \mathbb{R}$ be an increasing function. Then $f \in \mathcal{RS}(\alpha, I)$.

[Proof] If α is a constant function, then the proof is obvious. Hence we suppose that α is not a constant function. Let $\epsilon > 0$ be given. Put

$$\rho = \frac{\epsilon}{\alpha(b) - \alpha(a)}.$$

Since f is continuous on the closed and bounded interval $I = [a, b]$, f is uniformly continuous on I. Therefore, there exists $\delta > 0$ such that for any $x, y \in I$ with $|x - y| < \delta$,

$$|f(x) - f(y)| < \rho.$$

Consider the partition

$$\mathcal{P} : x_0 = a < x_1 = a + \delta < \cdots < x_{n-1} = a + (n-1)\delta < x_n = b,$$

where $n \in \mathbb{N}$ with $a + n\delta \geq b$. Then we obtain that

$$\begin{aligned} U(f, \alpha, \mathcal{P}) - L(f, \alpha, \mathcal{P}) &= \sum_{k=1}^{n} (M_k - m_k)\Delta\alpha_k \\ &< \rho \sum_{k=1}^{n} \Delta\alpha_k = \rho(\alpha(b) - \alpha(a)) \\ &= \epsilon. \end{aligned}$$

Therefore, we see that

$$\sup_{\mathcal{P}}\{L(f, \alpha, \mathcal{P})\} = \inf_{\mathcal{P}}\{U(f, \alpha, \mathcal{P})\},$$

and so f is Riemann-Stieltjes integrable with respect to α over I. □

Let α be of bounded variation. Then by Jordan's theorem, there exist two increasing functions α_1 and α_2 such that $\alpha = \alpha_1 - \alpha_2$. Let $f : I \to \mathbb{R}$ be a function such that f is Riemann-Stieltjes integrable with respect to all increasing functions α over I. Then we define the Riemann-Stieltjes integral of f with respect to the bounded variation function α over I by

$$\int_I f d\alpha = \int_I f d\alpha_1 - \int_I f d\alpha_2. \tag{6.23}$$

The Riemann-Stieltjes integral defined as in (6.23) of f with respect to the bounded variation function α over I does not depend on the representation of α. In fact, if $\alpha = \alpha_3 - \alpha_4$ for two increasing functions α_1 and α_2, then we have

$$\alpha_1 - \alpha_2 = \alpha = \alpha_3 - \alpha_4,$$

which implies that

$$\alpha_1 + \alpha_4 = \alpha_2 + \alpha_3,$$

and by Proposition 6.7.3,

$$\int_I f d\alpha_1 + \int_I f d\alpha_4 = \int_I f d\alpha_2 + \int_I f d\alpha_3.$$

Therefore, we have

$$\int_I f d\alpha_1 - \int_I f d\alpha_2 = \int_I f d\alpha_3 - \int_I f d\alpha_4.$$

Proposition 6.7.6 Let $f : I \to \mathbb{R}$ be a continuous function and $\alpha : I \to \mathbb{R}$ be of bounded variation. Then $f \in \mathcal{RS}(\alpha, I)$. Moreover, we have

$$\left|\int_a^b f d\alpha\right| \leq \int_a^b |f| \, d|\alpha| = \int_a^b |f| \, dTV(\alpha), \tag{6.24}$$

where $d|\alpha|(x) = dTV(\alpha)(x) = dTV\left(\alpha_{[a,x]}\right)$.

[Proof] From Proposition 6.7.5 and the definition of the Riemann-Stieltjes integral, it is obvious that $f \in \mathcal{RS}(\alpha, I)$. By definition, there exists a countable family $\{\mathcal{P}_n'\}_{n=1}^{\infty}$ of partitions $\mathcal{P}_n'$ of I such that $\mathcal{P}_n'$ is finer than $\mathcal{P}_m'$ for any $n \geq m$ and

$$\lim_{n\to\infty} L(f, \alpha, \mathcal{P}_n') = \int_I f d\alpha = \lim_{n\to\infty} U(f, \alpha, \mathcal{P}_n').$$

On the other hand, since $|f|$ is continuous and $TV(\alpha)$ is an increasing function, $|f| \in \mathcal{RS}(TV(\alpha))$ and there exists a countable family $\{\mathcal{P}_n''\}_{n=1}^{\infty}$ of partitions $\mathcal{P}_n''$ of I such that $\mathcal{P}_n''$ is finer than $\mathcal{P}_m''$ for any $n \geq m$ and

$$\begin{aligned}\lim_{n\to\infty} L(|f|, TV(\alpha), \mathcal{P}_n'') &= \int_I |f| \, dTV(\alpha) \\ &= \lim_{n\to\infty} U(|f|, TV(\alpha), \mathcal{P}_n'').\end{aligned}$$

Let $\mathcal{P}_n = \mathcal{P}_n' \cup \mathcal{P}_n''$ for all $n \in \mathbb{N}$. Then we obtain that

$$\begin{aligned}\left|\int_I f d\alpha\right| &= \lim_{n\to\infty} |L(f, \alpha, \mathcal{P}_n')| \\ &\leq \lim_{n\to\infty} |U(|f|, TV(\alpha), \mathcal{P}_n)| \\ &= \int_I |f| dTV(\alpha),\end{aligned}$$

which implies (6.24). □

Corollary 6.7.7 Let $f : I \to \mathbb{R}$ be a continuous function and $\alpha : I \to \mathbb{R}$ be of bounded variation. Then we have

$$\left|\int_I f d\alpha\right| \leq \sup_{x\in I}\{|f(x)|\} TV(f_I).$$

[Proof] The proof is immediate from Proposition 6.7.6. □

Let $f, g : I \to \mathbb{R}$ be bounded functions. Then we say that f, g satisfy the **property** $(\mathcal{RS})$ if there exists a number L so that for any $\epsilon > 0$ there exists a partition $\mathcal{P}_\epsilon$ of I such that for any partition $\mathcal{P} : a = x_0 < x_1 < \cdots < x_n = b$ finer than $\mathcal{P}_\epsilon$ and any $t_k \in [x_{k-1}, x_k]$,

$$\left| \sum_{k=1}^{n} f(t_k)\Delta g_k - L \right| < \epsilon.$$

In that case, the number L is unique.

Theorem 6.7.8 Let $f : I \to \mathbb{R}$ be bounded and $g : I \to \mathbb{R}$ be increasing. Then The following are equivalent:

(i) $f \in \mathcal{RS}(g, I)$,

(ii) f and g satisfy the property $(\mathcal{RS})$.

[Proof] (i) $\Rightarrow$ (ii). Put $L = \int_I f dg$. Let $\epsilon > 0$ be given. Since $f \in \mathcal{RS}(g, I)$, there exists a partition $\mathcal{P}_\epsilon$ of I such that

$$U(f, g, \mathcal{P}_\epsilon) - L < \epsilon, \qquad L - L(f, g, \mathcal{P}_\epsilon) < \epsilon.$$

Therefore, for any partition $\mathcal{P} : a = x_0 < x_1 < \cdots < x_n = b$ of I which is finer than $\mathcal{P}_\epsilon$,

$$\begin{aligned} L(f, g, \mathcal{P}_\epsilon) &\leq L(f, g, \mathcal{P}) \\ &\leq \sum_{k=1}^{n} f(t_k)\Delta g_k \\ &\leq U(f, g, \mathcal{P}) \\ &\leq U(f, g, \mathcal{P}_\epsilon), \end{aligned}$$

which implies that

$$\left| \sum_{k=1}^{n} f(t_k)\Delta g_k - L \right| < \epsilon$$

for any $t_k \in [x_{k-1}, x_k]$.

(ii) $\Rightarrow$ (i). Let $\epsilon > 0$ be given and let

$$\mathcal{P}_\epsilon : a = x_0 < x_1 < \cdots < x_n = b$$

be a partition of I corresponding $\epsilon/4$ in the property $(\mathcal{RS})$. Note that

$$M_k - m_k = \sup \{f(x) - f(y) \,|\, x, y \in [x_{k-1}, x_k]\}.$$

Therefore, there exist $t_k, t'_k \in [x_{k-1}, x_k]$ such that

$$M_k - m_k \leq f(t_k) - f(t'_k) + \frac{\epsilon}{2(g(b) - g(a) + 1)}.$$

Hence we obtain that

$$\begin{aligned} U(f,g,\mathcal{P}_\epsilon) - L(f,g,\mathcal{P}_\epsilon) &= \sum_{k=1}^{n}(M_k - m_k)\Delta g_k \\ &\le \sum_{k=1}^{n}(f(t_k) - f(t'_k))\Delta g_k + \frac{\epsilon}{2} \\ &= \sum_{k=1}^{n}(f(t_k) - L)\Delta g_k + \sum_{k=1}^{n}(L - f(t'_k))\Delta g_k + \frac{\epsilon}{2} \\ &< \epsilon, \end{aligned}$$

which implies that f is Riemann-Stieltjes integrable with respect to g over I. □

Theorem 6.7.9 Let $f, g : I \to \mathbb{R}$ be bounded functions. Then $f \in \mathcal{RS}(g,I)$ if and only if $g \in \mathcal{RS}(f,I)$. In this case, we have

$$\int_I fdg + \int_I gdf = f(b)g(b) - f(a)g(a). \tag{6.25}$$

[Proof] Suppose that $f \in \mathcal{RS}(g,I)$. Let $\epsilon > 0$ be given. Then for some $L = \int_I fdg \in \mathbb{R}$, there exists a partition $\mathcal{P}_\epsilon$ of I which is corresponding to ϵ in the property $(\mathcal{RS})$ such that for all partitions $\mathcal{P} : a = x_0 < x_1 < \cdots < x_n = b$ finer than $\mathcal{P}_{\epsilon_1}$ and for any $t_k \in [x_{k-1}, x_k]$

$$\left|\sum_{k=1}^{n} f(t_k)\Delta g_k - L\right| < \epsilon_1$$

for any $t_k \in [x_{k-1}, x_k]$. Put

$$M = -L + f(b)g(b) - f(a)g(a).$$

Then we obtain that

$$\begin{aligned} M - \sum_{k=1}^{n} g(t_k)\Delta f_k &= -L + \sum_{k=1}^{n}(f(x_k)g(x_k) - f(x_{k-1})g(x_{k-1})) \\ &\quad - \sum_{k=1}^{n} g(t_k)(f(x_k) - f(x_{k-1})) \\ &= -L + \sum_{k=1}^{n} f(x_k)(g(x_k) - g(t_k)) + \sum_{k=1}^{n} f(x_{k-1})(g(t_k) - g(x_{k-1})). \end{aligned}$$

On the other hand, by taking

$$\{y_0, y_1, \cdots, y_l\} = \{x_1, \cdots, x_n\} \cup \{t_1, \cdots, t_n\},$$

where we assume that $a = y_0 < y_1 < \cdots < y_l = b$, we have

$$\sum_{k=1}^{n} f(x_k)(g(x_k) - g(t_k)) + \sum_{k=1}^{n} f(x_{k-1})(g(t_k) - g(x_{k-1})) = \sum_{k=1}^{l} f(s_l)(g(y_k) - g(y_{k-1})),$$

where $\{s_1, \cdots, s_l\} = \{x_0, \cdots, x_n\}$. Then for the partition $\mathcal{Q} : a = y_0 < y_1 < \cdots < y_l = b$ finer than $\mathcal{P}$, we obtain that

$$\left| M - \sum_{k=1}^{n} g(t_k)\Delta f_k \right| = \left| -L + \sum_{k=1}^{l} f(s_l)(g(y_k) - g(y_{k-1})) \right| < \epsilon,$$

which implies that $g \in \mathcal{RS}(f, I)$ and $\int_I gdf = M$. Therefore, we have

$$\int_I fdg + \int_I gdf = L + M = f(b)g(b) - f(a)g(a),$$

which implies (6.25). □

6.8 Convex Functions

Definition 6.8.1 A function φ defined on an open interval (a, b) is said to be **convex** if for each $x, y \in (a, b)$ and any $0 \leq \lambda \leq 1$, it holds that

$$\varphi(\lambda x + (1 - \lambda)y) \leq \lambda\varphi(x) + (1 - \lambda)\varphi(y).$$

Lemma 6.8.2 (Chordal Slope Lemma) Let $\varphi : (a, b) \to \mathbb{R}$ be a convex function. Then for any $x_1, x_2 \in (a, b)$ with $x_1 < x_2$ and any $x_1 < x < x_2$, it holds that

$$\frac{\varphi(x) - \varphi(x_1)}{x - x_1} \leq \frac{\varphi(x_2) - \varphi(x_1)}{x_2 - x_1} \leq \frac{\varphi(x_2) - \varphi(x)}{x_2 - x}.$$

[Proof] Let $x_1, x_2, x \in (a, b)$ with $x_1 < x_2$ and $x_1 < x < x_2$ be given. Then there exists a $0 < \lambda < 1$ such that $x = \lambda x_1 + (1 - \lambda)x_2 \in (x_1, x_2)$, in fact, $\lambda = \frac{x - x_2}{x_1 - x_2}$. Then by the convexity of φ, we have

$$\begin{aligned}\varphi(x) &\leq \lambda\varphi(x_1) + (1 - \lambda)\varphi(x_2) \\ &= \frac{x - x_2}{x_1 - x_2}\varphi(x_1) + \frac{x_1 - x}{x_1 - x_2}\varphi(x_2),\end{aligned}$$

which implies that

$$\frac{\varphi(x) - \varphi(x_1)}{x - x_1} \leq \frac{\varphi(x_2) - \varphi(x_1)}{x_2 - x_1} \leq \frac{\varphi(x_2) - \varphi(x)}{x_2 - x}.$$

□

Let f be a real-valued function and x an interior point of its domain. Then we define the **upper left and right derivatives** $\overline{D}^{\mp}f(x)$ and the **lower left and right derivative** $\underline{D}^{\mp}f(x)$ of f at x by

$$\overline{D}^{-}f(x) = \lim_{h\to 0+}\left\{\sup_{0<t\le h}\frac{f(x)-f(x-t)}{t}\right\},$$

$$\overline{D}^{+}f(x) = \lim_{h\to 0+}\left\{\sup_{0<t\le h}\frac{f(x+t)-f(x)}{t}\right\},$$

$$\underline{D}^{-}f(x) = \lim_{h\to 0+}\left\{\inf_{0<t\le h}\frac{f(x)-f(x-t)}{t}\right\},$$

$$\underline{D}^{+}f(x) = \lim_{h\to 0+}\left\{\inf_{0<t\le h}\frac{f(x+t)-f(x)}{t}\right\}.$$

For each interior point x of the domain of f, if $\overline{D}^{-}f(x) = \underline{D}^{-}f(x)$ as finite, then we say that f is **differentiable on the left at** x and the common value is denoted by $f'(x-)$ and called the **left derivative at** x. Similarly, we say that f is differentiable on the right at x if $\overline{D}^{+}f(x) = \underline{D}_{+}f(x)$ as finite and the **right derivative at** x is denoted by $f'(x+)$.

Remark 6.8.3 Let f be a real-valued function and x an interior point of its domain. Then the derivatives $\overline{D}^{\mp}f(x)$ and $\underline{D}^{\mp}f(x)$ are sometimes called the **Dini derivatives** of f at x.

Proposition 6.8.4 Let φ be a convex function on (a, b).

(i) The right and left derivatives of φ exist at each point of (a, b) and at each point the left derivative is less than or equal to the right derivative.

(ii) The left and right derivatives are monotone increasing functions.

(iii) The right and left derivatives of φ are equal to each other except on a countable set.

[Proof] (i) Let $x_0 \in (a, b)$. Then by Lemma 6.8.2, the function

$$(a, x_0) \ni x \longmapsto \frac{\varphi(x)-\varphi(x_0)}{x-x_0}$$

is an increasing function, and bounded by $\frac{\varphi(x_1)-\varphi(x_0)}{x_1-x_0}$ for any $x_0 < x_1$, and the function

$$(x_0, b) \ni x \longmapsto \frac{\varphi(x)-\varphi(x_0)}{x-x_0}$$

is also an increasing function, and has a lower bound $\dfrac{\varphi(x_0)-\varphi(x_1)}{x_0-x_1}$ for $x_1 < x_0$, and so the limits

$$\varphi'(x_0-) := \lim_{x\uparrow x_0}\frac{\varphi(x)-\varphi(x_0)}{x-x_0},$$

$$\varphi'(x_0+) := \lim_{x \downarrow x_0} \frac{\varphi(x) - \varphi(x_0)}{x - x_0}$$

exist and are finite. Thus φ is differentiable on the right and on the left at each point. Moreover, it holds that

$$\varphi'(x_0-) \leq \varphi'(x_0+),$$

i.e., the left derivative is less than or equal to the right derivative.

(ii) If $x_0 < y_0$, $x < y_0$, and $x_0 < y$, then

$$\frac{\varphi(x) - \varphi(x_0)}{x - x_0} \leq \frac{\varphi(y) - \varphi(y_0)}{y - y_0},$$

and either derivative at x_0 is less than or equal to either derivative at y_0. Consequently, each derivative is monotone, i.e.,

$$\varphi'(x_0-) \leq \varphi'(x_0+) \leq \varphi'(y_0-) \leq \varphi'(y_0+). \tag{6.26}$$

(iii) By Theorem 6.4.1, a monotone function can have only a countable number of discontinuities. Therefore, there exists a countable subset D of (a, b) such that $\varphi'(x-)$ is continuous on $(a, b) \setminus D$. If $x \in (a, b) \setminus D$, then by (6.26), $\varphi'(x-) = \varphi'(x+)$. □

Proposition 6.8.5 Let φ be a continuous function on (a, b). If one of four derivatives of φ exists and is nondecreasing, then φ is convex.

[Proof] Suppose that $\overline{D}^+\varphi$ exists and is increasing. Let $x, y \in (a, b)$ with $x < y$ be given. Then we define a function ψ on $[0, 1]$ by

$$\psi(t) = t\varphi(x) + (1 - t)\varphi(y) - \varphi[tx + (1 - t)y].$$

We now prove that ψ is nonnegative on $[0, 1]$. Note that ψ is continuous, and $\psi(0) = \psi(1) = 0$. Moreover, it holds that

$$\overline{D}^+\psi(t) = (\varphi(x) - \varphi(y)) + (y - x)\overline{D}^+\varphi[tx + (1 - t)y],$$

and so $\overline{D}^+\psi$ is nonincreasing on $[0, 1)$ since $\overline{D}^+\varphi$ is nondecreasing.

Let $\gamma \in [0, 1]$ such that $\psi(\gamma) = \min\{\psi(t) \mid t \in [0, 1]\}$. If $\gamma = 1$, then $\psi(t) \geq \psi(1) = 0$ on $[0, 1]$. Hence suppose that $\gamma \in [0, 1)$. Since ψ has a local minimum at $t = \gamma$, we have $\overline{D}^+\psi(\gamma) \geq 0$. But $\overline{D}^+\psi$ is nonincreasing, and so $\overline{D}^+\psi \geq 0$ on $[0, \gamma]$. Consequently, ψ is nondecreasing on $[0, \gamma]$, and hence $\psi(\gamma) \geq \psi(0) = 0$. Thus the minimum of ψ on $[0, 1]$ is nonnegative, and so $\psi \geq 0$ on $[0, 1]$, which means that φ is convex. □

Corollary 6.8.6 Let φ have a second derivative at each point of (a, b). Then φ is convex on (a, b) if and only if $\varphi''(x) \geq 0$ for each $x \in (a, b)$.

Let φ be a convex function on (a, b) and $x_0 \in (a, b)$. The line $y = \alpha(x - x_0) + \varphi(x_0)$ through the point $(x_0, \varphi(x_0))$ is called a **supporting line** of φ at x_0 for any $x \in (a, b)$,

$$\varphi(x) \geq \alpha(x - x_0) + \varphi(x_0).$$

Remark 6.8.7 It follows from Lemma 6.8.2 that the line $y = \alpha(x - x_0) + \varphi(x_0)$ is a supporting line if and only if its slope α lies between the left and right derivatives at x_0. Thus, in particular, there is always at least one supporting line at each point.

Proposition 6.8.8 [**Jensen's Inequality**] Let $(X, \mathcal{M}, \mu)$ be a probability space, i.e., $(X, \mathcal{M}, \mu)$ is a measure space with $\mu(X) = 1$. Let $f : X \to (a, b)$ be an integrable function over X and φ be a convex function on (a, b). Then we have

$$\int_X \varphi \circ f \, d\mu \geq \varphi \left[\int_X f \, d\mu \right]. \tag{6.27}$$

[Proof] Let $m = \int_X f \, d\mu$. Then it holds that

$$a = \int_X a \, d\mu \leq m = \int_X f \, d\mu \leq \int_X b \, d\mu = b.$$

Let $y = \alpha(t - m) + \varphi(m)$ be the equation of a supporting line at m for the graph of φ. Then we have

$$\varphi(f(x)) \geq \alpha(f(x) - m) + \varphi(m), \qquad x \in X$$

and so we have

$$\begin{aligned} \int_X \varphi(f(x)) \, d\mu(x) &\geq \int_X [\alpha(f(x) - m) + \varphi(m)] \, d\mu(x) \\ &= \varphi(m) \\ &= \varphi \left(\int_X f \, d\mu(x) \right), \end{aligned}$$

which gives the proof of (6.27). $\square$

Exercises

6.1 Let $(X, \mathcal{M}, \mu)$ be a measure space. Let $\mathcal{F}$ and $\mathcal{G}$ be two families of integrable functions on X which are uniformly integrable over X. Prove that

(1) for each $\alpha \in \mathbb{R}$, $\alpha\mathcal{F} = \{\alpha f \mid f \in \mathcal{F}\}$ is uniformly integrable over X,

(2) $\mathcal{F} + \mathcal{G} = \{f + g \mid f \in \mathcal{F},\ \ g \in \mathcal{G}\}$ is uniformly integrable over X,

(3) $\mathcal{F} \cup \mathcal{G} = \{h \mid h \in \mathcal{F} \text{ or } h \in \mathcal{G}\}$ is uniformly integrable.

6.2 Let $(X, \mathcal{M}, \mu)$ be a measure space. Let $\mathcal{F}$ and $\mathcal{G}$ be two families of measurable functions on X which are tight over X. Prove that

(1) for each $\alpha \in \mathbb{R}$, $\alpha\mathcal{F} = \{\alpha f \mid f \in \mathcal{F}\}$ is tight over X,

(2) $\mathcal{F} + \mathcal{G} = \{f + g \mid f \in \mathcal{F},\ \ g \in \mathcal{G}\}$ is tight over X,

(3) $\mathcal{F} \cup \mathcal{G} = \{h \mid h \in \mathcal{F} \text{ or } h \in \mathcal{G}\}$ is tight over X.

6.3 Let $(X, \mathcal{M}, \mu)$ be a measure space and f a nonnegative integrable function over X. Prove that the set function ν on $\mathcal{M}$ defined by

$$\nu(E) = \int_E f d\mu, \qquad E \in \mathcal{M}$$

is a finite measure on $\mathcal{M}$.

6.4 Prove Theorem 6.2.10.

6.5 Let μ, ν, λ be σ-finite measures on a measurable space $(X, \mathcal{M})$. Prove that

(1) if $\mu \ll \nu$ and $\nu \ll \lambda$, then $\mu \ll \lambda$ and

$$\frac{d\mu}{d\lambda} = \frac{d\mu}{d\nu} \cdot \frac{d\nu}{d\lambda} \text{ a.e. } \lambda,$$

(2) if $\mu \ll \nu$ and $\nu \ll \mu$, then

$$\frac{d\mu}{d\nu} \cdot \frac{d\nu}{d\mu} = 1 \text{ a.e. } \mu,$$

(3) if $\mu \ll \lambda$ and $\nu \ll \lambda$, then $\mu + \nu \ll \lambda$ and

$$\frac{d(\mu+\nu)}{d\lambda} = \frac{d\mu}{d\lambda} + \frac{d\nu}{d\lambda} \text{ a.e. } \lambda.$$

6.6 Let $(X, \mathcal{M}, \mu)$ be a finite measure space and f an integrable function over X. Prove that for each sub σ-algebra $\mathcal{G}$ of $\mathcal{M}$, there exists a unique (up to sets of μ-measure zero) integrable function g over X such that for any $E \in \mathcal{G}$,

$$\int_E f\, d\mu = \int_E g\, d\mu.$$

The integrable function g is called the **conditional expectation** of f given (sub σ-algebra) $\mathcal{G}$ and denoted by $E[f|\mathcal{G}]$.

6.7 Let $I = [a, b]$ be a bounded closed interval and $f : I \to \mathbb{R}$ be of bounded variation. Prove the following.

(1) f is bounded and measurable,

(2) f is continuous except at the points of a countable subset of $[a, b]$,

(3) $\lim_{y\to x+} f(y)$ and $\lim_{y\to x-} f(y)$ exist for any $x \in (a, b)$.

6.8 Let h be an integrable function on $\mathbb{R}$ with respect to the Lebesgue measure m. Define a function f on $[0, \infty)$ by

$$f(x) = \int_0^x h(s)\, dm(s), \qquad x \geq 0.$$

Prove that for any $\epsilon > 0$, there exists $\delta > 0$ such that for any finite disjoint collection $\{(a_k, b_k)\}_{k=1}^n$ of open intervals in $[0, \infty)$,

$$\text{if } \sum_{k=1}^{\infty}(b_k - a_k) < \delta, \text{ then } \sum_{k=1}^{n} |f(b_k) - f(a_k)| < \epsilon.$$

6.9 Let E be a subset of $\mathbb{R}$ and $\mathcal{F}$ a collection of bounded closed intervals which is a Vitali covering of E. Prove that for each $\delta > 0$,

$$\mathcal{F}_\delta = \{I \in \mathcal{F} \mid \ell(I) < \delta\}$$

is a Vitali covering of E.

6.10 Every finite linear combination of absolutely continuous functions on the bounded closed interval $[a, b]$ is absolutely continuous on $[a, b]$.

6.11 Let f be a continuous function defined on the bounded closed interval $[a, b]$. Then f is absolutely continuous on $[a, b]$ if and only if $\{\mathrm{D}_h f\}_{0<h\leq 1}$ is uniformly integrable over $[a, b]$.

6.12 Let φ be a convex function on (a, b). Prove that φ is absolutely continuous on any closed subinterval $[c, d]$ of (a, b).

6.13 Prove that the Lebesgue decomposition of a bounded variation function f is unique up to a constant, i.e., if g_1 and g_2 are absolutely continuous parts of f, then there exists a constant c such that $g_2 = g_1 + c$.

6.14 Let $\{f_n\}_{n=1}^{\infty}$ be a sequence of continuous functions converges uniformly to f on $I = [a, b]$ and let $\alpha : I \to \mathbb{R}$ be of bounded variation. Prove that $f \in \mathcal{RS}(\alpha, I)$ and

$$\lim_{n\to\infty} \int_I f_n d\alpha = \int_I d\alpha.$$

6.15 Let $f : I = [a, b] \to \mathbb{R}$ be a continuous function and $\alpha : I \to \mathbb{R}$ is absolutely continuous. Prove that $f \in \mathcal{RS}(\alpha, I)$ and it holds that

$$\int_I f\, d\alpha = \int_I f\alpha' dm,$$

where m is the Lebesgue measure.

6.16 Let φ be a convex function on $(-\infty, \infty)$ and f an integrable function over $[0, 1]$. Prove that

$$\int_0^1 \varphi(f(t))dt \geq \varphi\left[\int_0^1 f(t)dt\right].$$

Chapter 7

The $L^p(X)$ Spaces

7.1 Seminormed and Normed Spaces

A $\mathbb{R}$-valued map $\|\cdot\| : V \to \mathbb{R}$ defined on a linear space (or vector space) V is called a **seminorm** if

(n1) $\|v\| \geq 0$ for any $v \in V$,

(n2) $\|u+v\| \leq \|u\| + \|v\|$ for any $u, v \in V$,

(n3) $\|\alpha v\| = |\alpha| \, \|v\|$ for any $\alpha \in \mathbf{C}$ and $v \in V$.

A linear space $(V, \|\cdot\|)$ with a seminorm $\|\cdot\|$ is called a **seminormed linear space** or simply **seminormed space**. A seminorm $\|\cdot\|$ defined on a linear space satisfying the following condition **(n1′)** instead of **(n1)**:

(n1′) $\|v\| \geq 0$ for any $v \in V$, and $\|v\| = 0$ if and only if $v = 0$

is called a **norm**, and a linear space $(V, \|\cdot\|)$ with a norm $\|\cdot\|$ is called a **normed linear space** or simply **normed space**.

Definition 7.1.1 Let $(V, \|\cdot\|)$ be a normed space.

(i) A sequence $\{v_n\}_{n=1}^{\infty} \subset V$ is said to be **Cauchy** if for any $\epsilon > 0$ there exists $N \in \mathbb{N}$ such that for any $m, n > N$,

$$\|v_n - v_m\| < \epsilon.$$

(ii) A sequence $\{v_n\}_{n=1}^{\infty} \subset V$ is called a **convergent sequence** to v in V if $v \in V$ and for any $\epsilon > 0$ there exists $N \in \mathbb{N}$ such that for any $n > N$,

$$\|v_n - v\| < \epsilon.$$

Proposition 7.1.2 Let $(V, \|\cdot\|)$ be a normed space. Let $\{v_n\}_{n=1}^{\infty} \subset V$ be a sequence and $v \in V$. If $\{v_n\}_{n=1}^{\infty}$ converges to v in V, then $\{\|v_n\|\}_{n=1}^{\infty}$ converges to $\|v\|$ in $\mathbb{R}$, i.e.,

$$\lim_{n\to\infty} \|v_n\| = \|v\|.$$

[Proof] By the triangle inequality, we can easily check that

$$\|v_n\| \le \|v_n - v\| + \|v\|, \qquad \|v\| \le \|v - v_n\| + \|v_n\|,$$

which imply that

$$\|v_n\| - \|v\| \le \|v_n - v\|, \qquad \|v\| - \|v_n\| \le \|v - v_n\|.$$

Therefore, it holds that

$$|\|v_n\| - \|v\|| \le \|v_n - v\|,$$

which implies that

$$\lim_{n\to\infty} |\|v_n\| - \|v\|| \le \lim_{n\to\infty} \|v_n - v\| = 0.$$

Hence $\{\|v_n\|\}_{n=1}^{\infty}$ converges to $\|v\|$ in $\mathbb{R}$. □

Proposition 7.1.3 Every convergent sequence $\{v_n\}_{n=1}^{\infty}$ in a normed space is bounded, i.e., there exists a $M \ge 0$ such that $\|v_n\| \le M$ for all $n \in \mathbb{N}$.

[Proof] Suppose that $\{v_n\}_{n=1}^{\infty}$ converges to a vector v in a normed space $(V, \|\cdot\|)$. Then by Proposition 7.1.2, $\{\|v_n\|\}_{n=1}^{\infty}$ converges to $\|v\|$ in $\mathbb{R}$, and so $\{\|v_n\|\}_{n=1}^{\infty}$ is a bounded sequence. Hence $\{v_n\}_{n=1}^{\infty}$ is bounded. □

Theorem 7.1.4 Let $(V, \|\cdot\|)$ be a normed space. Then it holds that

(i) every convergent sequence is Cauchy,

(ii) a Cauchy sequence converges if it has a convergent subsequence.

[Proof] (i) Let $\{v_n\}_{n=1}^{\infty} \subset V$ be a convergent sequence to $v \in V$ in a normed space $(V, \|\cdot\|)$. Then for any $\epsilon > 0$, there exists a $N \in \mathbb{N}$ such that for any $n > N$

$$\| v_n - v \| < \frac{\epsilon}{2}.$$

Therefore, for any $m, n > N$, it holds that

$$\| v_n - v_m \| \leq \| v_n - v \| + \| v_m - v \| < \frac{\epsilon}{2} + \frac{\epsilon}{2} = \epsilon.$$

Hence $\{v_n\}_{n=1}^{\infty}$ is a Cauchy sequence.

(ii) Let $\{v_n\}_{n=1}^{\infty}$ be a Cauchy sequence which has a convergent subsequence $\{v_{n_k}\}_{k=1}^{\infty}$. Suppose that

$$\lim_{k \to \infty} v_{n_k} = v.$$

Then for any given $\epsilon > 0$, there exists $N \in \mathbb{N}$ such that for any $m, n > N$ and $k > N$

$$\| v_n - v_m \| < \frac{\epsilon}{2} \quad \text{and} \quad \| v_{n_k} - v \| < \frac{\epsilon}{2}.$$

Therefore, for any $n > N$, since $n_k \geq k > N$, we have

$$\begin{aligned} \| v_n - v \| &\leq \| v_n - v_{n_k} \| + \| v_{n_k} - v \| \\ &< \frac{\epsilon}{2} + \frac{\epsilon}{2} = \epsilon, \end{aligned}$$

which implies that $\{v_n\}_{n=1}^{\infty}$ converges to v. □

Definition 7.1.5 A normed space $(V, \| \cdot \|)$ is said to be **complete** if every Cauchy sequence in V is convergent. A complete normed space is called a **Banach space**.

Definition 7.1.6 Let V be a complex linear (or vector) space (a linear space over the complex field $\mathbb{C}$). A map $\langle \cdot | \cdot \rangle : V \times V \to \mathbb{C}$ is called a **inner product** if it satisfies the following conditions:

(i) for any $u \in V$, $\langle u | u \rangle \geq 0$ and $\langle u | u \rangle = 0$ if and only if $u = 0$,

(ii) for any $u, v \in V$, $\langle u | v \rangle = \overline{\langle v | u \rangle}$,

(iii) for any $u, v, w \in V$ and $\alpha, \beta \in \mathbb{C}$,

$$\langle \alpha u + \beta v | w \rangle = \alpha \langle u | w \rangle + \beta \langle v | w \rangle.$$

Let $(V, \langle \cdot | \cdot \rangle)$ be an inner product space. Then one can define a canonical norm by

$$\| u \| = \sqrt{\langle u | u \rangle}, \qquad u \in V,$$

which is called the norm induced by the inner product $\langle \cdot | \cdot \rangle$.

Definition 7.1.7 An inner product space $(V, \langle \cdot | \cdot \rangle)$ is called a **Hilbert space** if V is complete with respect to the norm induced by the inner product, i.e., every Cauchy sequence in V converges in V (with respect to the norm).

Theorem 7.1.8 [**Cauchy-Schwarz Inequality**] Let $(H, \langle \cdot | \cdot \rangle)$ be a Hilbert space. Then for any $\xi, \eta \in H$,

$$|\langle \xi | \eta \rangle| \leq \|\xi\| \, \|\eta\|. \tag{7.1}$$

Moreover, the equality in (7.1) holds if and only if ξ and η are linearly dependent.

[Proof] If $\xi = 0$ or $\eta = 0$, then (7.1) is immediate. Suppose that $\eta \neq 0$. Then for any $\alpha \in \mathbb{C}$,

$$\begin{aligned} 0 &\leq \left\langle \xi + \alpha \frac{\eta}{\|\eta\|} \,\middle|\, \xi + \alpha \frac{\eta}{\|\eta\|} \right\rangle \\ &= \|\xi\|^2 + \overline{\alpha} \frac{\langle \xi | \eta \rangle}{\|\eta\|} + \alpha \frac{\overline{\langle \xi | \eta \rangle}}{\|\eta\|} + \alpha \overline{\alpha} \\ &= \|\xi\|^2 - \frac{|\langle \xi | \eta \rangle|^2}{\|\eta\|^2} + \left| \frac{\langle \xi | \eta \rangle}{\|\eta\|} + \alpha \right|^2. \end{aligned} \tag{7.2}$$

In particular, for $\alpha = -\frac{\langle \xi | \eta \rangle}{\|\eta\|}$, we have

$$\|\xi\|^2 - \frac{|\langle \xi | \eta \rangle|^2}{\|\eta\|^2} \geq 0,$$

which implies (7.1).

If the equality in (7.1) holds, then (7.2) holds with $\alpha = -\frac{\langle \xi | \eta \rangle}{\|\eta\|}$ and so $\xi + \alpha \frac{\eta}{\|\eta\|} = 0$ with $\alpha = -\frac{\langle \xi | \eta \rangle}{\|\eta\|}$, which implies that ξ and η are linearly dependent. The converse is straightforward. □

Corollary 7.1.9 Let $(H, \langle \cdot | \cdot \rangle)$ be a Hilbert space. The inner product $\langle \cdot | \cdot \rangle$ is continuous in the first and second variables.

[Proof] The proof is immediate from Theorem 7.1. In fact, if $\xi_n \to \xi$ and $\eta_n \to \eta$ in H, then

$$|\langle \xi_n | \eta_m \rangle - \langle \xi | \eta \rangle| \leq |\langle \xi_n | \eta_m \rangle - \langle \xi_n | \eta \rangle| + |\langle \xi_n | \eta \rangle - \langle \xi | \eta \rangle|$$

$$\leq \|\xi_n\|\|\eta_m - \eta\| + \|\xi_n - \xi\|\|\eta\|,$$

which converges to 0 as $n \to \infty$ and $m \to \infty$, since $\{\xi_n\}_{n=1}^{\infty}$ is bounded by Proposition 7.1.3. □

Theorem 7.1.10 **[Parallelogram Law]** Let $(H, \langle \cdot | \cdot \rangle)$ be a Hilbert space. Then for any $\xi, \eta \in H$, we have

$$\|\xi + \eta\|^2 + \|\xi - \eta\|^2 = 2\left(\|\xi\|^2 + \|\eta\|^2\right). \tag{7.3}$$

[Proof] For any $\xi, \eta \in H$, we obtain that

$$\begin{aligned}\|\xi + \eta\|^2 + \|\xi - \eta\|^2 &= \langle \xi + \eta | \xi + \eta \rangle + \langle \xi - \eta | \xi - \eta \rangle \\ &= 2\langle \xi | \xi \rangle + 2\langle \eta | \eta \rangle \\ &= 2\left(\|\xi\|^2 + \|\eta\|^2\right),\end{aligned}$$

which gives the proof of (7.3). □

Let S be a subset of a vector space V. Then S is said to be **convex** if for any $u, v \in S$ and $0 \leq \lambda \leq 1$,

$$\lambda u + (1 - \lambda)v \in V.$$

Theorem 7.1.11 **[Minimizing Vector]** Let S be a closed convex subset of a Hilbert space H. Then for any $\xi \in H$, there exists a unique vector $\xi_0 \in S$ such that

$$\|\xi - \xi_0\| = \inf\{\|\xi - \eta\| \mid \eta \in S\}. \tag{7.4}$$

[Proof] Let $\xi \in H$. Put

$$a = \inf\{\|\xi - \eta\| \mid \eta \in S\}.$$

Then there exists a sequence $\{\eta_n\}_{n-1}^{\infty} \subset S$ such that

$$a = \lim_{n \to \infty} \|\xi - \eta_n\| \quad \text{and} \quad a^2 = \lim_{n \to \infty} \|\xi - \eta_n\|^2$$

and then $\{\eta_n\}_{n=1}^{\infty}$ is a Cauchy sequence in S. In fact, for any $\epsilon > 0$, there exists $N \in \mathbb{N}$ such that for any $n \geq N$,

$$\|\xi - \eta_n\|^2 - a^2 = |\|\xi - \eta_n\|^2 - a^2| < \frac{\epsilon}{4},$$

from which for any $m, n \geq N$, by the Parallelogram Law (Theorem 7.1.10), it holds that

$$\begin{aligned}\|\eta_n - \eta_m\|^2 &= \|(\eta_n - \xi) - (\eta_m - \xi)\|^2 \\ &= 2\|\eta_n - \xi\|^2 + 2\|\xi - \eta_m\|^2 - \|(\eta_n - \xi) + (\eta_m - \xi)\|^2\end{aligned}$$

$$
\begin{aligned}
&= 2\|\eta_n - \xi\|^2 + 2\|\xi - \eta_m\|^2 - 4\left\|\frac{\eta_n + \eta_m}{2} - \xi\right\|^2 \\
&\leq 2\|\eta_n - \xi\|^2 + 2\|\xi - \eta_m\|^2 - 4a^2 \quad \left(\text{since } \frac{\eta_n + \eta_m}{2} \in S\right) \\
&< \frac{\epsilon}{2} + \frac{\epsilon}{2} = \epsilon.
\end{aligned}
$$

Since S is closed, there exists a unique vector $\xi_0 \in S$ such that $\{\eta_n\}_{n=1}^{\infty}$ converges to ξ_0 in H. Therefore, by the continuity of the Hilbertian norm, it holds that

$$\|\xi - \xi_0\| = \lim_{n\to\infty} \|\xi - \eta_n\| = a = \inf\{\|\xi - \eta\| \mid \eta \in S\}.$$

Suppose that there exists a vector $\zeta \in S$ such that

$$\|\xi - \zeta\| = \inf\{\|\xi - \eta\| \mid \eta \in S\}.$$

Then by the Parallelogram Law again, we have

$$\|\xi_0 - \zeta\|^2 = 2\|\xi_0 - \xi\|^2 + 2\|\xi - \zeta\|^2 - 4\left\|\frac{\xi_0 + \zeta}{2} - \xi\right\|^2 \leq 4a^2 - 4a^2 = 0,$$

which implies that $\xi_0 = \zeta$. Therefore, there exists a unique vector $\xi_0 \in S$ such that (7.4) holds. □

Corollary 7.1.12 [**Minimizing Vector**] Let S be a closed linear subspace of a Hilbert space H. Then for any $\xi \in H$, there exists a unique vector $\xi_0 \in S$ such that

$$\|\xi - \xi_0\| = \inf\{\|\xi - \eta\| \mid \eta \in S\}. \tag{7.5}$$

[Proof] Since linear subspace of a vector space is convex, the proof is immediate from Theorem 7.1.11. □

Theorem 7.1.13 Let $S, U \subset H$ be subspaces of a Hilbert space H such that S and U are orthogonal, i.e., for any $\xi \in S$ and $\eta \in U$, $\langle \xi | \eta \rangle = 0$. Then for any $z \in S + U$ there exist unique vectors $\xi \in S$ and $\eta \in U$ such that

$$z = \xi + \eta. \tag{7.6}$$

In this case, we write $S \oplus U$ for $S + U$.

[Proof] Let $z \in S + U$. Then it is obvious tat there exist $\xi \in S$ and $\eta \in U$ such that $z = \xi + \eta$. If $z = \xi' + \eta'$ for $\xi' \in S$ and $\eta' \in U$. Then

$$\xi + \eta = z = \xi' + \eta' \quad \Longrightarrow \quad \xi - \xi' = \eta' - \eta.$$

Since $\xi - \xi' \in S$ and $\eta' - \eta \in U$,

$$\xi - \xi', \;\; \eta' - \eta \in S \cap U.$$

On the other hand, if $\zeta \in S \cap U$, then $\langle \zeta | \, \zeta \rangle = 0$ and so $\zeta = 0$. Therefore, $\xi = \xi'$ and $\eta = \eta'$ and so the decomposition of z given as in (7.6) is unique. □

Let H be a Hilbert space. For each nonempty subset $S \subset H$, put

$$S^{\perp} = \{\xi \in H \mid \langle \xi | \, \eta \rangle = 0 \text{ for all } \eta \in S\},$$

which is called the **orthogonal complement** of S.

Theorem 7.1.14 Let $S, U \subset H$ be nonempty subsets of a Hilbert space H. Then we have

(i) $S^{\perp}$ is a closed subspace of H,

(ii) if $S \subset U$, then $U^{\perp} \subset S^{\perp}$,

(iii) $S \subset S^{\perp\perp} = (S^{\perp})^{\perp}$.

[Proof] (i) Let $\xi, \eta \in S^{\perp}$ and $\alpha, \beta \in \mathbb{C}$. Then for any $\zeta \in S$, we have

$$\langle \alpha\xi + \beta\eta | \, \zeta \rangle = \alpha \langle \xi | \, \zeta \rangle + \beta \langle \eta | \, \zeta \rangle = 0,$$

which implies that $\alpha\xi + \beta\eta \in S^{\perp}$ and so $S^{\perp}$ is a linear subspace of H.

Let $\{\xi_n\}_{n=1}^{\infty} \subset S^{\perp}$ such that $\{\xi_n\}_{n=1}^{\infty}$ converges to ξ in H. Then for any $\zeta \in S$, by Proposition 7.1.9, we have

$$\langle \xi | \, \zeta \rangle = \lim_{n \to \infty} \langle \xi_n | \, \zeta \rangle = 0,$$

which implies that $\xi \in S^{\perp}$. Therefore, $S^{\perp}$ is closed. Hence $S^{\perp}$ is a closed subspace of H.

(ii) If $\xi \in U^{\perp}$, then for any $\zeta \in U$, $\langle \xi | \, \zeta \rangle = 0$ and so, since $S \subset U$, for any $\zeta \in S$, $\langle \xi | \, \zeta \rangle = 0$ if and only if $\xi \in S^{\perp}$. Thus $U^{\perp} \subset S^{\perp}$.

(iii) Let $\xi \in S$. Then for any $\zeta \in S^{\perp}$, $\langle \xi | \, \zeta \rangle = 0$, which implies that $\xi \in (S^{\perp})^{\perp} = S^{\perp\perp}$. Therefore, $S \subset S^{\perp\perp}$. □

Theorem 7.1.15 Let S be a nonempty subset of a Hilbert space H. If S is a closed subspace of H, then $S = S^{\perp\perp}$ and $H = S \oplus S^{\perp}$.

[Proof] For any $\xi \in H$, by Theorem 7.1.11, there exists a unique $\xi_0 \in S$ such that

$$\|\xi - \xi_0\| = \inf\{\|\xi - \eta\| \mid \eta \in S\}.$$

Put $\zeta = \xi - \xi_0$. Then $\xi = \xi_0 + \zeta$ and for any $\eta \in S$ with $\|\eta\| = 1$ and $\lambda \in \mathbb{C}$,

$$\begin{aligned} 0 \le \|\zeta + \lambda\eta\|^2 &= \langle \zeta + \lambda\eta | \, \zeta + \lambda\eta \rangle \\ &= \|\zeta\|^2 + \overline{\lambda} \langle \zeta | \, \eta \rangle + \lambda \langle \eta | \, \zeta \rangle + \lambda\overline{\lambda} \\ &= \|\zeta\|^2 - | \langle \zeta | \, \eta \rangle |^2 + |\langle \zeta | \, \eta \rangle + \lambda|^2, \end{aligned}$$

which implies that for $\lambda = -\langle \zeta | \, \eta \rangle$,

$$0 \le \|\zeta\|^2 - | \langle \zeta | \, \eta \rangle |^2 \le \|\zeta\|^2.$$

On the other hand, we have $\zeta + \lambda\eta = \xi - \xi_0 + \lambda\eta = \xi - (\xi_0 - \lambda\eta)$, where $\xi_0 - \lambda\eta \in S$ and for $\lambda = -\langle \zeta | \, \eta \rangle$,

$$\|\zeta\|^2 = \|\xi - \xi_0\|^2 \le \|\xi - (\xi_0 - \lambda\eta)\|^2 = \|\zeta + \lambda\eta\|^2 = \|\zeta\|^2 - | \langle \zeta | \, \eta \rangle |^2 \le \|\zeta\|^2,$$

which implies that $\langle \zeta | \, \eta \rangle = 0$, i.e., $\zeta \in S^\perp$. Therefore, by Theorem 7.1.13, $H = S \oplus S^\perp$.

By (ii) in Theorem 7.1.14, $S \subset S^{\perp\perp}$. Let $\xi \in S^{\perp\perp} \subset H$. Then there exist $\xi_0 \in S$ and $\zeta \in S^\perp$ such that $\xi = \xi_0 + \zeta$ and so

$$0 = \langle \xi | \, \zeta \rangle = \langle \xi_0 | \, \zeta \rangle + \langle \zeta | \, \zeta \rangle,$$

which implies that $\zeta = 0$ (since $\langle \xi_0 | \, \zeta \rangle = 0$ and so $\langle \zeta | \, \zeta \rangle = 0$) and so $\xi = \xi_0 \in S$. Hence $S^{\perp\perp} \subset S$ and so $S^{\perp\perp} = S$. □

Theorem 7.1.16 Let $S, U \subset H$ be nonempty subsets of a Hilbert space H. Then we have

(i) $S^{\perp\perp}$ is the smallest closed subspace containing S,

(ii) if S is a subspace of H, then $\overline{S} = S^{\perp\perp}$,

(iii) $(S^{\perp\perp})^\perp = S^\perp$.

[Proof] (i) By (i) and (ii) in Theorem 7.1.14, $S^{\perp\perp}$ is a closed subspace of H containing S. Let M be a closed subspace of H containing S. Then by (ii) and Theorem 7.1.15, we have

$$S \subset M \implies M^\perp \subset S^\perp \implies S^{\perp\perp} \subset M^{\perp\perp} = M.$$

Therefore, $S^{\perp\perp}$ is the smallest closed subspace containing S.

(ii) It is easy to see that $\overline{S}$ is a closed subspace. Let M be a closed subspace of H containing S. For any $\xi \in \overline{S}$, there exists a sequence $\{\xi_n\}_{n=1}^\infty \subset S$ such that $\{\xi_n\}_{n=1}^\infty$ converges to ξ, since $\{\xi_n\}_{n=1}^\infty \subset S \subset M$ and M is closed, $\xi \in M$ and so $\overline{S} \subset M$. Therefore, $\overline{S}$ is the smallest closed subspace containing S. Also, by (i), $S^{\perp\perp}$ is the smallest closed subspace containing S. Hence $\overline{S} = S^{\perp\perp}$.

(iii) The proof follows from (i) in Theorem 7.1.14, Theorem 7.1.15. In fact, the following:

$$(S^{\perp\perp})^{\perp} = \left((S^{\perp})^{\perp}\right)^{\perp} = (S^{\perp})^{\perp\perp} = S^{\perp}$$

holds. □

Let $(V, \|\cdot\|)$ be a seminormed space. Define a relation $\sim$ on V by

$$u \sim v \iff \|u - v\| = 0, \qquad u, v \in V. \tag{7.7}$$

Then $\sim$ is an equivalence relation, i.e.,

(i) $\sim$ is reflexive, i.e., for any $u \in V$, $u \sim u$

(ii) $\sim$ is symmetric, i.e., if $u \sim v$ for $u, v \in V$, then $v \sim u$,

(iii) $\sim$ is transitive, i.e., if $u \sim v$ and $v \sim w$ for $u, v, w \in V$, then $u \sim w$.

For each $u \in V$, the equivalence class of u is denoted by $[u]$, i.e.,

$$[u] = \{v \in V \mid v \sim u, \text{ i.e., } \|u - v\| = 0\}$$

and the collection of all equivalence classes of V is denoted by $V/\sim$ which is called the **quotient space**. Then $(V/\sim, \|\cdot\|)$ becomes a normed space, where the norm $\|\cdot\|$ is defined by

$$\|[u]\| = \|u\|, \qquad u \in V.$$

Remark 7.1.17 Let $(V, \|\cdot\|)$ be a seminormed space. Let $\mathcal{N} = \{v \in V \mid \|v\| = 0\}$. Then we can easily see that $\mathcal{N}$ is a closed subspace of V. In fact, for any scalars α, β and $v, w \in \mathcal{N}$, we have

$$0 \le \|\alpha v + \beta w\| \le |\alpha|\,\|v\| + |\beta|\,\|w\| = 0,$$

which implies that $\alpha v + \beta w \in \mathcal{N}$, i.e., $\mathcal{N}$ is a linear space. If $\{u_n\}_{n=1}^{\infty} \subset \mathcal{N}$ is a convergent sequence to u in V, then we have

$$\|u\| = \lim_{n\to\infty} \|u_n\| = 0,$$

and so $u \in \mathcal{N}$ and $\mathcal{N}$ is closed. The equivalence relation $\sim$ given as in (7.7) is equivalent to that

$$u \sim v \iff u - v \in \mathcal{N}.$$

Then the quotient space $V/\sim$ is sometimes denoted by $V/\mathcal{N}$ and the quotient norm $\|[u]\|$ for $[u] = u + \mathcal{N}$ with $u \in V$ is represented by

$$\|[u]\| = \inf\{\|u + v\| \mid v \in \mathcal{N}\}.$$

In fact, for any $u \in V$ and $v \in \mathcal{N}$, we obtain that

$$\|u\| \le \|u + v\| + \|-v\| \le \|u + v\| \le \|u\| + \|v\| = \|u\|,$$

which implies that

$$\|[u]\| = \inf\{\|u + v\| \mid v \in \mathcal{N}\} = \|u\|.$$

Example 7.1.18 Let $\ell^1 = \{f : \mathbb{N} \to \mathbb{C} \mid \|f\|_1 = \sum_{n=1}^{\infty} |f(n)| < \infty\}$, i.e., ℓ^1 is the set of all absolutely summation sequences of complex numbers. Then ℓ^1 is a linear space with addition of vectors and scalar multiplication defined as following: for any $\alpha \in \mathbb{C}$ and $f, g \in \ell^1$, it holds that

$$\sum_{n=1}^{\infty} |f(n) + g(n)| \le \sum_{n=1}^{\infty} |f(n)| + \sum_{n=1}^{\infty} |g(n)| = \|f\|_1 + \|g\|_1,$$

therefore, the addition $f + g$ of f and g is well-defined by

$$(f + g)(n) = f(n) + g(n),$$

and then we have $\|f + g\|_1 \le \|f\|_1 + \|g\|_1$, also we obtain that

$$\sum_{n=1}^{\infty} |\alpha f(n)| = |\alpha| \sum_{n=1}^{\infty} |f(n)| = |\alpha| \|f\|_1,$$

and so the scalar multiplication αf is well-defined by

$$(\alpha f)(n) = \alpha f(n)$$

and then $\|\alpha f\|_1 = |\alpha| \|f\|_1$. Moreover, for $f \in \ell^1$, if $0 = \|f\|_1 = \sum_{n=1}^{\infty} |f(n)|$, then for any $n \in \mathbb{N}$, $f(n) = 0$ and so $f = 0$. Therefore, $(\ell^1, \|\cdot\|_1)$ is a normed space.

Let $\{f_l\}_{l=1}^{\infty}$ be a Cauchy sequence in ℓ^1. Then for any $\epsilon > 0$, there exists $N \in \mathbb{N}$ such that for any $l, m > N$,

$$\sum_{n=1}^{\infty} |f_m(n) - f_l(n)| = \|f_m - f_l\|_1 < \frac{\epsilon}{2}, \tag{7.8}$$

which implies that for each $n \in \mathbb{N}$, $\{f_l(n)\}_{l=1}^{\infty}$ is a Cauchy sequence in $\mathbb{C}$. Since $\mathbb{C}$ is complete, $\{f_l(n)\}_{l=1}^{\infty}$ converges in $\mathbb{C}$ of which the limit is denoted by $f(n)$. Then we have a sequence $f : \mathbb{N} \to \mathbb{C}$. On the other hand, for any $k \in \mathbb{N}$ and $m > N$, from (7.8) we have

$$\sum_{n=1}^{k} |f_m(n) - f(n)| = \sum_{n=1}^{k} \lim_{l \to \infty} |f_m(n) - f_l(n)| = \lim_{l \to \infty} \sum_{n=1}^{k} |f_m(n) - f_l(n)| \le \frac{\epsilon}{2},$$

which implies that for any $m > N$,

$$\sum_{n=1}^{\infty} |f_m(n) - f(n)| = \lim_{k \to \infty} \sum_{n=1}^{k} |f_m(n) - f(n)| \le \frac{\epsilon}{2} < \epsilon, \tag{7.9}$$

and so for any $m > N$, $f - f_m \in \ell^1$. Therefore, we have

$$f = (f - f_m) + f_m \in \ell^1 \quad \text{for } m > N.$$

Also, from (7.9), we see that $\{f_l\}_{l=1}^{\infty}$ converges to f in ℓ^1. Therefore, $(\ell^1, \|\cdot\|_1)$ is a complete normed space, i.e., $(\ell^1, \|\cdot\|_1)$ is a Banach space.

Example 7.1.19 Let $\ell^\infty = \left\{ f : \mathbb{N} \to \mathbb{C} \;\middle|\; \|f\|_\infty = \sup_{n\in\mathbb{N}}\{|f(n)|\} < \infty \right\}$, i.e., ℓ^∞ is the set of all bounded sequences of complex numbers. Then ℓ^∞ is a linear space with addition of vectors and scalar multiplication defined as following: for any $\alpha \in \mathbb{C}$ and $f, g \in \ell^\infty$, we have

$$\sup_{n\in\mathbb{N}}\{|f(n)+g(n)|\} \le \sup_{n\in\mathbb{N}}\{|f(n)|\} + \sup_{n\in\mathbb{N}}\{|g(n)|\},$$

therefore, the addition $f+g \in \ell^\infty$ of f and g is well-defined by

$$(f+g)(n) = f(n) + g(n),$$

and then we have $\|f+g\|_\infty \le \|f\|_\infty + \|g\|_\infty$, also we obtain that

$$\sup_{n\in\mathbb{N}}\{|\alpha f(n)|\} = |\alpha| \sup_{n\in\mathbb{N}}\{|f(n)|\} = |\alpha|\|f\|_\infty,$$

and so the scalar multiplication $\alpha f \in \ell^\infty$ is well-defined by

$$(\alpha f)(n) = \alpha f(n)$$

and then $\|\alpha f\|_\infty = |\alpha|\|f\|_\infty$. Moreover, for $f \in \ell^\infty$, if $0 = \|f\|_\infty = \sup_{n\in\mathbb{N}}\{|f(n)|\}$, then for any $n \in \mathbb{N}$, $f(n) = 0$ and so $f = 0$. Therefore, $(\ell^\infty, \|\cdot\|_\infty)$ is a normed space.

Let $\{f_l\}_{l=1}^\infty$ be a Cauchy sequence in ℓ^∞. Then for any $\epsilon > 0$, there exists $N \in \mathbb{N}$ such that for any $l, m > N$,

$$\sup_{n\in\mathbb{N}}\{|f_m(n) - f_l(n)|\} = \|f_m - f_l\|_\infty < \frac{\epsilon}{2}, \tag{7.10}$$

which implies that for each $n \in \mathbb{N}$, $\{f_l(n)\}_{l=1}^\infty$ is a Cauchy sequence in $\mathbb{C}$. Since $\mathbb{C}$ is complete, $\{f_l(n)\}_{l=1}^\infty$ converges in $\mathbb{C}$ of which the limit is denoted by $f(n)$. Then we have a sequence $f : \mathbb{N} \to \mathbb{C}$. On the other hand, for any $m > N$ for any $n \in \mathbb{N}$, from (7.10) it holds that

$$|f_m(n) - f(n)| = \lim_{l\to\infty} |f_m(n) - f_l(n)| \le \frac{\epsilon}{2},$$

which implies that for any $m > N$,

$$\sup_{n\in\mathbb{N}}\{|f_m(n) - f(n)|\} \le \frac{\epsilon}{2} < \epsilon, \tag{7.11}$$

and so for any $m > N$, $f - f_m \in \ell^\infty$. Therefore, we have

$$f = (f - f_m) + f_m \in \ell^\infty \quad \text{for } m > N.$$

Also, from (7.11), it holds that $\{f_l\}_{l=1}^\infty$ converges to f in ℓ^∞. Therefore, $(\ell^\infty, \|\cdot\|_\infty)$ is a complete normed space, i.e., $(\ell^\infty, \|\cdot\|_\infty)$ is a Banach space.

Theorem 7.1.20 A normed space $(V, \|\cdot\|)$ is complete if and only if every absolutely convergent series converges.

[Proof] $\Rightarrow$) If $(V, \|\cdot\|)$ is complete and $\sum_{n=1}^{\infty} \| u_n \| < \infty$, then for any $\epsilon > 0$ there exists $N \in \mathbb{N}$ such that for any $n > N$,

$$\sum_{k=n}^{\infty} \| u_k \| < \epsilon.$$

Therefore, for any $n > N$ and $m \in \mathbb{N}$, we have

$$\left\| \sum_{k=1}^{n+m} u_k - \sum_{k=1}^{n} u_k \right\| \leq \left\| \sum_{k=n+1}^{n+m} u_k \right\| \leq \sum_{k=n+1}^{n+m} \| u_k \| \leq \sum_{k=n+1}^{\infty} \| u_k \| < \epsilon,$$

which implies that $\{\sum_{k=1}^{n} u_k\}_{n=1}^{\infty}$ is a Cauchy sequence in $(V, \|\cdot\|)$. Since $(V, \|\cdot\|)$ is complete, the series $\sum_{k=1}^{\infty} u_k$ converges.

$\Leftarrow$) Suppose that every absolutely convergent series converges. Let $\{u_n\}_{n=1}^{\infty}$ be a Cauchy sequence in $(V, \|\cdot\|)$. For any $k \in \mathbb{N}$, there exists a $N_k \in \mathbb{N}$ such that for any $m, n > N_k$,

$$\| u_m - u_n \| < 2^{-k}.$$

Choose a $n_k \in \mathbb{N}$ as $n_k > N_k$ and $n_1 < n_2 < \cdots$, then we have a subsequence $\{u_{n_k}\}_{k=1}^{\infty}$ of $\{u_n\}_{n=1}^{\infty}$ and

$$\sum_{k=1}^{\infty} \| u_{n_{k+1}} - u_{n_k} \| \leq \sum_{k=1}^{\infty} 2^{-k} = 1$$

and so by the assumption, the series $\sum_{k=1}^{\infty} \left(u_{n_{k+1}} - u_{n_k}\right) + u_{n_1}$ converges to a u in $(V, \|\cdot\|)$, i.e.,

$$\lim_{m\to\infty} \left(\sum_{k=1}^{m} \left(u_{n_{k+1}} - u_{n_k}\right) + u_{n_1} \right) = u,$$

which implies that

$$u = \lim_{m\to\infty} \sum_{k=1}^{m} \left(u_{n_{k+1}} - u_{n_k}\right) + u_{n_1} = \lim_{m\to\infty} u_{n_{m+1}}$$

and so the subsequence $\{u_{n_k}\}_{k=1}^{\infty}$ converges to u. Since $\{u_n\}_{n=1}^{\infty}$ is a Cauchy sequence, by (ii) in Theorem 7.1.4, $\{u_n\}_{n=1}^{\infty}$ converges to the same limit u. □

Definition 7.1.21 Let $(V, \|\cdot\|)$ be a normed space and let $Z \subset W \subset V$. Then Z is said to be **dense** in W if for any $w \in W$ and $\epsilon > 0$, there exists a $z \in Z$ such that $\| w - z \| < \epsilon$.

Remark 7.1.22 Let $(V, \|\cdot\|)$ be a normed space and let $Z \subset W \subset V$. Let $w \in W$ be a given vector. If Z is dense in W, then for each $n \in \mathbb{N}$, there exists a $z_n \in Z$ such that $\| w - z_n \| < 1/n$ and so we have a sequence $\{z_n\}_{n=1}^{\infty} \subset Z$ such that $\{z_n\}_{n=1}^{\infty}$ converges to

w in V. In fact, for any given $\epsilon > 0$, there exists $N \in \mathbb{N}$ such that $1/N < \epsilon$, then for any $n > N$, we have

$$\| z_n - w \| < \frac{1}{n} < \frac{1}{N} < \epsilon.$$

Thus we proved that if Z is dense in W, then for each $w \in W$, there exists a sequence $\{z_n\}_{n=1}^{\infty} \subset Z$ such that $\{z_n\}_{n=1}^{\infty}$ converges to w in $(V, \| \cdot \|)$.

Remark 7.1.23 Let $(V, \| \cdot \|)$ be a normed space and let $Z \subset W \subset U \subset V$. If Z is dense in W and W is dense in U, then for any $u \in U$ and $\epsilon > 0$, there exists a $w \in W$ such that $\| w - u \| < \epsilon/2$, and there exists a $z \in Z$ such that $\| z - w \| < \epsilon/2$. Therefore, we have

$$\| u - z \| \leq \| u - w \| + \| w - z \| < \frac{\epsilon}{2} + \frac{\epsilon}{2} = \epsilon.$$

Thus Z is dense in U.

Proposition 7.1.24 Let

$$\mathcal{F} = \{f \in \ell^1 \mid f(n) = 0 \text{ for all except finitely many values of } n \in \mathbb{N}\}.$$

Then we have

(i) $\mathcal{F}$ is a subspace of ℓ^1,

(ii) $\mathcal{F}$ is dense in ℓ^1.

[Proof] The proof is an exercise (see Exercise **7.3**). □

Definition 7.1.25 A normed space $(V, \| \cdot \|)$ is said to be **separable** if there exists a countable subset C of V such that C is dense in V.

Theorem 7.1.26 The normed space $(\ell^1, \| \cdot \|_1)$ is separable.

[Proof] The proof is immediate from Proposition 7.1.24. In fact, for

$$\mathcal{F}_{\mathbb{Q}} = \{f \in \ell^1 \mid f(\mathbb{N}) \subset \mathbb{Q}+i\mathbb{Q} \text{ and } f(n) = 0 \text{ for all except finitely many values of } n \in \mathbb{N}\},$$

$\mathcal{F}_{\mathbb{Q}}$ is countable and dense in ℓ^1, i.e., $\overline{\mathcal{F}_{\mathbb{Q}}} = \ell^1$. □

Remark 7.1.27 Suppose that a countable sequence $\{f_n\}_{n=1}^{\infty} \subset \ell^{\infty}$ is dense in ℓ^{∞}. Then for each $A \subset \mathbb{N}$, there exists a $n_A \in \mathbb{N}$ such that

$$\| \mathbf{1}_A - f_{n_A} \|_{\infty} < \frac{1}{2},$$

where $\mathbf{1}_A \in \ell^{\infty}$ is defined by

$$\mathbf{1}_A(i) = \begin{cases} 1, & i \in A, \\ 0, & i \notin A. \end{cases}$$

Note that for any subsets A, B of $\mathbb{N}$ with $A \neq B$, $f_{n_A} \neq f_{n_B}$. In fact, if $f_{n_A} = f_{n_B}$ for $A, B \subset \mathbb{N}$ with $A \neq B$, then

$$\begin{aligned} 1 &= \| \mathbf{1}_A - \mathbf{1}_B \|_{\infty} \\ &\leq \| \mathbf{1}_A - f_{n_A} \|_{\infty} + \| \mathbf{1}_B - f_{n_B} \|_{\infty} \\ &< \frac{1}{2} + \frac{1}{2} = 1, \end{aligned}$$

which is a contradiction. Therefore, for any subsets A, B of $\mathbb{N}$ with $A \neq B$, $f_{n_A} \neq f_{n_B}$. Hence, the map $n_{(\cdot)} : \mathcal{P}(\mathbb{N}) \ni A \mapsto n_A \in \mathbb{N}$ is an one-to-one map and onto a subset of $\mathbb{N}$. Since any subset of $\mathbb{N}$ is countable, $\mathcal{P}(\mathbb{N})$ is countable. But $\mathcal{P}(\mathbb{N})$ is a uncountable set. Therefore, it is a contradiction. Thus ℓ^{∞} is not separable.

7.2 Bounded Linear Operators on Normed Spaces

In this section, let V and W be two normed spaces over the complex field $\mathbb{C}$. A function T from a vector space U_1 into another vector space U_2, i.e., $T : U_1 \ni v \mapsto T(v) \in U_2$, is called an **operator**. A function T from a vector space U into the scalar field $\mathbb{C}$, i.e., $T : V \ni v \mapsto T(v) \in \mathbb{C}$, is called a **functional**. For an operator T, we usually denote Tv for $T(v)$.

Definition 7.2.1 Let V and W be two normed spaces. If T is continuous at v for all $v \in V$, then $T : V \to W$ is said to be **continuous** on V.

Remark 7.2.2 An operator $T : V \to W$ is said to be **sequentially continuous** at $v \in V$ if for any sequence $\{v_n\}_{n=1}^{\infty} \subset V$ with $\lim_{n\to\infty} v_n = v$,

$$\lim_{n\to\infty} Tv_n = Tv \quad \text{in } W. \tag{7.12}$$

Note that it is obvious that every continuous operator is sequentially continuous (see Exercise **7.9**), but, in general, the converse is not true. However, if for each point v (equivalently, $0 \in V$) in V there exists a countable family $\mathcal{B}(v)$ of open subsets containing v such that for any open subset U containing v, there exists an open subset $O \in \mathcal{B}(v)$ with $O \subset U$ (V is said to be **first countable**), then the sequentially continuity implies continuity (see Exercise **7.10**). In particular, if V is metric space, then every sequentially continuous function $f : V \to W$ is continuous.

Remark 7.2.3 Let (X, τ) be a topological space. Then X is said **second countable** if there exists a countable family $\mathcal{B}$ of open subsets in τ such that every open subset O of X can be written as a union of a subfamily of $\mathcal{B}$. Note that every second countable topological space is first countable (see Exercise **7.11**).

Proposition 7.2.4 Let V and W be normed spaces and $S, T : V \to W$ bounded linear operators. If there exists a dense subset D in V such that $S(\xi) = T(\xi)$ for any $\xi \in D$, then $S = T$.

[Proof] Let $\eta \in V$. Then there exists a sequence $\{\eta_n\}_{n=1}^{\infty} \subset D$ such that $\{\eta_n\}_{n=1}^{\infty}$ converges to η in V. Since S and T are continuous, we have

$$S(\eta) = \lim_{n\to\infty} S(\eta_n) = \lim_{n\to\infty} T(\eta_n) = T(\eta),$$

which implies that $S = T$. □

Definition 7.2.5 An operator $L : V \to W$ is said to be **bounded** if for any $r > 0$, there exists a $R > 0$ such that

$$\|Tv\|_W \leq R \quad \text{for any } v \in V \text{ with } \|v\|_V \leq r. \tag{7.13}$$

Remark 7.2.6 An operator $L : V \to W$ is bounded if and only if for any bounded subset $B \subset V$, i.e., $\sup_{v\in B}\{\|v\|_V\} < \infty$,

$$\sup_{v\in B}\{\|Tv\|_W\} < \infty. \tag{7.14}$$

[Proof] $\Rightarrow$) Let B be a bounded subset of V, i.e., $\sup_{v\in B}\{\|v\|_V\} < \infty$. There exists a $r > 0$ such that $\sup_{v\in B}\{\|v\|_V\} < r$. Since T is bounded, there exists $R > 0$ such that for any $\|v\|_V \leq r$, $\|Tv\|_W \leq R$

$$\sup_{v\in B}\{\|Tv\|_W\} \leq \sup_{\|v\|_V\leq r}\{\|Tv\|_W\} \leq R < \infty.$$

$\Leftarrow$) The proof is obvious. In fact, for any $r > 0$, put $B_r = \{v \in V \mid \|v\| \leq r\}$. Then B_r is bounded and so by assumption,

$$R := \sup_{v\in B_r} \|Tv\|_W < \infty,$$

which implies (7.13). Therefore, T is bounded. □

Definition 7.2.7 Let V and W be two normed spaces over the complex field $\mathbb{C}$. An operator $L : V \to W$ is said to be **linear** if

$$L(\alpha u + \beta v) = \alpha L(u) + \beta L(v) \tag{7.15}$$

for any $\alpha, \beta \in \mathbb{C}$ and $u, v \in V$.

Let $L : V \to W$ be a linear operator. We put

$$\|L\| := \sup_{\|v\|_V \leq 1} \{\|Lv\|_W\}. \tag{7.16}$$

In this case, $\|L\|$ can be infinity, but it is always finite for bounded operator T.

Theorem 7.2.8 Let V be a vector space over $\mathbb{C}$ and $f : V \to \mathbb{C}$ be a complex linear functional. Put $r = \Re(f)$, where $\Re(f)$ is the real part of f. Then r is a real linear functional and $f(v) = r(v) - ir(iv)$ for all $v \in V$. Conversely, if $r : V \to \mathbb{R}$ is a real linear functional, then the functional $f : V \to \mathbb{C}$ defined by $f(v) = r(v) - ir(iv)$ for all $v \in V$ is complex linear. In this case, if $(V, \|\cdot\|)$ is a normed space, then it holds that $\|f\| = \|r\|$.

[Proof] For any $u, v \in V$ and $\alpha, \beta \in \mathbb{R}$, we obtain that

$$r(\alpha u + \beta v) = \Re\left(f(\alpha u + \beta v)\right) = \alpha\Re(f(u)) + \beta\Re(f(v)) = \alpha r(u) + \beta r(v),$$

which implies that r is a real linear functional. On the other hand, we have

$$\Im(f(v)) = -\Re(if(v)) = -\Re(f(iv)), \qquad v \in V.$$

Therefore, we have

$$f(v) = \Re(f(v)) + i\Im(f(v)) = \Re(f(v)) - i\Re(f(iv)) = r(v) - ir(iv), \qquad v \in V.$$

Suppose that $r : V \to \mathbb{R}$ is real linear and $f(v) = r(v) - ir(iv)$ for all $v \in V$. Then it is clear that f is real linear and $f(iv) = r(iv) - ir(-v) = r(iv) + ir(v) = i(r(v) - ir(iv)) = if(v)$. Therefore, f is complex linear.

Suppose that $(V, \|\cdot\|)$ is a normed space. Then for any $v \in V$, $|r(v)| = |\Re(f(v))| \leq |f(v)|$ and so $\|r\| \leq \|f\|$. On the other hand, for any $v \in V$ with $f(v) \neq 0$, we have

$$\mathbb{R} \ni |f(v)| = \frac{|f(v)|}{f(v)} f(v) = f\left(\frac{|f(v)|}{f(v)} v\right),$$

and so we obtain that,

$$|f(v)| = f\left(\frac{|f(v)|}{f(v)} v\right) = r\left(\frac{|f(v)|}{f(v)} v\right) \leq \|r\| \left\|\frac{|f(v)|}{f(v)} v\right\| = \|r\|\|v\|,$$

which implies that $\|f\| \leq \|r\|$. Hence $\|f\| = \|r\|$. □

Theorem 7.2.9 Let $L : V \to W$ be a linear operator. Suppose that L is bounded. Then we have

(i) if $V \neq \{0\}$
$$\|L\| = \sup_{\|v\|_V=1} \{\|Lv\|_W\} < \infty,$$

(ii) for any $v \in V$, $\|Lv\|_W \leq \|L\|\|v\|_V$.

[Proof] (i) Since L is bounded,
$$M_1 := \sup_{\|v\|_V\leq 1} \{\|Lv\|_W\} < \infty, \qquad M_2 := \sup_{\|v\|_V=1} \{\|Lv\|_W\} < \infty.$$

Then it is obvious that $M_2 \leq M_1$. Let $v \in V$ with $v \neq 0$ and $\|v\|_V \leq 1$. Then for $u := v/\|v\|_V$, $\|u\| = 1$ and so
$$\frac{\|Lv\|_W}{\|v\|_V} = \|Lu\|_W \leq M_2 \quad \Longleftrightarrow \quad \|Lv\|_W \leq M_2\|v\|_V, \tag{7.17}$$

which implies that
$$M_1 = \sup_{\|v\|_V\leq 1} \{\|Lv\|_W\} \leq \sup_{\|v\|_V\leq 1} \{M_2\|v\|_V\} \leq M_2$$

and so $M_1 \leq M_2$. Therefore, $M_1 = M_2$.

(ii) The proof is immediate from (7.17) and (i). □

Theorem 7.2.10 Let $L : V \to W$ be a linear operator. Then

(i) L is bounded if and only if there exists a $M \geq 0$ such that for all $v \in V$,
$$\|Lv\|_W \leq M\|v\|_V,$$

(ii) the operator norm $\|L\|$ of L is given by
$$\|L\| = \inf\{M \geq 0 \,|\, \|Lv\|_W \leq M\|v\|_V \text{ for all } v \in V\}. \tag{7.18}$$

[Proof] (i) $\Rightarrow$) The proof is immediate from Theorem 7.2.9.

$\Leftarrow$) For any given $r > 0$, put $R = (M + 1)r$. Then for any $v \in V$ with $\|v\|_V \leq r$, we have
$$\|Lv\|_W \leq M\|v\|_V \leq (M + 1)r = R.$$

Therefore, L is bounded.

(ii) By (ii) in Theorem 7.2.9, the following inequality:

$$\|L\| \geq \inf\{M \geq 0 \,|\, \|Lv\|_W \leq M\|v\|_V \text{ for all } v \in V\}$$

holds. On the other hand, if $\|Lv\|_W \leq M\|v\|_V$ for some $M \geq 0$, then we have

$$\|L\| = \sup_{\|v\|_V \leq 1} \{\|Lv\|_W\} \leq \sup_{\|v\|_V \leq 1} \{M\|v\|_V\} = M.$$

Therefore, we have

$$\|L\| \leq \inf\{M \geq 0 \,|\, \|Lv\|_W \leq M\|v\|_V \text{ for all } v \in V\}.$$

Finally, we proved (7.18). □

Theorem 7.2.11 Let $L : V \to W$ be a linear operator. Then the following are equivalent:

(i) L is continuous at $0 \in V$,

(ii) L is continuous on V,

(iii) L is bounded.

[Proof] (i) $\Rightarrow$ (ii) Let U be an open subset of W. Suppose that $L^{-1}(U)$ is nonempty and $v \in L^{-1}(U)$. Then $Lv \in U$ and U is an open subset containing Lv. Therefore, $U - Lv$ is an open subset of W containing 0, and so since L is continuous at 0, there exists an open subset O of V containing 0 such that $L(O) \subset U - Lv$, equivallently, $L(O + v) \subset U$. Therefore, $v \in O + v \subset L^{-1}(U)$ and v is an interior point of $L^{-1}(U)$ and so $L^{-1}(U)$ is open. Hence L is continuous.

(ii) $\Rightarrow$ (i) It is obvious.

(i) $\Rightarrow$ (iii) Suppose that L is not bounded, i.e., there exists $r > 0$ such that

$$\sup_{\|v\|_V \leq r} \{\|Lv\|_W\} = \infty,$$

which means that there exists a sequence $\{v_n\}_{n=1}^{\infty} \subset V$ such that $\|v_n\|_V \leq r$ and $\|Lv_n\|_W \geq n$ for all $n = 1, 2, \cdots$. Then, the sequence $\{v_n/n\}_{n=1}^{\infty}$ converges to 0 in V. In fact,

$$\lim_{n\to\infty} \left\| \frac{v_n}{n} \right\|_V \leq \lim_{n\to\infty} \frac{r}{n} = 0.$$

But by the linearity of L, $\|L(v_n/n)\|_W = \|L(v_n)\|_W/n \geq n/n = 1$, which implies that $L(v_n/n)$ does not converge to 0 in W. Therefore, L is not sequentially continuous at 0, and so L is not continuous at 0 which is in contradiction to the assumption that L is continuous at 0.

(iii) $\Rightarrow$ (i) Let O be an open subset of W containing 0. Then there exists $R > 0$ such that

$$\{w \in W \,|\, \|w\|_W < R\} \subset O.$$

On the other hand, since L is bounded, it holds that

$$\|Lv\|_W \le \|L\| \|v\|_V, \qquad v \in V.$$

Therefore, for $r = R/(\|L\|+1)$, we put

$$B = \{v \in V \mid \|v\|_V < r\}.$$

Then for any $v \in B$, it holds that

$$\|Lv\|_W \le \|L\| \|v\|_V < \|L\| \frac{R}{\|L\|+1} < R,$$

which implies that

$$L(B) \subset \{w \in W \mid \|w\|_W < R\} \subset O,$$

and so $B \subset L^{-1}(O)$. Hence L is continuous at $0 \in V$. □

Remark 7.2.12 Every normed space is a metric space and so first countable. Therefore, on a normed space, the continuity and the sequentially continuity are equivalent. Hence we can prove more easily the equivalence of (i) and (ii) in Theorem 7.2.11 by using the sequentially continuity. In fact, $\{v_n\}_{n=1}^{\infty}$ converges to v if and only if $\{v_n - v\}_{n=1}^{\infty}$ converges to 0. Also, $\{Tv_n\}_{n=1}^{\infty}$ converges to Tv if and only if $\{T(v_n - v)\}_{n=1}^{\infty}$ converges to 0. Therefore, L is continuous at $0 \in V$ if and only if L is continuous at v for every $v \in V$.

Let $\mathcal{L}(V, W)$ be the set of all bounded linear operators from V into W. For each $S, T \in \mathcal{L}(V, W)$, we define an operator $S+T$ from V into W by

$$(S+T)(v) = Sv + Tv, \qquad v \in V.$$

Then it is easy to see that $S+T$ is linear. For any $r > 0$, there exist $R_1, R_2 > 0$ such that for any $v \in V$ with $\|v\|_V \le r$

$$\|Sv\|_W \le R_1, \qquad \|Tv\|_W \le R_2,$$

and so

$$\|(S+T)(v))\|_W \le \|Sv\|_W + \|Tv\|_W \le R_1 + R_2.$$

Therefore, $S+T$ is bounded. Moreover, we obtain that

$$\begin{aligned} \|S+T\| &= \sup_{\|v\|_V \le 1} \{\|(S+T)v\|_W\} \le \sup_{\|v\|_V \le 1} \{\|Sv\|_W\} + \sup_{\|v\|_V \le 1} \{\|Tv\|_W\} \\ &= \|S\| + \|T\|. \end{aligned} \tag{7.19}$$

For each $\alpha \in \mathbb{C}$ and $T \in \mathcal{L}(V, W)$, define a linear operator αT by

$$(\alpha T)(v) = \alpha T v, \qquad v \in V.$$

Then we have

$$\sup_{\|v\|_V \le 1} \{\|\alpha T v\|_W\} = |\alpha| \sup_{\|v\|_V \le 1} \{\|Tv\|_W\} = |\alpha| \|T\| < \infty,$$

which implies that αT is bounded and

$$\|\alpha T\| = |\alpha|\|T\|. \tag{7.20}$$

Therefore, $\mathcal{L}(V, W)$ is linear (vector) space. Moreover, if $\|T\| = 0$, then by (7.16), for any vector $v \in V$, $\|T(v)\|_V = 0$, which implies that $Tv = 0$. Therefore, for any $v \in V$, $Tv = 0$, i.e., $T = 0$ and so by (7.19) and (7.20), $\|\cdot\|$ is a norm on $\mathcal{L}(V, W)$, i.e., $(\mathcal{L}(V, W), \|\cdot\|)$ is a normed space.

Definition 7.2.13 For a bounded linear operator $L : V \to W$, $\|L\|$ given as in (7.16) is called the **operator norm** of L.

The normed space $\mathcal{L}(V, \mathbb{C})$ of all bounded linear functional on V is called the **dual space** of V and denoted by V^*.

Definition 7.2.14 For a bounded linear functional $f : V \to \mathbb{C}$, $\|f\|$ defined by

$$\|f\| := \sup_{\|v\|_V \leq 1} \{|f(v)|\} \tag{7.21}$$

is called the **functional norm** of f.

Theorem 7.2.15 If W is complete, then $\mathcal{L}(V, W)$ is complete with respect to the operator norm.

[Proof] Let $\{L_n\}_{n=1}^{\infty} \subset \mathcal{L}(V, W)$ such that

$$\sum_{n=1}^{\infty} \|L_n\| < \infty.$$

Then for any $v \in V$,

$$\sum_{n=1}^{\infty} \|L_n v\|_W \leq \sum_{n=1}^{\infty} \|L_n\|\|v\|_V = \left(\sum_{n=1}^{\infty} \|L_n\|\right) \|v\|_V < \infty,$$

which means that the series $\sum_{n=1}^{\infty} L_n v$ is absolutely convergent in W. Since W is complete, by Theorem 7.1.20, the series $\sum_{n=1}^{\infty} L_n v$ converges in W. Define a linear operator L on V by

$$L : V \ni v \longmapsto L(v) = \sum_{n=1}^{\infty} L_n v \in W.$$

Then we see that

$$\|L\| = \sup_{\|v\|_V \le 1} \{\|Lv\|_W\} \le \sup_{\|v\|_V \le 1} \left\{ \sum_{n=1}^{\infty} \|L_n\| \|v\|_V \right\} = \sum_{n=1}^{\infty} \|L_n\|,$$

and so L is bounded and

$$\left\| L - \sum_{k=1}^{n} L_k \right\| = \sup_{\|v\|_V \le 1} \left\{ \left\| \left(L - \sum_{k=1}^{n} L_k \right) v \right\|_W \right\} = \sup_{\|v\|_V \le 1} \left\{ \left\| \sum_{k=n+1}^{\infty} L_k v \right\|_W \right\}$$
$$\le \sum_{k=n+1}^{\infty} \|L_k\|$$

which converges to 0 as $n \to \infty$. Therefore, $\sum_{n=1}^{\infty} L_n$ converges to L in $\mathcal{L}(V, W)$. Hence, by Theorem 7.1.20, $\mathcal{L}(V, W)$ is complete with respect to the operator norm. □

Corollary 7.2.16 Let $(V, \|\cdot\|)$ is a normed space. Then V^* is complete with respect to the functional norm $\|\cdot\|$.

[Proof] Since the complex field $\mathbb{C}$ is complete, the proof is immediate from Theorem 7.2.15 □

Proposition 7.2.17 Let H be a Hilbert space. Let $\xi \in H$. Then for the continuous linear functional f_ξ defined by

$$f_\xi(\eta) = \langle \eta | \, \xi \rangle, \qquad \eta \in H,$$

it holds that

$$\|f_\xi\| = \|\xi\|.$$

[Proof] For any $\eta \in H$, by the Cauchy-Schwarz Inequality (Theorem 7.1), it holds that

$$|f_\xi(\eta)| \le \|\xi\| \|\eta\|,$$

which implies that $\|f_\xi\| \le \|\xi\|$. On the other hand, if $\xi \neq 0$, then we have

$$\left| f_\xi \left(\frac{\xi}{\|\xi\|} \right) \right| = \left| \left\langle \xi \left| \frac{\xi}{\|\xi\|} \right. \right\rangle \right| = \|\xi\|,$$

which implies that $\|f_\xi\| \ge \|\xi\|$. □

Theorem 7.2.18 Let $(V, \|\cdot\|)$ be a normed space. Let $f \in V^*$. Then $\mathrm{Ker}(f) = \{\xi \in V \,|\, f(\xi) = 0\}$, which is called the **Kernel** (or **null space**) of f, is a closed subspace of V.

[Proof] Let $\{\xi_n\}_{n=1}^{\infty} \subset \mathrm{Ker}(f)$ such that $\{\xi_n\}_{n=1}^{\infty}$ converges to $\xi \in V$. Then by the continuity of f, we see that

$$f(\xi) = \lim_{n\to\infty} f(\xi_n) = 0,$$

which implies that $\xi \in \mathrm{Ker}(f)$ and so $\mathrm{Ker}(f)$ is closed. Moreover, for any $\xi, \eta \in \mathrm{Ker}(f)$ and $\alpha, \beta \in \mathbb{C}$, by the linearity of f, we have

$$f(\alpha\xi + \beta\eta) = \alpha f(\xi) + \beta f(\eta) = 0,$$

which implies that $\alpha\xi + \beta\eta \in \mathrm{Ker}(f)$ and so $\mathrm{Ker}(f)$ is linear space. Therefore, $\mathrm{Ker}(f)$ is a closed subspace of V. □

Theorem 7.2.19 [**Riesz-Frechet Theorem**] For any $f \in H^*$, there exists a unique vector $\xi_f \in H$ such that $\|\xi_f\| = \|f\|$ and

$$f(\eta) = \langle \eta |\, \xi_f \rangle, \qquad \eta \in H. \tag{7.22}$$

[Proof] If $f = 0$, then the proof is immediate with the vector $\xi_f = 0$. Suppose that $f \neq 0$. Then $\{0\} \neq \mathrm{Ker}(f) \subsetneqq H$ (nontrivial proper subspace) and so there exists a nonzero vector $\xi \in \mathrm{Ker}(f)^{\perp} \subsetneqq H$ and then for any $\eta \in H$,

$$f\left(\eta - f(\eta)\frac{\xi}{f(\xi)}\right) = 0,$$

i.e.,

$$\eta - f(\eta)\frac{\xi}{f(\xi)} \in \mathrm{Ker}(f).$$

Therefore,

$$0 = \left\langle \eta - f(\eta)\frac{\xi}{f(\xi)} \middle| \xi \right\rangle = \langle \eta |\, \xi \rangle - \frac{f(\eta)}{f(\xi)} \langle \xi |\, \xi \rangle,$$

which implies that

$$f(\eta) = \left\langle \eta \middle| \frac{\overline{f(\xi)}}{\|\xi\|^2}\xi \right\rangle.$$

Therefore, for the vector

$$\xi_f = \frac{\overline{f(\xi)}}{\|\xi\|^2}\xi,$$

we have

$$f(\eta) = \langle \eta | \, \xi_f \rangle \,, \qquad \eta \in H.$$

Suppose that for $\zeta \in H$, $f(\eta) = \langle \eta | \, \zeta \rangle$ for all $\eta \in H$. Then for all $\eta \in H$,

$$\langle \eta | \, \xi_f \rangle = \langle \eta | \, \zeta \rangle \quad \Longleftrightarrow \quad \langle \eta | \, \xi_f - \zeta \rangle = 0,$$

in particular, $\|\xi_f - \zeta\|^2 = 0$ and so $\zeta = \xi_f$. Therefore, there exists a unique vector $\xi_f \in H$ such that (7.22) holds. Then by Proposition 7.2.17 $\|f\| = \|\xi_f\|$. □

7.3 L^p Spaces

Throughout this section, let $(X, \mathcal{M}, \mu)$ be a measure space. Let $1 \le p < \infty$ and $f : X \to \overline{\mathbb{R}}$ be a measurable function. Put

$$\|f\|_p = \left(\int_X |f|^p d\mu \right)^{1/p}$$

and

$$L^p(X) = \{ f : X \to \mathbb{R} \text{ a measurable function} | \, \|f\|_p < \infty \} \,.$$

Let $f : X \to \overline{\mathbb{R}}$ be a measurable function. If there exists a $0 \le M < \infty$ such that $|f| \le M$ a.e. on X, then we say that f is **essentially bounded** on X. For the essentially bounded function f on X, we put

$$\|f\|_\infty = \inf \{ M \, | \, |f| \le M \text{ a.e on } X \} \,.$$

Remark 7.3.1 Let $f \in L^\infty(X)$. By definition, for each $n \in \mathbb{N}$, $|f| \le \|f\|_\infty + 1/n$ a.e. on X, i.e., there exists a measurable set $E_n \subset X$ with $\mu(E_n) = 0$ such that $|f| \le \|f\|_\infty + 1/n$ on $X \setminus E_n$. Therefore, for $E = \cup_{n=1}^\infty E_n$, $|f| \le \|f\|_\infty$ on $X \setminus E$ and $\mu(E) \le \sum_{n=1}^\infty \mu(E_n) = 0$ and so $|f| \le \|f\|_\infty$ a.e. on X.

Let $L^0(X)$ denote the collection of all extended real-valued measurable functions defined on X which are finite a.e. on X. Define a relation $\sim$ on $L^0(X)$ as following: for $f, g \in L^0(X)$,

$$f \sim g \quad \Longleftrightarrow \quad f = g \text{ a.e. on } X.$$

Then $\sim$ is an **equivalence relation**, i.e.,

(i) [**reflexive**] for any $f \in L^0(X)$, $f \sim f$,

(ii) [**symmetric**] if $f \sim g$ for $f, g \in L^0(X)$, then $g \sim f$,

(iii) [**transitive**] if $f \sim g$ and $g \sim h$ for $f, g, h \in L^0(X)$, then $f \sim h$.

For $f \in L^0(X)$, the equivalence class of f is denoted by $[f]$, i.e.,

$$[f] = \{ g \in L^0(X) \, | \, g \sim f \}$$

and the collection of all equivalence classes of $L^0(X)$ is denoted by $L^0(X)/\sim$ which is called a **quotient space**. If there is no confusion, then for notational convenience, we use f and $L^0(X)$ again for $[f]$ and $L^0(X)/\sim$, respectively. Similarly, we use the notation $L^p(X)$ again for $L^p(X)/\sim$ if there is no confusion.

Theorem 7.3.2 Let $1 \le p \le \infty$. $L^p(X)$ is a linear space (or vector space).

[Proof] (1) The case of $p = 1$. For any $\alpha, \beta \in \mathbb{R}$ and $f, g \in L^1(X)$, we obtain that

$$\int_X |\alpha f + \beta g| d\mu \le |\alpha| \int_X |f| d\mu + |\beta| \int_X |g| d\mu < \infty,$$

which implies that $\alpha f + \beta g \in L^1(X)$ and so $L^1(X)$ is a linear space.

(2) The case of $p = \infty$. For any $\alpha, \beta \in \mathbb{R}$ and $f, g \in L^\infty(X)$, by Remark 7.3.1 we obtain that

$$\begin{aligned} |\alpha f + \beta g| &\le |\alpha||f| + |\beta||g| \\ &\le |\alpha|\|f\|_\infty + |\beta|\|g\|_\infty \\ &< \infty \end{aligned}$$

a.e. on X, which implies that $\alpha f + \beta g \in L^\infty(X)$ and so $L^\infty(X)$ is a linear space.

(3) The case of $1 < p < \infty$. Let $f, g \in L^p(X)$. Consider the convex function $\varphi(t) = t^p$, $t \ge 0$. Then by Jensen's inequality, we have

$$\left| \frac{a+b}{2} \right|^p \le \frac{1}{2} \left(a^p + b^p \right), \qquad a, b \ge 0,$$

which implies that

$$|f + g|^p \le (|f| + |g|)^p \le 2^{p-1}(|f|^p + |g|^p),$$

and

$$\begin{aligned} \int_X |f + g|^p d\mu &\le 2^p \left(\int_X |f|^p d\mu + \int_X |g|^p d\mu \right) \\ &< \infty \end{aligned}$$

and so $f + g \in L^p(X)$. Moreover, for any $\alpha \in \mathbb{R}$ and $f \in L^p(X)$, we have

$$\int_X |\alpha f|^p d\mu = |\alpha|^p \int_X |f|^p d\mu < \infty$$

and so $\alpha f \in L^p(X)$. Therefore, $L^p(X)$ is a linear space. □

7.4 Inequalities

Definition 7.4.1 For each $1 < p < \infty$, the number $1 < q < \infty$ satisfying that

$$\frac{1}{p} + \frac{1}{q} = 1 \tag{7.23}$$

is called the **conjugate number of** p. The **conjugate number of** 1 is defined to be ∞ and the **conjugate number of** ∞ is 1.

Lemma 7.4.2 Let $0 < \lambda < 1$. Then for any nonnegative real numbers $a, b \geq 0$, we have

$$a^\lambda b^{1-\lambda} \leq \lambda a + (1-\lambda) b. \tag{7.24}$$

Moreover, the equality in (7.24) holds if and only if $a = b$.

[Proof] If $a = 0$ or $b = 0$, then the inequality (7.24) is obvious. Suppose that $a \neq 0$ and $b \neq 0$. Since exponential function is convex (or by applying the general Jensen's Inequality (see Theorem 6.8.8)), we obtain that

$$a^\lambda b^{1-\lambda} = e^{\lambda \ln a + (1-\lambda)\ln b} \leq \lambda e^{\ln a} + (1-\lambda) e^{\ln b} = \lambda a + (1-\lambda) b, \tag{7.25}$$

which gives the proof. If $a = b$, then the equality in (7.24) holds. Conversely, suppose that the equality in (7.24) holds. If $a = 0$, then we can easily see that $b = 0$. Similarly, if $b = 0$, then $a = 0$. Suppose that $a \neq 0$ and $b \neq 0$. Then we obtain that

$$a^\lambda b^{1-\lambda} = \lambda a + (1-\lambda) b \iff 1 = \lambda \left(\frac{a}{b}\right)^{1-\lambda} + (1-\lambda)\left(\frac{a}{b}\right)^{-\lambda}. \tag{7.26}$$

Consider the function f defined on $(0, \infty)$ by

$$f(t) = \lambda t^{1-\lambda} + (1-\lambda) t^{-\lambda} - 1, \qquad t \in (0, \infty). \tag{7.27}$$

Then we obtain that

$$f'(t) = \lambda(1-\lambda) t^{-\lambda} - \lambda(1-\lambda) t^{-(1+\lambda)} = \lambda(1-\lambda) t^{-\lambda}\left(1 - t^{-1}\right)$$

and so $f'(t) < 0$ for $0 < t < 1$, $f'(t) > 0$ for $t > 1$ and $f'(1) = 0$. Therefore, f has the minimum value $f(1) = 0$ only at $t = 1$, i.e., the equation $f(t) = 0$ have a unique solution at $t = 1$ and by (7.26), $\frac{a}{b}$ is a solution of $f(t) = 0$. Therefore, $\frac{a}{b} = 1$, i.e., $a = b$. □

Remark 7.4.3 In the proof of Lemma 7.4.2, the function given as in (7.27) is nonnegative. In particular, we have

$$0 \leq f\left(\frac{a}{b}\right) = \lambda\left(\frac{a}{b}\right)^{1-\lambda} + (1-\lambda)\left(\frac{a}{b}\right)^{-\lambda} - 1,$$

which implies (7.24) by (7.26). It is a direct proof of (7.24).

Theorem 7.4.4 [**Young's Inequality**] Let $1 < p, q < \infty$ be conjugate numbers of each other. Then for any positive numbers $a, b \geq 0$, it holds that

$$ab \leq \frac{a^p}{p} + \frac{b^q}{q}. \tag{7.28}$$

Moreover, the equality given as in (7.28) holds if and only if $a^p = b^q$.

[Proof] In equation (7.24), by applying a^p, b^q and $\lambda = \dfrac{1}{p}$ instead of a, b and λ, respectively, we obtain that

$$\begin{aligned} ab = (a^p)^{1/p}(b^q)^{1/q} &= (a^p)^{1/p}(b^q)^{1-1/p} \leq \frac{1}{p}a^p + \left(1 - \frac{1}{p}\right)b^{1-1/p} \\ &= \frac{a^p}{p} + \frac{b^q}{q}, \end{aligned}$$

which implies the proof. □

Remark 7.4.5 Now, we consider an another proof of the Young's Inequality. If $a = 0$ or $b = 0$, then the inequality is trivial. Suppose that $a \neq 0$ and $b \neq 0$. Consider the function f defined on $(0, \infty)$ by

$$f(t) = \frac{1}{p} + \frac{t^q}{q} - t.$$

Then

$$f'(t) = t^{q-1} - 1 \begin{cases} < 0, \ 0 < t < 1 \\ > 0, \ 1 < t < \infty \end{cases}, \qquad f(1) = 0,$$

which implies that $f(t) \geq 0$. In particular, for $t_0 = \dfrac{b}{a^{p-1}}$, we have

$$\frac{1}{p} + \frac{\left(\dfrac{b}{a^{p-1}}\right)^q}{q} - \frac{b}{a^{p-1}} \geq 0 \iff ab \leq \frac{a^p}{p} + \frac{b^q}{q}.$$

Then from the previous equation and the fact that $f(t) = 0$ if and only if $t = 1$, we obtain that the equality given as in (7.28) holds if and only if $t_0 = 1$, i.e., $\dfrac{b}{a^{p-1}} = 1$ if and only if $b = a^{p-1}$ if and only if $b^q = a^{q(p-1)} = a^p$.

Theorem 7.4.6 Let $1 \leq p \leq \infty$ and q be the conjugate number of p. If $f \in L^p(X)$ and $g \in L^q(X)$, then

(i) fg is integrable,

(ii) **[Hölder's Inequality]**

$$\int_X |fg| d\mu \leq \left(\int_X |f|^p d\mu\right)^{1/p} \left(\int_X |g|^q d\mu\right)^{1/q} = \|f\|_p \|g\|_q, \tag{7.29}$$

Moreover, the equality in (7.29) holds if and only if $\alpha|f|^p = \beta|g|^q$ for some constants α, β with $(\alpha, \beta) \neq (0,0)$.

(iii) if $1 \leq p < \infty$ and $f \neq 0$ (equivalently, $\|f\|_p \neq 0$), then the function $f^* = \|f\|_p^{1-p} \operatorname{sgn}(f) |f|^{p-1} \in L^q(X)$ and

$$\int_X f f^* d\mu = \|f\|_p, \qquad \|f^*\|_q = 1.$$

In this case, the function f^* is called the **conjugate function** of f.

[Proof] (i) The proof is immediate from (ii)

(ii) If $f = 0$ or $g = 0$, then the inequality is trivial as the left and right hand sides are zero. Suppose that $f \neq 0$ and $g \neq 0$.

(1) The case of $p = 1$. We obtain that

$$\int_X |fg| d\mu \leq \|g\|_\infty \int_X |f| d\mu = \|f\|_1 \|g\|_\infty.$$

(2) The case of $p = \infty$. We obtain that

$$\int_X |fg| d\mu \leq \|f\|_\infty \int_X |g| d\mu = \|f\|_\infty \|g\|_1.$$

(3) The case of $1 < p < \infty$. Put

$$f' = \frac{f}{\|f\|_p}, \qquad g' = \frac{g}{\|g\|_q}.$$

Then $\|f'\|_p = \|g'\|_q = 1$ and by apply Young's Inequality (Theorem 7.4.4), we obtain that

$$\begin{aligned} \frac{1}{\|f\|_p \|g\|_q} \int_X |fg| d\mu = \int_X |f'g'| d\mu &\leq \frac{1}{p} \int_X |f'|^p d\mu + \frac{1}{q} \int_X |g'|^q d\mu \\ &= \frac{1}{p} \|f'\|_p^p + \frac{1}{q} \|g'\|_q^q = \frac{1}{p} + \frac{1}{q} \\ &= 1, \end{aligned}$$

which implies the Hölder's inequality. The equality in (7.29) holds if and only if $|f'|^p = |g'|^q$ a.e on X if and only if $\|g\|_q^q|f|^p = \|f\|_p^p|g|^q$ a.e. on X.

(iii) We first note that $f \cdot \text{sgn}(f) = |f|$. Therefore, we obtain that

$$\int_X ff^* = \int_X \|f\|_p^{1-p}|f|^p d\mu = \|f\|_p^{1-p}\|f\|_p^p = \|f\|_p.$$

If $p = 1$, then $f^* = \text{sgn}(f)$ and $f \neq 0$. Therefore, there exists $x \in X$ such that $f(x) \neq 0$ and so $\text{sgn}(f)(x) = 1$ or -1. Hence $\|f^*\|_\infty = 1$. If $1 < p < \infty$, then for the conjugate number q of p, we obtain that $p = q(p-1)$ and

$$\begin{aligned}\|f^*\|_q &= \|f\|_p^{1-p}\left(\int_X |f|^{q(p-1)} d\mu\right)^{1/q} = \|f\|_p^{1-p}\left(\int_X |f|^p d\mu\right)^{1/q} \\ &= \|f\|_p^{1-p}\|f\|_p^{p/q} = \|f\|_p^{1-p}\|f\|_p^{p-1} \\ &= 1.\end{aligned}$$

We complete the proof. □

Corollary 7.4.7 Let $f, g \in L^2(X)$. Then

(i) fg is integrable,

(ii) [**Cauchy-Schwarz Inequality**]

$$\int_X |fg| d\mu \leq \left(\int_X |f|^2 d\mu\right)^{1/2}\left(\int_X |g|^2 d\mu\right)^{1/2} = \|f\|_2\|g\|_2. \tag{7.30}$$

[Proof] The proof is immediate from the Hölder's Inequality ((ii) in Theorem 7.4.6) by taking $p = q = 2$. □

Theorem 7.4.8 [**Minkowski's Inequality**] Let $1 \leq p \leq \infty$. Then for any $f, g \in L^p(X)$, $f + g \in L^p(X)$ and, moreover, we have

$$\|f + g\|_p \leq \|f\|_p + \|g\|_p. \tag{7.31}$$

[Proof] (1) The case of $p = 1$. For any $f, g \in L^1(X)$, we obtain that

$$\|f + g\|_1 = \int_X |f + g| d\mu \leq \int_X |f| d\mu + \int_X |g| d\mu = \|f\|_1 + \|g\|_1.$$

(2) The case of $p = \infty$. For any $f, g \in L^\infty(X)$, we obtain that

$$|f + g| \leq |f| + |g| \leq \|f\|_\infty + \|g\|_\infty \text{ a.e.on } X,$$

which implies that

$$\|f+g\|_\infty \le \|f\|_\infty + \|g\|_\infty.$$

(3) The case of $1<p<\infty$. Let q be the conjugate number of p and $f, g \in L^p(X)$. If $f+g=0$, then the proof is obvious. Suppose that $f+g\neq 0$. Then by Theorem 7.3.2, $f+g\in L^p(X)$ and, by (iii) in Theorem 7.4.6 and the Hölder's Inequality ((ii) in Theorem 7.4.6), we obtain that

$$\begin{aligned}
\|f+g\|_p &= \int_X (f+g)(f+g)^* d\mu \\
&= \int_X f(f+g)^* d\mu + \int_X g(f+g)^* d\mu \\
&\le \|f\|_p \|(f+g)^*\|_q + \|g\|_p \|(f+g)^*\|_q \\
&= \|f\|_p + \|g\|_p \quad \text{since } \|(f+g)^*\|_q = 1,
\end{aligned}$$

which gives the proof. □

Theorem 7.4.9 Let $1\le p\le\infty$. Then $(L^p(X), \|\cdot\|_p)$ is a normed space.

[Proof] From Theorem 7.3.2, $L^p(X)$ is a vector space. By definition, for any $f\in L^p(X)$, $\|f\|_p\ge 0$ and $\|f\|_p=0$ if and only if $|f|=0$ a.e. if and only if $f=0$ a.e. Let $\alpha\in\mathbb{R}$ and $f\in L^p(X)$. Also, from the definition of $\|\cdot\|_p$, it is obvious that

$$\|\alpha f\|_p = |\alpha|\|f\|_p.$$

Finally, by the Minkowski's Inequality (Theorem 7.4.8), $\|\cdot\|_p$ satisfies the triangle inequality. Therefore, $\|\cdot\|_p$ is a norm on $L^p(X)$, and so $(L^p(X), \|\cdot\|_p)$ is a normed space. □

Corollary 7.4.10 Let $1<p\le\infty$ and $\mathcal{F}$ be a family of functions in $L^p(X)$. Suppose that $\mathcal{F}$ is bounded, i.e.,

$$\sup\{\|f\|_p \mid f\in\mathcal{F}\} < \infty.$$

Then $\mathcal{F}$ is uniformly integrable on X, i.e., for any $\epsilon>0$ there exists $\delta>0$ such that for any $f\in\mathcal{F}$ and any measurable subset $A\subset X$ with $\mu(A)<\delta$,

$$\int_A |f| < \epsilon.$$

[Proof] Let $\epsilon > 0$ be given and q be the conjugate number of p. Suppose that

$$\sup\{\|f\|_p \mid f \in \mathcal{F}\} < M$$

for some $0 < M < \infty$. Then for any $f \in \mathcal{F}$ and any measurable subset $A \subset X$ with $\mu(A) < \infty$, by applying the Hölder's Inequality ((ii) in Theorem 7.4.6), for $1 < p < \infty$, we obtain that

$$\begin{aligned}\int_A |f| d\mu = \int_A |f \cdot 1| d\mu &\leq \left(\int_A |f|^p d\mu\right)^{1/p} \left(\int_A 1^q d\mu\right)^{1/q} = \|f\|_p \, (\mu(A))^{1/q} \\ &\leq M \, (\mu(A))^{1/q}.\end{aligned}$$

If $p = \infty$, then $\int_A |f| d\mu \leq \|f\|_\infty \mu(A) \leq M\mu(A)$. Therefore, by taking

$$\delta = \left(\frac{\epsilon}{M}\right)^q,$$

for any $f \in \mathcal{F}$ and any measurable subset $A \subset X$ with $\mu(A) < \delta$

$$\int_A |f| < \epsilon,$$

which implies that $\mathcal{F}$ is uniformly integrable on X. □

Corollary 7.4.11 Let $(X, \mathcal{M}, \mu)$ be a finite measure space and $1 \leq p < q \leq \infty$. Then $L^q(X) \subseteq L^p(X)$.

[Proof] (1) The case of $q < \infty$. Let $f \in L^q(X)$. Then by applying the Hölder's Inequality ((ii) in Theorem 7.4.6), we obtain that

$$\begin{aligned}\|f\|_p = \left(\int_X |f|^p \cdot 1 d\mu\right)^{1/p} &\leq \left(\left(\int_X (|f|^p)^{q/p} d\mu\right)^{p/q}\right)^{1/p} \left(\left(\int_X 1^{q/(q-p)} d\mu\right)^{(q-p)/q}\right)^{1/p} \\ &= \mu(X)^{\frac{q-p}{pq}} \|f\|_q,\end{aligned}$$

which implies that $f \in L^p(X)$.

(2) The case of $q = \infty$. Let $f \in L^q(X)$. Then we obtain that

$$\|f\|_p = \left(\int_X |f|^p \cdot 1 d\mu\right)^{1/p} \leq \|f\|_\infty \left(\int_X 1 d\mu\right)^{1/p} = \|f\|_\infty \mu(X)^{1/p} = \mu(X)^{1/p} \|f\|_\infty,$$

which implies that $f \in L^p(X)$. □

Remark 7.4.12 Let $(X, \mathcal{M}, \mu)$ be a finite measure space and $1 \leq p < q \leq \infty$. Then in the proof of Corollary 7.4.11, we proved that for any $f \in L^q(X)$, there exists a constant $c \geq 0$ such that

$$\|f\|_p \leq c\|f\|_q.$$

Moreover, the constant c is given by

$$c = \begin{cases} \mu(X)^{\frac{q-p}{pq}}, & q < \infty, \\ \mu(X)^{1/p}, & q = \infty. \end{cases}$$

Example 7.4.13 Let $X = (0,1]$, $\gamma > 0$ and f_γ be a function defined on X by $f_\gamma(x) = x^{-\gamma}$ for $x \in X$. For any $\gamma > 0$ and $M > 0$, put

$$A_{\gamma,M} = \{x \in (0,1] \,|\, |f_\gamma(x)| \le M\} = \left\{x \in (0,1] \,|\, x \ge (1/M)^{1/\gamma}\right\}.$$

Then the Lebesgue measure of $A^{\mathrm{c}}_{\gamma,M}$ is given by

$$m\left(A^{\mathrm{c}}_{\gamma,M}\right) = (1/M)^{1/\gamma}$$

and so $f_\gamma \notin L^\infty((0,1])$. Let $1 \le p < \infty$. Then we obtain that

$$\begin{aligned}\int_0^1 |f_\gamma|^p &= \int_0^1 x^{-\gamma p} dx \\ &= \begin{cases} [\ln x]_0^1, & \gamma p = 1 \\ \left[\frac{1}{1-\gamma p} x^{1-\gamma p}\right]_0^1, & \gamma p \ne 1 \end{cases} \\ &= \begin{cases} \infty, & \gamma p = 1 \\ \frac{1}{1-\gamma p}, & \gamma < 1/p \\ \infty, & \gamma > 1/p \end{cases}\end{aligned}$$

and so for fixed $0 < \gamma < 1$, $f_\gamma \in L^p((0,1])$ for $p < 1/\gamma$, but $f_\gamma \notin L^q((0,1])$ for $1/\gamma < q$. Therefore, for $1 \le p < q \le \infty$, $L^q((0,1])$ is a proper subspace of $L^p((0,1])$.

Example 7.4.14 Let $X = [1,\infty)$, $\gamma > 0$ and f_γ be a function defined on X by $f_\gamma(x) = x^{-\gamma}$ for $x \in X$. For any $\gamma > 0$ and $M > 0$, put

$$A_{\gamma,M} = \{x \in [1,\infty)] \,|\, |f_\gamma(x)| \le M\} = \left\{x \in [1,\infty) \,|\, x \ge (1/M)^{1/\gamma}\right\}.$$

Then for any $M \ge 1$, $(1/M)^{1/\gamma} \le 1$, and so $A_{\gamma,M} = [1,\infty)$ and

$$m\left(A^{\mathrm{c}}_{\gamma,M}\right) = m(\emptyset) = 0.$$

Therefore, $f_\gamma \in L^\infty([1,\infty))$ for any $\gamma > 0$. Let $1 \le p < \infty$. Then we obtain that

$$\begin{aligned}\int_0^1 |f_\gamma|^p &= \int_1^\infty x^{-\gamma p} dx \\ &= \begin{cases} [\ln x]_1^\infty, & \gamma p = 1 \\ \left[\frac{1}{1-\gamma p} x^{1-\gamma p}\right]_1^\infty, & \gamma p \ne 1 \end{cases} \\ &= \begin{cases} \infty, & \gamma p = 1 \\ \frac{1}{\gamma p-1}, & \gamma > 1/p \\ \infty, & \gamma < 1/p \end{cases}\end{aligned}$$

and so for fixed $0 < \gamma < 1$, $f_\gamma \in L^p([1,\infty))$ for $p > 1/\gamma$, but $f_\gamma \notin L^q([1,\infty))$ for $1/\gamma > q$. Therefore, for $1 \le p < q \le \infty$, $L^q([1,\infty)) \nsubseteq L^p([1,\infty))$.

7.5 Completeness of $L^p(X)$

Theorem 7.5.1 **[Riesz-Fischer Theorem]** Let $(X, \mathcal{M}, \mu)$ be a measure space. Let $1 \le p \le \infty$. Then $(L^p(X), \|\cdot\|_p)$ is complete.

[Proof] (i) The case of $1 \le p < \infty$. To apply Theorem 7.1.20, we suppose that $\{f_n\}_{n=1}^{\infty}$ is a sequence in $L^p(X)$ such that

$$\sum_{k=1}^{\infty} \| f_k \|_p = M < \infty.$$

For each $n \in \mathbb{N}$, put

$$g_n = \sum_{k=1}^{n} |f_k|, \qquad g = \sum_{k=1}^{\infty} |f_k|.$$

Then we have

$$\| g_n \|_p = \left\| \sum_{k=1}^{n} |f_k| \right\|_p \le \sum_{k=1}^{n} \| f_k \|_p \le M, \tag{7.32}$$

and by applying the monotone convergence theorem (Theorem 5.1.15) and (7.32), we obtain that

$$\int_X g^p d\mu = \lim_{n\to\infty} \int_X g_n^p d\mu = \lim_{n\to\infty} \| g_n \|_p^p \le M^p,$$

which implies that $g \in L^p(X)$. Therefore, g is finite a.e. on X and so the series

$$\sum_{k=1}^{\infty} f_k$$

converges a.e. on X. Put

$$f = \begin{cases} \displaystyle\sum_{k=1}^{\infty} f_k, & \text{if the series converges,} \\ 0, & \text{otherwise.} \end{cases}$$

Then f is measurable and we have

$$|f| \le \sum_{k=1}^{\infty} |f_k| = g \in L^p(X),$$

which implies that $f \in L^p(X)$. Moreover, we have

$$\left| f - \sum_{k=1}^{n} f_k \right|^p \le g^p \in L^1(X)$$

and so by applying the dominated convergence theorem, we have

$$\lim_{n\to\infty}\left\| f-\sum_{k=1}^{n} f_k\right\|_p^p = \lim_{n\to\infty}\int_X \left| f-\sum_{k=1}^{n} f_k\right|^p d\mu$$
$$= \int_X \lim_{n\to\infty}\left| f-\sum_{k=1}^{n} f_k\right|^p d\mu$$
$$= 0,$$

which implies that the series

$$\sum_{k=1}^{\infty} f_k$$

converges in $L^p(X)$. Therefore, by Theorem 7.1.20, $L^p(X)$ is complete.

(ii) The case of $p=\infty$. We suppose that $\{f_n\}_{n=1}^{\infty}$ is a sequence in $L^\infty(X)$ such that

$$\sum_{k=1}^{\infty} \| f_k \|_\infty = M < \infty.$$

For each $k \in \mathbb{N}$, there exists a measurable subset $D_k \subset X$ with $\mu(D_k) = 0$ such that $|f_k(x)| \leq \|f_k\|_\infty$ for all $x \in X \setminus D_k$. Put

$$D = \bigcup_{k=1}^{\infty} D_k.$$

Then $\mu(D) = 0$ and for any $x \in X \setminus D$ and any $k \in \mathbb{N}$, $|f_k(x)| \leq \|f_k\|_\infty$. Therefore, for any $x \in X \setminus D$, the series $\sum_{k=1}^{\infty} |f_k|$ converges a.e. on X, in fact, for any $x \in X \setminus D$,

$$\sum_{k=1}^{\infty} |f_k(x)| \leq \sum_{k=1}^{\infty} \| f_k \|_\infty = M < \infty,$$

and so the series $\sum_{k=1}^{\infty} f_k$ converges a.e. on X. Put

$$f = \begin{cases} \sum_{k=1}^{\infty} f_k, & \text{on } X \setminus D, \\ 0, & \text{on } D. \end{cases}$$

Then for any $x \in X \setminus D$, we have

$$\left| f(x) - \sum_{k=1}^{n} f_k(x)\right| = \left|\sum_{k=n+1}^{\infty} f_k(x)\right| \leq \sum_{k=n+1}^{\infty} |f_k(x)| \leq \sum_{k=n+1}^{\infty} \|f_k\|_\infty,$$

which implies that

$$\left| f - \sum_{k=1}^{n} f_k\right| \leq \sum_{k=n+1}^{\infty} \|f_k\|_\infty \quad \text{a.e. on } X$$

and so

$$\left\| f - \sum_{k=1}^{n} f_k\right\|_\infty \leq \sum_{k=n+1}^{\infty} \|f_k\|_\infty$$

and

$$\lim_{n\to\infty}\left\| f - \sum_{k=1}^{n} f_k \right\|_\infty \le \lim_{n\to\infty} \sum_{k=n+1}^{\infty} \|f_k\|_\infty = 0,$$

which implies that the series

$$\sum_{k=1}^{\infty} f_k$$

converges to f in $L^\infty(X)$. Therefore, by Theorem 7.1.20, $L^\infty(X)$ is complete. □

Corollary 7.5.2 Let $(X, \mathcal{M}, \mu)$ be a measure space. Let $1 \le p \le \infty$. Then $(L^p(X), \|\cdot\|_p)$ is a Banach space.

[Proof] The proof is immediate from Theorems 7.4.9 and 7.5.1. In fact, from Theorem 7.4.9, $(L^p(X), \|\cdot\|_p)$ is a normed space and it is complete from Theorem 7.5.1. Therefore, $(L^p(X), \|\cdot\|_p)$ is a Banach space. □

Theorem 7.5.3 Let $(X, \mathcal{M}, \mu)$ be a measure space. Let $1 \le p < \infty$ and $\{f_n\}_{n=1}^\infty \subset L^p(X)$ which converges pointwise a.e. on X to the function $f \in L^p(X)$. Then $\{f_n\}_{n=1}^\infty$ converges to f in $L^p(X)$ if and only if

$$\lim_{n\to\infty} \int_X |f_n|^p d\mu = \int_X |f|^p d\mu.$$

[Proof] $\Rightarrow$) It is obvious from Proposition 7.1.2.

$\Leftarrow$) Suppose that $\{f_n\}_{n=1}^\infty$ converges pointwise on $X \setminus N$ to f for some measurable set N with $\mu(N) = 0$ and

$$\lim_{n\to\infty} \int_X |f_n|^p d\mu = \int_X |f|^p d\mu.$$

Consider the function φ defined on $\mathbb{R}$ by $\varphi(t) = |t|^p$ for $t \in \mathbb{R}$. Then φ is increasing, convex on $(0, \infty)$ and $\varphi(-t) = \varphi(t)$ for all $t \in \mathbb{R}$. Therefore, φ is convex on $\mathbb{R}$. Therefore, for any $a, b \in \mathbb{R}$, we have

$$\left|\frac{a-b}{2}\right|^p = \varphi\left(\frac{a+(-b)}{2}\right) \le \frac{\varphi(a)+\varphi(-b)}{2} = \frac{|a|^p+|b|^p}{2},$$

which implies that

$$\frac{|a|^p+|b|^p}{2} - \left|\frac{a-b}{2}\right|^p \ge 0.$$

Now, for each $n \in \mathbb{N}$, we put

$$g_n = \begin{cases} \dfrac{|f_n|^p + |f|^p}{2} - \left|\dfrac{f_n - f}{2}\right|^p, & \text{on } X \setminus N \\ 0, & \text{on } N. \end{cases}$$

Then $\{g_n\}_{n=1}^{\infty}$ is a sequence of nonnegative measurable functions which is converges pointwise a.e. on X to $|f|^p$. Therefore, by applying the Fatou's Lemma (Corollary 5.1.19), we obtain that

$$\begin{aligned} \int_X |f|^p d\mu &\leq \liminf_{n\to\infty} \int_X g_n d\mu = \liminf_{n\to\infty} \int_X \left(\frac{|f_n|^p + |f|^p}{2} - \left|\frac{f_n - f}{2}\right|^p\right) d\mu \\ &= \int_X |f|^p d\mu - \limsup_{n\to\infty} \int_X \left|\frac{f_n - f}{2}\right|^p d\mu, \end{aligned}$$

which implies that

$$\limsup_{n\to\infty} \int_X \left|\frac{f_n - f}{2}\right|^p d\mu \leq 0$$

and so $\lim_{n\to\infty} \| f_n - f \|_p^p = 0$, equivalently, $\lim_{n\to\infty} \| f_n - f \|_p = 0$. Therefore, $\{f_n\}_{n=1}^{\infty}$ converges to f in $L^p(X)$. □

Corollary 7.5.4 Let $(\mathbb{R}, \mathfrak{M}, m)$ be the Lebesgue measure space. Let $1 \leq p < \infty$ and $\{f_n\}_{n=1}^{\infty} \subset L^p(\mathbb{R})$ which converges pointwise a.e. on $\mathbb{R}$ to the function $f \in L^p(\mathbb{R})$. Then $\{f_n\}_{n=1}^{\infty}$ converges to f in $L^p(\mathbb{R})$ if and only if

$$\lim_{n\to\infty} \int_{\mathbb{R}} |f_n|^p = \int_{\mathbb{R}} |f|^p.$$

[Proof] The proof is immediate from Theorem 7.5.3. □

Remark 7.5.5 Now, we recall Corollary 6.1.11. Let $\{f_n\}_{n=1}^{\infty}$ be a sequence of nonnegative integrable functions on X. Suppose $\{f_n\}$ converges to $f = 0$ pointwise on X. Then

$$\lim_{n\to\infty} \int_X f_n d\mu - 0$$

if and only if $\{f_n\}$ is uniform integrable and tight over X. □

Theorem 7.5.6 Let $(X, \mathcal{M}, \mu)$ be a measure space. Let $1 \leq p < \infty$ and $\{f_n\}_{n=1}^{\infty} \subset L^p(X)$ which converges pointwise a.e. on X to the function $f \in L^p(X)$. Then $\{f_n\}_{n=1}^{\infty}$ converges to f in $L^p(X)$ if and only if $\{|f_n|^p\}_{n=1}^{\infty}$ is uniformly integrable and tight over X.

[Proof] By Corollary 6.1.11, $\{f_n\}_{n=1}^{\infty}$ converges to f in $L^p(X)$ if and only if $\{|f_n - f|^p\}_{n=1}^{\infty}$ is uniformly integrable and tight over X. On the other hand, for any $n \in \mathbb{N}$, we have

$$|f_n - f|^p \leq 2^p \left(|f_n|^p + |f|^p\right), \qquad |f_n|^p \leq 2^p \left(|f_n - f|^p + |f|^p\right),$$

and since $|f|^p$ is integrable, $|f|^p$ is uniformly integrable and tight over X. Therefore, $\{|f_n - f|^p\}_{n=1}^{\infty}$ is uniformly integrable and tight over X if and only if $\{|f_n|^p\}_{n=1}^{\infty}$ is uniformly integrable and tight over X. □

Corollary 7.5.7 Let $(\mathbb{R}, \mathfrak{M}, m)$ be the Lebesgue measure space. Let $1 \leq p < \infty$ and $\{f_n\}_{n=1}^{\infty} \subset L^p(\mathbb{R})$ which converges pointwise a.e. on $\mathbb{R}$ to the function $f \in L^p(\mathbb{R})$. Then $\{f_n\}_{n=1}^{\infty}$ converges to f in $L^p(\mathbb{R})$ if and only if $\{|f_n|^p\}_{n=1}^{\infty}$ is uniformly integrable and tight over $\mathbb{R}$.

[Proof] The proof is immediate from Theorem 7.5.6. □

Remark 7.5.8 Let (X, τ) be a topological space. Then X is said to be

(1) **first countable** if there exists a countable neighborhood local basis at each point;

(2) **second countable** if there exists a countable neighborhood basis for the topology τ.

Example 7.5.9 Every metric space is first countable.

Proposition 7.5.10 Let (X, τ) be a first countable topological space, E a subset of X and x a point of X. Then $x \in \overline{E}$ if and only if there exists a sequence $\{x_n\}_{n=1}^{\infty} \subset E$ such that $\{x_n\}_{n=1}^{\infty}$ converges to x.

[Proof] The proof is an exercise (see Exercise **7.17**) □

Theorem 7.5.11 Let $(X, \mathcal{M}, \mu)$ be a measure space and let $1 \leq p < \infty$. The subspace of all simple functions ϕ on X with $\mu(\{x \in X | \phi(x) \neq 0\}) < \infty$ is dense in $L^p(X)$.

[Proof] Let $f \in L^p(X)$ be given. Then by Proposition 5.1.8, $\{x \in X \,|\, f(x) \neq 0\}$ is σ-finite. Therefore, we may assume that μ is σ-finite and so by the Simple Approximation Theorem ((iii) in Theorem 4.1.37), there exists a sequence $\{\phi_n\}_{n=1}^{\infty}$ of simple functions ϕ_n which vanishes outside a set of finite measure, i.e., $\mu(\{x \in X \,|\, \phi_n \neq 0\}) < \infty$, such that $|\phi_n| \leq |f|$ and ϕ_n converges pointwise on X to f. Then it holds that

$$|f - \phi_n|^p \leq 2^p |f|^p, \qquad n \in \mathbb{N}.$$

Since $|f|^p$ is integrable, by the Lebesgue Dominated Convergence Theorem (Theorem 5.1.26), ϕ_n converges to f in $L^p(X)$. □

7.6 Dual of $L^p(X)$

Let $(X, \mathcal{M}, \mu)$ be a measure space and $1 \leq p \leq \infty$. Let q be the conjugate number of p and $g \in L^q(X)$. Then for each $f \in L^p(X)$, by the Hölder's Inequality (Theorem 7.4.6), $fg \in L^1(X)$ and so the linear functional

$$T_g : L^p(X) \ni f \longmapsto T_g(f) = \int_X g \cdot f d\mu \tag{7.33}$$

is well-defined. Moreover, it holds that

$$\|T_g\| = \sup_{\|f\|_p \leq 1} \{|T_g(f)|\} \leq \|g\|_q.$$

In fact, we obtain that

$$|T_g(f)| \leq \int_X |g \cdot f| d\mu = \|f\|_p \|g\|_q,$$

and so T_g is a bounded linear functional on $L^p(X)$, i.e., $T_g \in (L^p(X))^*$.

Let $(X, \mathcal{M}, \mu)$ be a measure space. If for each $E \in \mathcal{M}$ with $\mu(E) = \infty$, there exists a measurable subset $F \subset E$ such that $0 < \mu(F) < \infty$, then μ is said to be ***semifinite***. Every σ-finite is semifinite.

Proposition 7.6.1 Let $(X, \mathcal{M}, \mu)$ be a measure space and $1 \leq p \leq \infty$. Let q be the conjugate number of p. Let $g \in L^q(X)$. Then the linear functional T_g defined on $L^p(X)$ as in (7.33) is bounded and $\|T_g\| = \|g\|_q$ for $1 \leq q < \infty$. In the case of $q = \infty$, if μ is semifinite, then $\|T_g\| = \|g\|_q$.

[Proof] We already proved that T_g is bounded and $\|T_g\| \leq \|g\|_q$. If $g = 0$, then the proof is obvious. Suppose that $g \neq 0$.

(i) The case of $1 \leq q < \infty$ $(1 < p \leq \infty)$. Consider the conjugate function g^* given by

$$g^* = \|g\|_q^{1-q} \operatorname{sgn}(g) \, |g|^{q-1}.$$

Then by (iii) in Theorem 7.4.6, we see that $g^* \in L^p(X)$ and

$$\int_X g^* g \, d\mu = \|g\|_q, \qquad \|g^*\|_p = 1.$$

Therefore, it holds that

$$\begin{aligned} \|T_g\| &= \sup_{\|f\|_p \leq 1} \{|T_g(f)|\} \\ &\geq |T_g(g^*)| = \left| \int_X g^* g \, d\mu \right| \\ &= \|g\|_q, \end{aligned}$$

and so $\|T_g\| = \|g\|_q$.

(ii) The case of $q = \infty$ $(p = 1)$. Let $\epsilon > 0$ be given. Put

$$A = \{x \in X \mid |g(x)| > \|g\|_\infty - \epsilon\}.$$

Then by the definition of $\|\cdot\|_\infty$, $\mu(A) > 0$, if $\mu(A) = \infty$, then since μ is semifinite, there exists a measurable subset $B \subset A$ such that $0 < \mu(B) < \infty$. If $\mu(A) < \infty$, then we take $B = A$. Put

$$f = \frac{1}{\mu(B)} \chi_B \operatorname{sgn}(g).$$

Then $f \in L^1(X)$ and $\|f\|_1 = 1$, and so

$$\begin{aligned} \|T_g\| &\geq \int_X fg \, d\mu \\ &= \frac{1}{\mu(B)} \int_B |g| \, d\mu \\ &\geq \|g\|_\infty - \epsilon. \end{aligned}$$

Since $\epsilon > 0$ is arbitrary, $\|T_g\| \geq \|g\|_\infty$ and so $\|T_g\| = \|g\|_\infty$. □

Example 7.6.2 Let X be a nonempty set and $\mathcal{M} = \{X, \emptyset\}$. Define a measure μ by $\mu(X) = \infty$ and $\mu(\emptyset) = 0$. Then $(X, \mathcal{M}, \mu)$ is not s semifinite measure space. Then we have $L^1(X) = \{0\}$ and $L^\infty(X) =$ the set of all constant functions on X Then for any constant function $g = a \neq 0$, $T_g = 0 \in (L^1(X))^*$ and $\|T_g\| = 0 < |a| = \|g\|_\infty$.

Lemma 7.6.3 Let $(X, \mathcal{M}, \mu)$ be a σ-finite measure space and $1 \leq p < \infty$. Let $g \in L^1(X)$ and there exists a $M \geq 0$ such that

$$\left| \int_X gf \, d\mu \right| \leq M \|f\|_p \tag{7.34}$$

for every simple functions f in $L^p(X)$ which vanishes outside a set of finite measure. Then $g \in L^q(X)$ and $\|g\|_q \leq M$, where q is the conjugate number of p.

[Proof] Since g is finite a.e. on X, without loss of generality we may assume that g is finite on X.

(i) The case of $1 < p < \infty$ $(1 < q < \infty)$. Since g is measurable function, by the Simple Approximation Theorem (Theorem 4.1.37) there exists a sequence $\{\psi_n\}_{n=1}^{\infty}$ of simple functions which vanishes outsides of a set of finite measure such that $\{\psi_n\}_{n=1}^{\infty}$ converges pointwise to $|g|$ and $0 \leq \psi_n \leq |g|$. Then we obtain that

$$\begin{aligned}\int_X \psi_n^q d\mu &= \int_X \psi_n \cdot \psi_n^{q-1} d\mu \\ &\leq \int_X |g| \cdot \psi_n^{q-1} d\mu \\ &= \int_X g \cdot \operatorname{sgn}(g) \cdot \psi_n^{q-1} d\mu,\end{aligned}$$

which implies that for the simple function $f_n = \operatorname{sgn}(g) \cdot \psi_n^{q-1}$,

$$\int_X \psi_n^q d\mu \leq \int_X g \cdot f_n d\mu,$$

where ψ_n is simple function which vanishes outsides a set of finite measure and so $f_n \in L^p(X)$. Therefore, by (7.34) it holds that

$$\begin{aligned}\int_X \psi_n^q d\mu \leq \int_X g \cdot f_n d\mu &\leq M \left(\int_X |f_n|^p d\mu \right)^{1/p} \\ &= M \left(\int_X \psi_n^q d\mu \right)^{1/p},\end{aligned}$$

which implies that

$$\left(\int_X \psi_n^q d\mu \right)^{1/q} = \left(\int_X \psi_n^q d\mu \right)^{1-1/p} \leq M$$

and so $\psi_n \in L^q(X)$ and then by the Fatou's Lemma, we see that

$$\int_X |g|^q d\mu \leq \liminf_{n\to\infty} \int_X \psi_n^q d\mu \leq M^q. \tag{7.35}$$

Therefore, $g \in L^q(X)$ and $\|g\|_q \leq M$.

(ii) The case of $p = 1$ $(q = \infty)$. Suppose not, i.e., $\|g\|_\infty > M$. There exists $\epsilon > 0$ such that $\|g\|_\infty > M + \epsilon > M$. Then for $A = \{x \in X \mid |g(x)| > M + \epsilon\}$, we have

$$\mu(A) > 0.$$

and so there exists a measurable set $B \subset A$ with $0 < \mu(B) < \infty$ such that

$$\int_X g \cdot \operatorname{sgn}(g)\chi_B \, d\mu \geq (M + \epsilon) \int_X \chi_B \, d\mu > M \int_X |\operatorname{sgn}(g)\chi_B| \, d\mu,$$

which is a contradiction to (7.34). □

Proposition 7.6.4 Let $(X, \mathcal{M}, \mu)$ be a finite measure space and $1 \le p < \infty$. Then for each $T \in (L^p(X))^*$, there exists a finite signed measure ν on $\mathcal{M}$, which is absolutely continuous with respect to μ, such that

$$\nu(E) = T(\chi_E), \qquad E \in \mathcal{M}. \tag{7.36}$$

[Proof] For each $E \in \mathcal{M}$, since μ is a finite measure, it holds that

$$\int_X |\chi_E|^p d\mu = \mu(E) < \infty,$$

and so the set function ν on $\mathcal{M}$ given as in (7.36) is well-defined. Let $\{E_k\}_{k=1}^{\infty}$ be a countable disjoint family of sets in $\mathcal{M}$. Put

$$E = \bigcup_{k=1}^{\infty} E_k$$

Then by the countable additivity of μ, it holds that

$$\sum_{k=1}^{\infty} \mu(E_k) = \mu(E) \le \mu(X) < \infty,$$

which implies that $\lim_{n\to\infty} \sum_{k=n+1}^{\infty} \mu(E_k) = 0$. Let $\{s_k\}_{k=1}^{\infty}$ be a sequence of real numbers such that $s_k = 1$ or -1 for all $k \in \mathbb{N}$. Then, for any $m, n \in \mathbb{N}$, it holds that

$$\begin{aligned} \left\| \sum_{k=m}^{m+n} s_k \chi_{E_k} \right\|_p^p &\le \int_X \left(\sum_{k=m}^{m+n} \chi_{E_k} \right)^p d\mu \\ &= \sum_{k=m}^{m+n} \mu(E_k), \end{aligned}$$

which implies that, since $L^p(X)$ is complete, the series $\sum_{k=1}^{\infty} s_k \chi_{E_k}$ converges in $L^p(X)$. Therefore, by the continuity and the linearity of T, it holds that

$$\begin{aligned} \nu(E) = T(\chi_E) &= \lim_{n\to\infty} T\left(\sum_{k=1}^{n} \chi_{E_k} \right) \\ &= \lim_{n\to\infty} \sum_{k=1}^{n} T(\chi_{E_k}) \\ &= \sum_{k=1}^{\infty} \nu(E_k). \end{aligned}$$

For each $k \in \mathbb{N}$, put

$$s_k = \operatorname{sgn}(T(\chi_{E_k})).$$

Then by the continuity and the linearity of T, we obtain that

$$\sum_{k=1}^{\infty} |\nu(E_k)| = \sum_{k=1}^{\infty} s_k T(\chi_{E_k})$$
$$= T\left(\sum_{k=1}^{\infty} s_k \chi_{E_k}\right),$$

which implies that the series $\sum_{k=1}^{\infty} \nu(E_k)$ converges absolutely. Thus ν is a signed measure. Let (P, N) be the Hahn decomposition of X. Then it holds that

$$|\nu|(X) = \nu(P) - \nu(N) = T(\chi_P) - T(\chi_N) < \infty,$$

and so ν is a finite signed measure. Also, if $E \in \mathcal{M}$ and $\mu(E) = 0$, then $\chi_E = 0$ in $L^p(X)$, which implies that $\nu(E) = T(\chi_E) = T(0) = 0$, and so ν is absolutely continuous with respect to μ. □

Theorem 7.6.5 **[Riesz Representation]** Let $(X, \mathcal{M}, \mu)$ be a σ-finite measure space and $1 \leq p < \infty$. Then for each $T \in (L^p(X))^*$, there exists a unique $g \in L^q(X)$, where q is the conjugate of p, such that

$$T_g = T \quad \text{and} \quad \|T\| = \|T_g\| = \|g\|_q,$$

where T_g is defined by

$$T_g(f) = \int_X g \cdot f \, d\mu \quad \text{for all } f \in L^p(X). \tag{7.37}$$

[Proof] (i) The case of $1 < p < \infty$ $(1 < q < \infty)$.

(1) Uniqueness. If there exist $g_1, g_2 \in L^q(X)$ such that $T_{g_1} = T$ and $T_{g_2} = T$. Then $T_{g_1} = T_{g_2}$ and $T_{g_1 - g_2} = T_{g_1} - T_{g_2} = 0$ and so by Proposition 7.6.1,

$$0 = \|T_{g_1 - g_2}\| = \|g_1 - g_2\|_q,$$

which implies that $g_1 = g_2$. In fact, $g_1 = g_2$ a.e. on X.

(2) Existence. Suppose that μ is finite. Let $T \in (L^p(X))^*$. Then by Proposition 7.6.4, there exists a finite signed measure ν on $\mathcal{M}$, which is absolutely continuous with respect to μ, such that (7.36) holds. Then by the Radon-Nikodym Theorem (Theorem 6.2.10), there exists a unique integrable function g such that

$$\nu(E) = \int_E g d\mu = \int_X \chi_E g d\mu, \qquad E \in \mathcal{M},$$

which, from (7.36), implies that for any $E \in \mathcal{M}$,

$$T(\chi_E) = \nu(E) = \int_X \chi_E g d\mu,$$

and so, from the linearities of T and the integral, for any simple function $f \in L^2(X)$, it holds that

$$T(f) = \int_X fg d\mu. \tag{7.38}$$

Since T is bounded on the Banach space $L^p(X)$, for any simple function f,

$$\begin{aligned}\left|\int_X fg d\mu\right| &= |T(f)| \\ &\leq \|T\|\|f\|_p.\end{aligned}$$

Therefore, by Lemma 7.6.3, $g \in L^q(X)$. Put

$$S = T - T_g.$$

Then $S \in (L^p(X))^*$ and for any simple function $f \in L^p(X)$, $S(f) = 0$. On the other hand, by Theorem 7.5.11, the linear space of all simple functions is dense in $L^p(X)$, $S = 0$, i.e., $T = T_g$.

Let μ is σ-finite. Then there exists an ascending sequence $\{X_n\}_{n=1}^{\infty}$ of measurable sets with finite measure such that $X = \cup_{n=1}^{\infty} X_n$. For each $n \in \mathbb{N}$, define a linear functional T_n on $L^p(X_n)$ by

$$T_n(f) = T(\widetilde{f}), \qquad f \in L^p(X_n),$$

where $\widetilde{f} \in L^p(X)$ such that $\widetilde{f} \restriction_{X_n} = f$ and $\widetilde{f} \restriction_{X_n^c} = 0$. Then we have

$$\begin{aligned}\|T_n\| &= \sup\{|T_n(f)| \mid f \in L^p(X_n); \|f\|_p = 1\} \\ &= \sup\{|T(\widetilde{f})| \mid f \in L^p(X_n); \|f\|_p = 1\} \\ &\leq \|T\|,\end{aligned}$$

and so there exists $g_n \in L^q(X_n)$ such that $T_n = T_{g_n}$ and $\|g_n\|_q = \|T_n\| \leq \|T\|$. Then for each $n \in \mathbb{N}$, $g_{n+1} \restriction_{X_n} \in L^q(X_n)$ and for any $f \in L^p(X_n)$, we have

$$\begin{aligned}T_{g_{n+1}\restriction X_n}(f) &= \int_{X_n} g_{n+1} \restriction_{X_n} \cdot f \, d\mu \\ &= \int_{X_{n+1}} g_{n+1} \cdot \widetilde{f} \restriction_{X_{n+1}} \, d\mu \\ &= T_{g_{n+1}}(\widetilde{f} \restriction_{X_{n+1}}) \\ &= T_{n+1}(\widetilde{f} \restriction_{X_{n+1}}) \\ &= T(\widetilde{f}) = T_n(f) \\ &= T_{g_n}(f),\end{aligned}$$

which, by the uniqueness, implies that $g_{n+1} \restriction_{X_n} = g_n$ a.e. on X_n. Therefore, define a measurable function g on X as following: for each $x \in X$, there exists $n \in \mathbb{N}$ such that $x \in X_n$ and then

$$g(x) = g_n(x).$$

Then $\{\widetilde{g_n}\}_{n=1}^{\infty}$ converges pointwise to g a.e. on X, by the fatou's Lemma we have

$$\int_X |g|^q \, d\mu \leq \liminf_{n\to\infty} \int_X |\widetilde{g_n}|^q \, d\mu$$

$$
\begin{aligned}
&= \liminf_{n\to\infty} \int_{X_n} |g_n|^q \, d\mu \\
&\leq \|T\|^q,
\end{aligned}
$$

which implies that $g \in L^q(X)$. The for any $f \in L^p(X)$ vanishing outside a bounded set, i.e., there exists a $n \in \mathbb{N}$ such that $f \restriction_{X_n^c} = 0$,

$$
\begin{aligned}
T_g(f) &= \int_X g \cdot f \, d\mu \\
&= \int_{X_n} g \cdot f \, d\mu \\
&= \int_{X_n} g_n \cdot f \restriction_{X_n} \, d\mu \\
&= T_n(f \restriction_{X_n}) = T(f),
\end{aligned}
$$

which means that $T_g(f) = T(f)$ for all $f \in D$, where D is the set of all $L^p(X)$ vanishing outside a bounded set. Since D is a dense subset of $L^p(X)$, $T_g = T$ and $\|T\| = \|T_g\| = \|g\|_q$.

(ii) The case of $p = 1$ ($q = \infty$) is left as an exercise (see Exercise **7.19**). □

Corollary 7.6.6 Let $(X, \mathcal{M}, \mu)$ be a σ-finite measure space and $1 \leq p < \infty$. Then $(L^p(X))^*$ and $L^q(X)$ are isometric isomorphic, where q is the conjugate of p.

[Proof] The proof is immediate from Theorem 7.6.5. In fact, by Theorem 7.6.5, the map

$$L^q(X) \ni g \longmapsto T_g \in (L^p(X))^*$$

is an isometric isomorphism. □

Theorem 7.6.7 Let $(\mathbb{R}, \mathfrak{M}, m)$ be the Lebesgue measure space and $1 \leq p < \infty$. Then $(L^p(\mathbb{R}))^*$ and $L^q(\mathbb{R})$ are isometric isomorphic, where q is the conjugate of p.

[Proof] Since Lebesgue measure m is σ-finite, the proof is immediate from Corollary 7.6.6. □

7.7 Separability of L^p

Proposition 7.7.1 Let $(X, \mathcal{M}, \mu)$ be a measure space and let $1 \leq p < \infty$. Then the subspace of simple functions in $L^p(X)$ is dense in $L^p(X)$.

[Proof] (i) The case of $1 \leq p < \infty$. Let $f \in L^p(X)$. Then since f is measurable, by the Simple Approximation Theorem (Theorem 4.1.37), there exists a sequence $\{\varphi_n\}_{n=1}^{\infty}$ of simple functions such that $\{\varphi\}_{n=1}^{\infty}$ converges pointwise to f on X and

$$|\varphi_n| \leq |f| \quad \text{on } X \quad \text{for all } n.$$

In this case, since $f \in L^p(X)$, $\{\varphi\}_{n=1}^{\infty} \subset L^p(X)$ and, by Lebesgue Dominated Convergence Theorem (Theorem 5.1.26), we have

$$\lim_{n\to\infty} \int_X |\varphi_n|^p \, d\mu = \int_X |f|^p \, d\mu$$

and so by Theorem 7.5.3, $\{\varphi\}_{n=1}^{\infty}$ converges to f in $L^p(X)$.

(ii) The case of $p = \infty$. Let $f \in L^\infty(X)$. There exists a measurable subset X_0 of X with $\mu(X_0) = 0$ and f is bounded on $X \setminus X_0$. Therefore, by by the Simple Approximation Theorem (Theorem 4.1.37), for any given $\epsilon > 0$ there are simple functions ϕ_ϵ and ψ_ϵ defined on $X \setminus X_0$ such that

$$\phi_\epsilon \leq f \leq \psi_\epsilon \text{ and } 0 \leq \psi_\epsilon - \phi_\epsilon < \epsilon \text{ on } X \setminus X_0,$$

and so $0 \leq f - \phi_\epsilon < \psi_\epsilon - \phi_\epsilon < \epsilon$, which implies that $\|f - \phi_\epsilon\|_\infty < \epsilon$ and gives the proof. □

Proposition 7.7.2 Let $[a, b]$ be a closed and bounded interval and $1 \leq p < \infty$. The subspace of all step functions on $[a, b]$ is dense in $L^p([a, b])$.

[Proof] Let $f \in L^p([a, b])$.

(i) Let $f = \chi_A$ for a measurable subset A of $[a, b]$ and $\epsilon > 0$ be given. Then since $m(A) \leq b - a < \infty$, Littlewood's First Principle (Theorem 3.2.19), there exists a finite disjoint family $\{I_k\}_{k=1}^{n}$ of open intervals such that for $O = \cup_{k=1}^{n} I_k$,

$$m((A \setminus O) \cup (O \setminus A)) < \epsilon^p.$$

Then χ_O is a step function and

$$\begin{aligned}\|f - \chi_O\|_p &\leq \left(\int_a^b |f - \chi_O|^p \right)^{1/p} \\ &\leq m((A \setminus O) \cup (O \setminus A))^{1/p} \\ &< \epsilon.\end{aligned}$$

(ii) Let f be a simple function in $L^p([a, b])$. Then there exist a finite sequence $\{a_n\}_{k=1}^{n}$ of nonzero real numbers and a finite sequence $\{A_k\}_{k=1}^{n}$ of measurable subsets of $[a, b]$ such that

$$f = \sum_{k=1}^{n} a_k \chi_{A_k}.$$

By (i), for each $k = 1, 2, \cdots, n$, there exists a step function φ_k such that

$$\|\chi_{A_k} - \varphi_k\|_p < \frac{\epsilon}{|a_k|n}.$$

Therefore, for the step function $\varphi = \sum_{k=1}^{n} a_k\varphi_k$, we have

$$\|f - \varphi\|_p \leq \sum_{k=1}^{n} |a_k|\|\chi_{A_k} - \varphi_k\|_p < \epsilon.$$

Therefore, the set of all step functions is dense in the space of all simple functions with respect to the norm $\|\cdot\|_p$.

(iii) For the general $f \in L^p([a,b])$, by Proposition 7.7.1 there exists a simple function $\psi \in L^p([a,b])$ and by (ii) a step function $\varphi \in L^p([a,b])$ such that

$$\|f - \psi\|_p < \frac{\epsilon}{2}, \qquad \|\psi - \varphi\|_p < \frac{\epsilon}{2}.$$

Therefore, we have

$$\begin{aligned} \|f - \varphi\|_p &\leq \|f - \psi\|_p + \|\psi - \varphi\|_p \\ &< \frac{\epsilon}{2} + \frac{\epsilon}{2} \\ &= \epsilon. \end{aligned}$$

Hence the subspace of all step functions on $[a,b]$ is dense in $L^p([a,b])$. □

Theorem 7.7.3 Let $1 \leq p < \infty$. Then the Banach space $L^p(\mathbb{R})$ is separable.

[Proof] For each $n \in \mathbb{N}$, let $\mathbf{S}_{\mathbb{Q}}[-n,n]$ be the set of all step functions φ on $[-n,n]$ of the form:

$$\varphi = \sum_{k=1}^{n} a_k \chi_{(x_k, y_k)},$$

where a_k, x_k, y_k are rational numbers and $(x_k, y_k) \subset [-n,n]$ for all $k = 1, 2, \cdots, n$. Then $\mathbf{S}_{\mathbb{Q}}[-n,n]$ is countable and, by Proposition 7.7.2, dense in $L^p([-n,n])$. For each $n \in \mathbb{N}$, put

$$\mathcal{C} = \bigcup_{n=1}^{\infty} \mathcal{C}_n, \qquad \mathcal{C}_n = \{\widetilde{\varphi} \mid \varphi \in \mathbf{S}_{\mathbb{Q}}[-n,n]\},$$

where for $\varphi \in \mathbf{S}_{\mathbb{Q}}[-n,n]$,

$$\widetilde{\varphi}(x) = \begin{cases} \varphi(x), & x \in [-n,n], \\ 0, & x \in [-n,n]^c. \end{cases}$$

Then $\mathcal{C}$ is countable and dense in $L^p(\mathbb{R})$. In fact, for any given $f \in L^p(\mathbb{R})$ and $\epsilon > 0$, by the Monotone Convergence Theorem (Theorem 5.1.15),

$$\lim_{n\to\infty} \int_{[-n,n]} |f|^p \, d\mu = \int_{\mathbb{R}} |f|^p \, d\mu,$$

there exist $n \in \mathbb{N}$ and $\varphi \in \mathbf{S}_{\mathbb{Q}}[-n,n]$ such that

$$\int_{[-n,n]^c} |f|^p \, d\mu < \frac{\epsilon^p}{2}, \qquad \int_{[-n,n]} |f-\varphi|^p \, d\mu < \frac{\epsilon^p}{2}.$$

Therefore, we have

$$\begin{aligned} \|f - \widetilde{\varphi}\|_p^p &\le \int_{[-n,n]} |f-\varphi|^p \, d\mu + \int_{[-n,n]^c} |f|^p \, d\mu \\ &< \epsilon^p, \end{aligned}$$

and so $\mathcal{C}$ is dense in $L^p(\mathbb{R})$. Hence $L^p(\mathbb{R})$ is separable. □

Corollary 7.7.4 Let E be a Lebesgue measurable set and $1 \le p < \infty$. Then the Banach space $L^p(E)$ is separable.

[Proof] Let $f \in L^p(E)$ and $\epsilon > 0$ be given. Since $L^p(\mathbb{R})$ is separable, there exists a countable family $\mathcal{C}$ of elements of $L^p(\mathbb{R})$ such that $\mathcal{C}$ is dense in $L^p(\mathbb{R})$. Let $\mathcal{C}(E)$ be the restriction of $\mathcal{C}$ to E, i.e.,

$$\mathcal{C}(E) = \{h \upharpoonright_E \mid h \in \mathcal{C}\}.$$

Then there exists $h \in \mathcal{C}$ such that

$$\|\widetilde{f} - h\|_p < \epsilon,$$

where $\widetilde{f}$ is defined on $\mathbb{R}$ such that $\widetilde{f} \upharpoonright_E = f$ and $\widetilde{f} \upharpoonright_{E^c} = 0$, and so

$$\begin{aligned} \|f - h \upharpoonright_E \|_p^p &= \int_E |f - h \upharpoonright_E |^p \, d\mu \le \int_E |f - h \upharpoonright_E |^p \, d\mu + \int_{E^c} |h|^p \, d\mu \\ &= \|\widetilde{f} - h\|_p^p < \epsilon^p, \end{aligned}$$

which implies that $L^p(E)$ is separable. □

Let $C_c(E)$ be the linear space of continuous real-valued functions on E which vanish outside a bounded set.

Theorem 7.7.5 Let E be a Lebesgue measurable set and $1 \le p < \infty$. Then $C_c(E)$ is dense in $L^p(E)$.

[Proof] Let $f \in L^p(E)$ and $\epsilon > 0$ be given. Then by the proof of Theorem 7.7.3, there exists a $h \in \mathcal{C}$ such that

$$\|\widetilde{f} - h\|_p < \frac{\epsilon}{2},$$

where $\widetilde{f}$ is defined on $\mathbb{R}$ such that $\widetilde{f} \upharpoonright_E = f$ and $\widetilde{f} \upharpoonright_{E^c} = 0$, and there exists a $n \in \mathbb{N}$ such that h vanishes outside $[-n,n]$. Since the restriction $h \upharpoonright_{[-n,n]}$ of h to $[-n,n]$ is a step function, there exists a continuous and piecewise linear function k that vanishes outside a bounded set such that

$$\|h-k\|_p < \frac{\epsilon}{2}.$$

In fact, for any nonzero real numbers α, β and $a<b<c<d$, consider the step function $h=\alpha\chi_{(a,b)}+\beta\chi_{(c,d)}$ and the continuous and piecewise linear function k defined as following: for sufficiently small $\delta>0$ such that $b+2\delta<c$,

$$k(x)=\begin{cases} 0; & x\le a-\delta,\\ \text{linear with } f(a-\delta)=0,\ \ f(a)=\alpha; & a-\delta\le x\le a,\\ h(x)=\alpha; & a<x<b,\\ \text{linear with } f(b)=\alpha,\ \ f(b+\delta)=0; & b\le x\le b+\delta,\\ 0; & b+\delta\le x\le c-\delta,\\ \text{linear with } f(c-\delta)=0,\ \ f(c)=\beta; & c-\delta\le x\le c,\\ h(x)=\beta; & c<x<d,\\ \text{linear with } f(d)=\beta,\ \ f(d+\delta)=0; & d\le x\le d+\delta,\\ 0; & x\ge d+\delta. \end{cases}$$

Then we have

$$\|h-k\|_p<\frac{\epsilon}{2}$$

for sufficiently small $\delta>0$. Then $k\restriction_E\in C_c(E)$ and

$$\|f-k\restriction_E\|_p^p\le\int_E|f-k\restriction_E|^p\,d\mu+\int_{E^c}|k|^p\,d\mu=\int_{\mathbb{R}}|\widetilde{f}-k|^p\,d\mu,$$

which implies

$$\|f-k\restriction_E\|_p\le\|\widetilde{f}-k\|_p\le\|\widetilde{f}-h\|_p+\|h-k\|_p.$$

Therefore, $C_c(E)$ is dense in $L^p(E)$. □

Example 7.7.6 Let E be a measurable measurable subset of $\mathbb{R}$. Consider a class $\mathcal{G}$ of measurable subsets of E satisfying the following conditions:

(i) $\mathcal{G}=\{A\,|\,A \text{ is a measurable subset of } E \text{ with } m(A)\neq 0\}$,

(ii) if $A,B\in\mathcal{G}$ are distinct, then $m(A\setminus B)\neq 0$ or $m(B\setminus A)\neq 0$.

If $\#\mathcal{G}=\infty$, then $L^\infty(E)$ is not separable. Suppose that $\{f_n\}_{n=1}^\infty$ is a countable family of functions in $L^\infty(E)$. In fact, for any distinct $A,B\in\mathcal{G}$, there exists a $n_A,n_B\in\mathbb{N}$ such that

$$\|\chi_A-f_{n_A}\|_\infty<\frac{1}{2},\qquad \|\chi_B-f_{n_B}\|_\infty<\frac{1}{2}.$$

Suppose that $m(A\setminus B)\neq 0$. If $n_A=n_B$, then we have

$$1=\|\chi_{A\setminus B}\|_\infty=\|\chi_A-\chi_B\|_\infty\le\|\chi_A-f_{n_A}\|_\infty+\|\chi_B-f_{n_B}\|_\infty<\frac{1}{2}+\frac{1}{2}=1.$$

Therefore, the map

$$n_{(\cdot)}:\mathcal{G}\ni A\longmapsto n_A\in\mathbb{N}$$

is one to one and onto on a subset of $\mathbb{N}$. But $\mathcal{G}$ is a uncountable set and every subset of $\mathbb{N}$ and so the existence of the map $n_{(\cdot)}$ is a contradiction. Therefore, $L^\infty(E)$ is not separable.

Example 7.7.7 From Example 7.7.6 it is obvious that for any $a<b$, $L^\infty([a,b])$ is not separable.

Exercises

7.1 Let $(V, \|\cdot\|)$ be a normed space. Prove that if a sequence $\{u_n\}_{n=1}^{\infty} \subset V$ converges to u in V, then it holds that

$$\lim_{n\to\infty} \|u_n\| = \|u\|.$$

7.2 Let $\|\cdot\|$ and $\|\cdot\|'$ be norms in a vector space V. Then we say that $\|\cdot\|$ and $\|\cdot\|'$ are **equivalent** if there exist constants C, C' with $0 < C \le C'$ such that

$$C\|v\| \le \|v\|' \le C'\|v\|, \qquad v \in V.$$

Let $\mathcal{N} = \{\|\cdot\| \mid \|\cdot\| \text{ is a norm on } V\}$ and define a relation $\sim$ by

$$\|\cdot\| \quad \sim \quad \|\cdot\|' \quad \Leftrightarrow \quad \|\cdot\| \text{ and } \|\cdot\|' \text{ are equivalent}$$

for $\|\cdot\|, \|\cdot\|' \in \mathcal{N}$. Prove that $\sim$ is an equivalent relation.

7.3 Prove Proposition 7.1.24.

7.4 Let V be a n-dimensional vector space over $\mathbb{C}$ and $\{e_1, e_2, \cdots, e_n\}$ is a basis of V.

(1) For each $v \in V$ and the coordinate representation $(\alpha_1, \alpha_2, \cdots, \alpha_n)$ of v corresponding to the basis $\{e_1, e_2, \cdots, e_n\}$, we define

$$\|v\|_1 = \sum_{k=1}^{n} |\alpha_k|.$$

Prove that $\|\cdot\|_1$ is a norm on V.

(2) Let $\|\cdot\|$ be a norm on V. Prove that $\|\cdot\|$ and $\|\cdot\|_1$ are equivalent, where $\|\cdot\|_1$ is the norm given as in **(1)**.

(3) Let $\|\cdot\|$ and $\|\cdot\|'$ be norms on V. Prove that $\|\cdot\|$ and $\|\cdot\|'$ are equivalent.

7.5 Let $(V, \langle\cdot,\cdot\rangle)$ be an inner product space over $\mathbb{C}$ and $\|\cdot\|$ be the norm induced by the inner product $\langle\cdot,\cdot\rangle$.

(1) Prove that if u and v are orthogonal, then

$$\|u+v\|^2 = \|u\|^2 + \|v\|^2,$$

which is called the **Pythagorean Theorem**.

(2) Prove that if $\langle w, u\rangle = \langle w, v\rangle$ for any $w \in V$, then $u = v$.

(3) Prove that

$$\langle u, v\rangle = \frac{1}{4}\left(\left(\|u+v\|^2 - \|u-v\|^2\right) + i\left(\|u+iv\| - \|u-iv\|\right)\right), \qquad u, v \in V,$$

which is called the **polarization identity**.

7.6 Let $(V, \|\cdot\|)$ be a Banach space. Prove that if the norm $\|\cdot\|$ satisfies the parallelogram raw, i.e.,

$$\|u+v\|^2 + \|u-v\|^2 = 2\left(\|u\|^2 + \|v\|^2\right), \qquad u, v \in V,$$

then $(V, \|\cdot\|)$ is a Hilbert space.

7.7 Let H be a Hilbert space and $S \subset H$ be a subset of H. We say that S is **total** if for $h \in H$, $\langle h, z\rangle = 0$ for all $z \in S$ implies that $h = 0$. Prove that if S is total, then

$$H = S^{\perp\perp}.$$

7.8 Let H be a Hilbert space and $\{e_n\}_{n=1}^{\infty} \subset H$ be a sequence of vectors in H. Then $\{e_n\}_{n=1}^{\infty}$ is called an **orthonormal basis** for the Hilbert space H if

(i) $\{e_n\}_{n=1}^{\infty}$ is orthonormal, i.e., for any $n \neq m$, $\langle e_n, e_m\rangle = 0$ and $\|e_n\| = 1$,

(ii) $\{e_n\}_{n=1}^{\infty}$ is total.

Prove that if $\{e_n\}_{n=1}^{\infty}$ is an orthonormal basis for a Hilbert space H, then every element $h \in H$ is represented by

$$h = \sum_{n=1}^{\infty} \langle h, e_n\rangle e_n,$$

which is called the **Fourier series**, and prove that

$$\|h\|^2 = \sum_{n=1}^{\infty} |\langle h, e_n\rangle|^2,$$

which is called the **Parseval's indentity**.

7.9 Let X and Y be topological spaces. Suppose that every point in X has countable neighborhood (X is first countable). Let $f : X \to Y$ be a continuous function. Prove that f is sequentially continuous.

7.10 Let X and Y be topological spaces. Suppose that X is first countable, i.e., for each point x in X there exists a countable family $\mathcal{B}(x)$ of open subsets containing x such that for any open subset U containing x, there exists an open subset $O \in \mathcal{B}(x)$ with $O \subset U$. In this case, $\mathcal{B}(x)$ is called a countable **open (neighborhood) basis** of x. Let $f : X \to Y$ be a sequentially continuous function. Prove that f is continuous.

7.11 Prove that every second countable topological space is first countable.

7.12 Let H and K be Hilbert spaces and $T \in \mathcal{L}(H, K)$. Prove that there exists a unique operator $T^* \in \mathcal{L}(K, H)$ such that

$$\langle h, T^*k\rangle_H = \langle Th, k\rangle_K, \qquad h \in H, \quad k \in K,$$

in which the operator T^* is called the **adjoint operator** of T.

7.13 Let H and K be Hilbert spaces and $T \in \mathcal{L}(H, K)$. Prove that

(1) $\|T^*\|_{\mathrm{OP}} = \|T\|_{\mathrm{OP}}$,

(2) $T^{**} := (T^*)^* = T$,

(3) for any $\alpha \in \mathbb{C}$, $(\alpha T)^* = \overline{\alpha} T^*$,

(4) for a Hilbert space L and any $S \in \mathcal{L}(K, L)$, $(ST)^* = T^* S^*$.

7.14 Let $(X, \mathcal{F}, \mu)$ be a finite measure space. Prove that every bounded family in $L^1(X, \mathcal{F}, \mu)$ is uniformly integrable.

7.15 Let $p \geq 1$ and let $\ell^p = \left\{ f : \mathbb{N} \to \mathbb{C} \mid \|f\|_p^p = \sum_{n=1}^{\infty} |f(n)|^p < \infty \right\}$. Prove that

(1) $(\ell^p, \|\cdot\|_p)$ is a Banach space,

(2) $(\ell^p, \|\cdot\|_p)$ is separable.

7.16 Let $(X, \mathcal{F}, \mu)$ be a measure space and $0 < p < 1$. Put

$$L^p(X, \mathcal{F}, \mu) = \left\{ f : X \to \mathbb{R} \text{ is } \mathcal{F}\text{-measurable} \;\middle|\; \|f\|_p = \left(\int_X |f|^p d\mu \right)^{1/p} < \infty \right\}.$$

Prove that

(1) for any measurable functions f, g and for $r = -p/(1-p)$, it holds that

$$\int_X |fg| \, d\mu \geq \left(\int_X |f|^p \, d\mu \right)^{1/p} \left(\int_X |g|^r \, d\mu \right)^{1/r},$$

where the equality holds only for $\alpha|fg| = \beta|g|^r$ with $(\alpha, \beta) \neq (0, 0)$,

(2) $\|\cdot\|_p$ is not a seminnorm.

7.17 Prove Proposition 7.5.10

7.18 Let $(X, \mathcal{M}, \mu)$ be a finite measure space and $1 \leq p < \infty$. A linear functional $T : L^p(X) \to \mathbb{R}$ is said to be **positive** if for any $f \in L^p(X)$ with $f \geq 0$, $T(f) \geq 0$. Prove that for each positive continuous linear functional $T \in (L^p(X))^*$, there exists a finite measure ν on $\mathcal{M}$, which is absolutely continuous with respect to μ, such that

$$\nu(E) = T(\chi_E), \qquad E \in \mathcal{M}.$$

7.19 Let $(X, \mathcal{M}, \mu)$ be a σ-finite measure space and let $T \in (L^1(X))^*$. Prove that there exists a unique $g \in L^\infty(X)$ such that

$$T_g = T \quad \text{and} \quad \|T\| = \|T_g\| = \|g\|_\infty,$$

where T_g is defined by

$$T_g(f) = \int_X g \cdot f \, d\mu \quad \text{for all } f \in L^1(X).$$

Chapter 8

Product Measures and Integration

8.1 Product Measures

For the sets X and Y, the **Cartesian product** $X \times Y$ of X and Y is defined by

$$X \times Y = \{(x, y) \mid x \in X, \quad y \in Y\}.$$

Let $(X, \mathcal{A})$ and $(Y, \mathcal{B})$ be measurable spaces. For measurable subsets $A \in \mathcal{A}$ and $B \in \mathcal{B}$, the subset $A \times B$ of $X \times Y$ is called a **measurable rectangle** or simply **rectangle**. Let $\mathfrak{M}_0$ be the set of all finite unions of measurable rectangles and put

$$\mathcal{A} \otimes \mathcal{B} = \sigma\left(\{A \times B \mid A \in \mathcal{A}, \quad B \in \mathcal{B}\}\right)$$

which is σ-algebra generated by all (measurable) rectangles, i.e., the smallest σ-algebra containing all rectangles, and called the **product σ-algebra**. We note that

(i) the intersection of measurable rectangles is a measurable rectangle, i.e.,

$$(A \times B) \cap (C \times D) = (A \cap C) \times (B \cap D), \qquad A, C \in \mathcal{A}, \quad B, D \in \mathcal{B},$$

(ii) the complement of a measurable rectangle is a finite union of measurable rectangles, i.e.,

$$(A \times B)^{\mathrm{c}} = (A^{\mathrm{c}} \times B) \cup (A \times B^{\mathrm{c}}) \cup (A^{\mathrm{c}} \times B^{\mathrm{c}}).$$

Then we can easily see that $\mathfrak{M}_0$ is an algebra of subsets of $X \times Y$, which is the smallest algebra containing all measurable rectangles of $X \times Y$ (see Exercise **8.1**). In fact, it is obvious that $\mathfrak{M}_0$ is closed under the finite union. Also, if $\cup_{i=1}^{n} A_i \times B_i \in \mathfrak{M}_0$, then we obtain that

$$\begin{aligned}
\left(\bigcup_{i=1}^{n} A_i \times B_i\right)^{\mathrm{c}} &= \bigcap_{i=1}^{n} (A_i \times B_i)^{\mathrm{c}} = \bigcap_{i=1}^{n} \left(\bigcup_{k=1}^{3} E_{ik}\right) \\
&= \bigcup_{\sigma \in \mathcal{F}} \left(\bigcap_{i=1}^{n} E_{i\sigma(i)}\right) \\
&\in \mathfrak{M}_0,
\end{aligned}$$

where

$$E_{i1} = A_i^c \times B_i, \quad E_{i2} = A_i \times B_i^c, \quad E_{i3} = A_i^c \times B_i^c$$

and

$$\mathcal{F} = \{\sigma : \{1, 2, \cdots, n\} \to \{1, 2, 3\} \text{ a function }\}.$$

Therefore, it holds that

$$\mathcal{A} \otimes \mathcal{B} = \sigma(\mathfrak{M}_0).$$

Moreover, every element in $\mathfrak{M}_0$ can be represented as a finite disjoint union of measurable rectangles, though not necessarily in a unique way (see Exercise **8.1**).

To construct a measure on the σ-algebra $\mathcal{A} \otimes \mathcal{B}$ by applying the Carathéodory-Hahn Theorem (Theorem 2.5.26), we need to define a premeasure on the algebra $\mathfrak{M}_0$ and so we start with a set function defined on the family of all (measurable) rectangles.

Definition 8.1.1 Let $(X, \mathcal{A}, \mu)$ and $(Y, \mathcal{B}, \nu)$ be measure spaces. The product set function λ on the set of all (measurable) rectangles is defined by

$$\lambda(A \times B) = \mu(A) \times \nu(B), \qquad A \in \mathcal{A}, \quad B \in \mathcal{B}.$$

Lemma 8.1.2 Let $(X, \mathcal{A}, \mu)$ and $(Y, \mathcal{B}, \nu)$ be measure spaces. Let $\{A_k \times B_k\}_{k=1}^{\infty}$ be a countable disjoint family of measurable rectangles of which the union $\cup_{k=1}^{\infty} A_k \times B_k$ is again a measurable rectangle, i.e., there exist $A \in \mathcal{A}$ and $B \in \mathcal{B}$ such that

$$A \times B = \bigcup_{k=1}^{\infty} A_k \times B_k.$$

Then we have

$$\lambda(A \times B) = \mu(A) \times \nu(B) = \sum_{k=1}^{\infty} \mu(A_k) \times \nu(B_k).$$

[Proof] Since

$$A \times B = \bigcup_{k=1}^{\infty} A_k \times B_k.$$

is a disjoint union, it holds that for $(x, y) \in X \times Y$, $\chi_{A \times B}(x, y) = 1$ if and only if $(x, y) \in A \times B$ if and only if there exists only one $k \in \mathbb{N}$ such that $(x, y) \in A_k \times B_k$, and so we have

$$\chi_{A \times B}(x, y) = \chi_A(x)\chi_B(y) = \sum_{k=1}^{\infty} \chi_{A_k}(x)\chi_{B_k}(y).$$

Then for each fixed $x \in X$, by applying the monotone convergence theorem, we obtain that

$$\chi_A(x)\nu(B) = \int_Y \chi_A(x)\chi_B(y)\,d\nu(y) = \int_Y \sum_{k=1}^{\infty} \chi_{A_k}(x)\chi_{B_k}(y)\,d\nu(y)$$
$$= \sum_{k=1}^{\infty} \chi_{A_k}(x)\nu(B_k),$$

form which, by taking the integration with respect to μ and applying the monotone convergence theorem again, we obtain that

$$\lambda(A \times B) = \mu(A) \times \nu(B) = \sum_{k=1}^{\infty} \mu(A_k) \times \nu(B_k),$$

which gives the proof. □

Proposition 8.1.3 Let $(X, \mathcal{A}, \mu)$ and $(Y, \mathcal{B}, \nu)$ be measure spaces. Define a set function π_* on $\mathfrak{M}_0$ by

$$\pi_*\left(\bigcup_{i=1}^{n} A_i \times B_i\right) = \sum_{i=1}^{n} \lambda(A_i \times B_i) = \sum_{i=1}^{n} \mu(A_i) \times \nu(B_i)$$

for any finite disjoint family $\{A_i \times B_i\}_{i=1}^{n} \subset \mathcal{A} \times \mathcal{B}$. Then π_* is a premeasure.

[Proof] It is obvious that $\pi_*(\emptyset) = 0$ and finitely additive. Let $F \in \mathfrak{M}_0$ and let $\{E_k\}_{k=1}^{\infty}$ be a countable family of elements in $\mathfrak{M}_0$ such that

$$F \subset \bigcup_{k=1}^{\infty} E_k.$$

Since $\mathfrak{M}_0$ is an algebra, we may assume that $\{E_k\}_{k=1}^{\infty}$ is a disjoint family. Also, since every element in $\mathfrak{M}_0$ can be represented as a finite disjoint union of measurable rectangles, we can assume that $\{E_k\}_{k=1}^{\infty}$ is a countable disjoint family of measurable rectangles. Also, F can be represented as a finite disjoint union of measurable rectangles, i.e.,

$$F = \bigcup_{i=1}^{\ell} F_i.$$

Then by applying Lemma 8.1.2, we obtain that

$$\pi_*(F) = \sum_{i=1}^{\ell} \lambda(F_i) = \sum_{i=1}^{\ell} \lambda\left(\bigcup_{k=1}^{\infty} F_i \cap E_k\right) = \sum_{i=1}^{\ell}\sum_{k=1}^{\infty} \lambda\left(F_i \cap E_k\right)$$
$$= \sum_{k=1}^{\infty}\sum_{i=1}^{\ell} \lambda\left(F_i \cap E_k\right) = \sum_{k=1}^{\infty} \pi_*\left(F \cap E_k\right)$$

$$\leq \sum_{k=1}^{\infty} \pi_* (E_k),$$

where the first equality is from the definition of π_* and the third equality is by Lemma 8.1.2, which implies that π_* is countably monotone. Hence π_* is a premeasure. □

Remark 8.1.4 In Proposition 8.1.3, if an element $E \in \mathfrak{M}_0$ has any two representations as finite disjoint unions of rectangles, then the two representations have common refinement such that each of the original rectangles is a disjoint union of rectangles in the refinement. Therefore, the set function π_* is well-defined.

Definition 8.1.5 Let $(X, \mathcal{A}, \mu)$ and $(Y, \mathcal{B}, \nu)$ be measure spaces. Then the restriction to $\mathcal{A} \otimes \mathcal{B}$ of the Carathéodory extension (measure) π (on the σ-algebra $\mathfrak{M}$ of all π_*^*-measurable subsets of $X \times Y$) of the premeasure π_* defined as in Proposition 8.1.3 is called the **product measure** and denoted by $\mu \times \nu$.

In fact, the product measure $\mu \times \nu$ of μ and ν is the restriction of the outer measure π_*^* generated by the premeasure π_* to $\mathcal{A} \otimes \mathcal{B}$. By Proposition 2.5.24, it holds that

$$\mathcal{A} \otimes \mathcal{B} = \sigma(\mathfrak{M}_0) \subset \mathfrak{M}.$$

If μ and ν are σ-finite, then π_* is σ-finite. In fact, if

$$X = \bigcup_{i=1}^{\infty} A_i, \qquad Y = \bigcup_{j=1}^{\infty} B_j$$

with $\mu(A_i) < \infty$ and $\nu(B_j) < \infty$, then

$$X \times Y = \bigcup_{i,j=1}^{\infty} A_i \times B_j$$

with $\pi_*(A_i \times B_j) = \mu(A_i)\nu(B_j) < \infty$.

Therefore, if μ and ν are σ-finite, then by (iii) in Theorem 2.5.27, the Carathéodory extension π of the premeasure π_* to the σ-algebra $\mathfrak{M}$ of all π_*^*-measurable subsets of $X \times Y$ defined as in Proposition 8.1.3 is the unique measure on the σ-algebra $\mathfrak{M}$. Moreover, by Corollary 2.5.28, the product measure $\mu \times \nu$ is the unique measure on the σ-algebra $\mathcal{A} \otimes \mathcal{B}$ such that

$$\mu \times \nu(A \times B) = \mu(A)\nu(B)$$

for all $A \in \mathcal{A}$ and $B \in \mathcal{B}$.

8.2 Integration on Product Measure Spaces

Let E be a subset of $X \times Y$. Then for each $x \in X$,

$$E_x = \{y \in Y \mid (x, y) \in E\}$$

is called the x**-section** of E. Similarly, for each $y \in Y$,

$$E^y = \{x \in X \mid (x, y) \in E\}$$

is called the y**-section** of E.

Lemma 8.2.1 Let $(X, \mathcal{A}, \mu)$ and $(Y, \mathcal{B}, \nu)$ be measure spaces. Let E be a subset of $X \times Y$. If $E \in \mathcal{A} \otimes \mathcal{B}$, then $E_x \in \mathcal{B}$ for all $x \in X$ and $E^y \in \mathcal{A}$ for all $y \in Y$.

[Proof] Put

$$\mathcal{E} = \{E \subset X \times Y \mid E_x \in \mathcal{B}, \quad x \in X\}.$$

Then every measurable rectangle belongs to $\mathcal{E}$, in fact, for $A \in \mathcal{A}$ and $B \in \mathcal{B}$, it holds that for any $x \in X$,

$$(A \times B)_x = \begin{cases} B, \, x \in A, \\ \emptyset, \;\; x \notin A \end{cases} \in \mathcal{B}.$$

Also, we can easily see that for any subset E of $X \times Y$ and any sequence $\{E_i\}_{i=1}^{\infty}$ of subsets of $X \times Y$, it holds that

$$(E^c)_x = (E_x)^c, \qquad \left(\bigcup_{i=1}^{\infty} E_i\right)_x = \bigcup_{i=1}^{\infty} (E_i)_x$$

for any $x \in X$. Therefore, $\mathcal{E}$ is a σ-algebra and so $\mathcal{A} \otimes \mathcal{B} \subset \mathcal{E}$ and so for given $E \in \mathcal{A} \otimes \mathcal{B}$, $E_x \in \mathcal{B}$ for all $x \in X$. Similarly, we prove that $E^y \in \mathcal{A}$ for all $y \in Y$. □

Let f be a extended real valued function defined on $X \times Y$. Then for each $x \in X$, the function

$$f_x : Y \ni y \longmapsto f_x(y) = f(x, y) \in \overline{\mathbb{R}}$$

is called the x**-section** of f. Similarly, for each $y \in Y$,

$$f^y : X \ni x \longmapsto f^y(x) = f(x, y) \in \overline{\mathbb{R}}$$

is called the y**-section** of f. Let E be a subset of $X \times Y$. Then it holds that

$$(\chi_E)_x = \chi_{E_x}, \qquad (\chi_E)^y = \chi_{E^y}$$

for any $x \in X$ and $y \in Y$.

Lemma 8.2.2 Let f be an extended real valued function defined on $X \times Y$. If f is $\mathcal{A} \otimes \mathcal{B}$-measurable, then for any $x \in X$, the x-section f_x of f is $\mathcal{B}$-measurable and for any $y \in Y$, the y-section f^y of f is $\mathcal{A}$-measurable.

[Proof] Let $x \in X$, $y \in Y$ and $B \in \mathcal{B}(\mathbb{R})$, Then by applying Lemma 8.2.1, we obtain that

$$(f_x)^{-1}(B) = \{y \in Y \mid f_x(y) \in B\} = \{y \in Y \mid f(x,y) \in B\} = \left(f^{-1}(B)\right)_x \in \mathcal{B},$$
$$(f^y)^{-1}(B) = \{x \in X \mid f^y(x) \in B\} = \{x \in X \mid f(x,y) \in B\} = \left(f^{-1}(B)\right)^y \in \mathcal{A},$$

which implies that f_x is $\mathcal{B}$-measurable and f^y is $\mathcal{A}$-measurable. □

A **monotone class** is a nonempty family $\mathcal{M}$ of subsets of a set X which contains the union of each ascending sequence in $\mathcal{M}$ and the intersection of each descending sequence in $\mathcal{M}$.

Let $\mathcal{C}$ be a family of subsets of X. Then there exists the smallest monotone class containing $\mathcal{C}$, which is called the **monotone class generated by $\mathcal{C}$** and denoted by $\mathcal{M}(\mathcal{C})$.

Theorem 8.2.3 [Monotone Class Theorem] Let $\mathcal{A}$ be an algebra of subsets of a set X. Then the monotone class $\mathcal{M}(\mathcal{A})$ generated by $\mathcal{A}$ coincides with the σ-algebra $\sigma(\mathcal{A})$ generated by $\mathcal{A}$.

[Proof] Since $\sigma(\mathcal{A})$ is a monotone class containing $\mathcal{A}$, $\mathcal{M}(\mathcal{A}) \subset \sigma(\mathcal{A})$ by the minimality of $\mathcal{M}(\mathcal{A})$ (minimal monotone class containing $\mathcal{A}$). We now prove that $\mathcal{M}(\mathcal{A})$ is a σ-algebra. Note that if the monotone class $\mathcal{M}(\mathcal{A})$ is an algebra, then $\mathcal{M}(\mathcal{A})$ is a σ-algebra. Therefore, we now prove that $\mathcal{M}(\mathcal{A})$ is an algebra. For each $E \in \mathcal{M}(\mathcal{A})$, define

$$\mathcal{M}(E) = \{F \in \mathcal{M}(\mathcal{A}) \mid E \setminus F, E \cap F \text{ and } F \setminus E \text{ belong to } \mathcal{M}(\mathcal{A})\}.$$

Then it is easy to see that $\mathcal{M}(E)$ is a monotone class. If $A \in \mathcal{A}$, then for any $B \in \mathcal{A}$,

$$A \cap B, \quad A \setminus B, \quad B \setminus A \in \mathcal{A} \subset \mathcal{M}(\mathcal{A}),$$

which implies that $\mathcal{A} \subset \mathcal{M}(A)$. Therefore, by the minimality of $\mathcal{M}(\mathcal{A})$, $\mathcal{M}(\mathcal{A}) \subset \mathcal{M}(A)$, and so $\mathcal{M}(\mathcal{A}) = \mathcal{M}(A)$ for any $A \in \mathcal{A}$. Hence, if $E \in \mathcal{M}(\mathcal{A})$, then $E \in \mathcal{M}(A)$ $(= \mathcal{M}(\mathcal{A}))$ for any $A \in \mathcal{A}$, and so for any $A \in \mathcal{A}$,

$$A \cap E, \quad A \setminus E, \quad E \setminus A \in \mathcal{M}(\mathcal{A})$$

which implies that $\mathcal{A} \subset \mathcal{M}(E)$. Again, by the minimality of $\mathcal{M}(\mathcal{A})$, we see that $\mathcal{M}(\mathcal{A}) = \mathcal{M}(E)$ for any $E \in \mathcal{M}(\mathcal{A})$. Since $\emptyset, X \in \mathcal{A}$, it is obvious that $\emptyset, X \in \mathcal{M}(\mathcal{A})$. If $E, F \in \mathcal{M}(\mathcal{A})$, then

$$F \in \mathcal{M}(\mathcal{A}) = \mathcal{M}(E),$$

and so

$$F \cap E, \quad F^c \cap E, \quad F \cap E^c \in \mathcal{M}(\mathcal{A}),$$

which implies that $\mathcal{M}(\mathcal{A})$ is an algebra. Hence, $\mathcal{M}(\mathcal{A})$ is a σ-algebra. □

Theorem 8.2.4 Let $(X, \mathcal{A}, \mu)$ and $(Y, \mathcal{B}, \nu)$ be σ-finite measure spaces. Let $E \in \mathcal{A} \otimes \mathcal{B}$ be given. Then we have that

(i) the functions

$$X \ni x \longmapsto \nu(E_x) \in \overline{\mathbb{R}}, \qquad Y \ni y \longmapsto \mu(E^y) \in \overline{\mathbb{R}}$$

are $\mathcal{A}$-measurable and $\mathcal{B}$-measurable, respectively,

(ii) and it holds that

$$\int_X \nu(E_x)\, d\mu(x) = \mu \times \nu(E) = \int_Y \mu(E^y)\, d\nu(y).$$

[Proof] Suppose that μ and ν are finite. Let

$$\mathcal{M} = \{E \in \mathcal{A} \otimes \mathcal{B} \mid \text{ for } E, \text{ (i) and (ii) are satisfied}\}.$$

Then it is obvious that for $E = A \times B$ with $A \in \mathcal{A}$ and $B \in \mathcal{B}$, $E \in \mathcal{M}$. In fact,

$$E_x = \begin{cases} B, & x \in A, \\ \emptyset, & x \notin A \end{cases} \in \mathcal{B},$$

which implies that the function

$$X \ni x \longmapsto \nu(E_x) = \begin{cases} \nu(B), & x \in A \\ 0, & x \notin A \end{cases}$$

coincides with $\nu(B)\chi_A$ and is $\mathcal{A}$-measurable, and

$$\int_X \nu(E_x)\, d\mu(x) = \mu(A)\nu(B) = \mu \times \nu(E).$$

Similarly, we prove that the function $Y \ni y \longmapsto \mu(E^y) \in \mathbb{R}$ is $\mathcal{B}$-measurable and $\mu \times \nu(E) = \int_Y \mu(E^y)\, d\nu(y)$. Therefore, $E = A \times B \in \mathcal{M}$. Since every element in $\mathfrak{M}_0$ can be represented as a finite disjoint union of (measurable) rectangles, $\mathfrak{M}_0 \subset \mathcal{M}$. We now prove that $\mathcal{M}$ is a monotone class. Let $\{E_n\}_{n=1}^{\infty}$ is an ascending sequence of elements in $\mathcal{M}$. For each $n \in \mathbb{N}$, put

$$f_n(x) = \nu((E_n)_x), \qquad g_n(y) = \mu((E_n)^y), \qquad x \in X, \quad y \in Y.$$

Then since $E_n \in \mathcal{M}$, f_n and g_n are measurable and

$$\mu \times \nu(E_n) = \int_X f_n(x)\, d\mu(x) = \int_Y g_n(y)\, d\nu(y).$$

Since $\{(E_n)_x\}_{n=1}^{\infty}$ is an ascending sequence (for each $x \in X$), by the continuity of measure (Theorem 2.3.15), we obtain that

$$\lim_{n\to\infty} f_n(x) = \lim_{n\to\infty} \nu((E_n)_x) = \nu\left(\bigcup_{n=1}^{\infty} (E_n)_x\right)$$

$$= \nu\left(\left(\bigcup_{n=1}^{\infty} E_n\right)_x\right),$$

which implies that the function $x \mapsto \nu\left(\left(\cup_{n=1}^{\infty} E_n\right)_x\right)$ is $\mathcal{A}$-measurable. Similarly, we see that the function $y \mapsto \mu\left(\left(\cup_{n=1}^{\infty} E_n\right)^y\right)$ is $\mathcal{B}$-measurable. Also, by applying the monotone convergence theorem (Theorem 5.1.15) and the continuity of measure (Theorem 2.3.15), we obtain that

$$\begin{aligned}\int_X \nu\left(\left(\bigcup_{n=1}^{\infty} E_n\right)_x\right) d\mu(x) &= \int_X \lim_{n\to\infty} \nu((E_n)_x)\, d\mu(x) = \lim_{n\to\infty} \int_X \nu((E_n)_x)\, d\mu(x)\\ &= \lim_{n\to\infty} \mu\times\nu(E_n)\\ &= \mu\times\nu\left(\bigcup_{n=1}^{\infty} E_n\right).\end{aligned}$$

By same arguments, we see that

$$\int_Y \mu\left(\left(\bigcup_{n=1}^{\infty} E_n\right)^y\right) d\nu(y) = \mu\times\nu\left(\bigcup_{n=1}^{\infty} E_n\right).$$

Therefore, $\cup_{n=1}^{\infty} E_n \in \mathcal{M}$. Similarly, we prove that $\mathcal{M}$ is closed under the intersection of descending sequence for which the finiteness (σ-finiteness) of μ and ν are essential. Hence $\mathcal{M}$ is a monotone class. Then by the monotone class theorem (Theorem 8.2.3), $\mathcal{M}$ is a σ-algebra and so $\mathcal{M} = \mathcal{A}\otimes\mathcal{B}$. Therefore, for any $E \in \mathcal{A}\otimes\mathcal{B}$, (i) and (ii) hold.

Finally, if μ and ν are σ-finite, then $\mu\times\nu$ is σ-finite. Therefore, there exists an increasing sequence $\{Z_n\}_{n=1}^{\infty}$ of rectangles with $\mu\times\nu(Z_n) < \infty$ such that $X\times Y = \cup_{n=1}^{\infty} Z_n$. Then for each $E \in \mathcal{A}\otimes\mathcal{B}$, and $x \in X$, $y \in Y$, by the continuity of measure (Theorem 2.3.15), we obtain that

$$\nu(E_x) = \nu\left(\left(\bigcup_{n=1}^{\infty} E\cap Z_n\right)_x\right) = \nu\left(\bigcup_{n=1}^{\infty} E_x\cap (Z_n)_x\right) = \lim_{n\to\infty} \nu\left(E_x\cap (Z_n)_x\right),$$

which implies that the function $x \mapsto \nu(E_x)$ is $\mathcal{A}$-measurable. Similarly, we see that the function $y \mapsto \mu(E^y)$ is $\mathcal{B}$-measurable. Also, by applying the monotone convergence theorem (Theorem 5.1.15) and the continuity of measure (Theorem 2.3.15), we obtain that

$$\begin{aligned}\int_X \nu(E_x)\, d\mu(x) &= \int_X \lim_{n\to\infty} \nu\left(E_x\cap (Z_n)_x\right) d\mu(x) = \lim_{n\to\infty} \int_X \nu\left(E_x\cap (Z_n)_x\right) d\mu(x)\\ &= \lim_{n\to\infty} \mu\times\nu(E\cap Z_n)\\ &= \mu\times\nu(E).\end{aligned}$$

By same arguments, we see that

$$\int_Y \mu(E^y)\, d\nu(y) = \mu\times\nu(E).$$

Therefore, we complete the proof. □

Example 8.2.5 Let $X = Y = [0,1]$ and $\mathcal{A} = \mathcal{B} = \mathcal{B}([0,1])$. Let μ be the Lebesgue measure on $\mathcal{A}$ and ν is the counting measure on $\mathcal{B}$. Let ϕ be a function on $X \times Y$ into $\mathbb{R}$ defined by

$$\phi(x,y) = x - y.$$

Then ϕ is continuous and so

$$D = \{(x,y) \in X \times Y \,|\, x = y\} = \phi^{-1}(\{0\})$$

is a measurable set, i.e., a Borel set. Then we obtain that

$$\begin{aligned}\int_X \nu(D_x)\, d\mu(x) &= \int_X 1\, d\mu(x) = 1\\ &\neq 0 = \int_X 0\, d\nu(y) = \int_Y \mu(D_y)\, d\nu(y).\end{aligned}$$

Therefore, Theorem 8.2.4 fails if we drop the hypothesis that μ and ν are σ-finite.

Theorem 8.2.6 **[Tonelli's Theorem]** Let $(X, \mathcal{A}, \mu)$ and $(Y, \mathcal{B}, \nu)$ be σ-finite measure spaces and let $f : X \times Y \to \overline{\mathbb{R}}$ a nonnegative $\mathcal{A} \otimes \mathcal{B}$-measurable function. Then the functions g and h defined on X and Y, respectively, defined by

$$g(x) = \int_Y f_x(y)\, d\nu(y), \qquad h(y) = \int_X f^y(x)\, d\mu(x)$$

are measurable and

$$\int_X g\, d\mu = \int_{X\times Y} f\, d(\mu \times \nu) = \int_Y h\, d\nu.$$

[Proof] By Lemma 8.2.2, f_x and f^y are measurable and so the functions g and h are well-defined. If f is a characteristic function, then the assertion follows from Theorem 8.2.4. Also, if f is a simple function, then, by the linearity, the assertion holds. If f is an arbitrary nonnegative function on $X \times Y$ to $\overline{\mathbb{R}}$, then, by the simple approximation theorem (Theorem 4.1.37), there exists an increasing sequence $\{\phi_n\}$ of nonnegative simple functions defined on X such that $\{\phi_n\}$ converges to f pointwise on X. For each $n \in \mathbb{N}$, put

$$\begin{aligned} g_n(x) &= \int_Y (\phi_n)_x(y)\, d\nu(y), \qquad x \in X,\\ h_n(y) &= \int_X (\phi_n)^y(x)\, d\mu(x), \qquad y \in Y.\end{aligned}$$

Then $\{g_n\}_{n=1}^{\infty}$ and $\{h_n\}_{n=1}^{\infty}$ are increasing sequence of nonnegative measurable functions. Therefore, by applying the monotone convergence theorem (Theorem 5.1.15), we see that $\{g_n\}$ and $\{h_n\}$ converges to g and h pointwise on X and Y, respectively. Therefore, g and

h are measurable and by applying the monotone convergence theorem (Theorem 5.1.15) again, we obtain that

$$\begin{aligned}\int_X g(x)\,d\mu(x) &= \lim_{n\to\infty}\int_X g_n(x)\,d\mu(x) = \lim_{n\to\infty}\int_{X\times Y}\phi_n(x,y)\,d(\mu\times\nu)(x,y)\\ &= \lim_{n\to\infty}\int_Y h_n(y)\,d\nu(y) = \int_X h(y)\,d\nu(y).\end{aligned}$$

Also, by applying the monotone convergence theorem (Theorem 5.1.15), we prove that

$$\lim_{n\to\infty}\int_{X\times Y}\phi_n(x,y)\,d(\mu\times\nu)(x,y) = \int_{X\times Y} f\,d\mu\times\nu.$$

Hence we complete the proof. □

Let $(X,\mathcal{A},\mu)$ and $(Y,\mathcal{B},\nu)$ be σ-finite measure spaces and let $f : X\times Y \to \overline{\mathbb{R}}$ an integrable function over $X\times Y$ with respect to product measure $\mu\times\nu$. Then by definition, it holds that $\int_{X\times Y}|f|\,d\mu\times\nu < \infty$, and by Tonelli's Theorem, it holds that

$$\begin{aligned}\int_X\left(\int_Y |f(x,y)|\,d\nu(y)\right)d\mu(x) &= \int_{X\times Y}|f|\,d(\mu\times\nu)\\ &= \int_Y\left(\int_X |f(x,y)|\,d\mu(x)\right)d\nu(y).\end{aligned}$$

Therefore, for almost all $x\in X$ and almost all $y\in Y$, the integrals

$$\int_Y |f(x,y)|\,d\nu(y), \qquad \int_X |f(x,y)|\,d\mu(x)$$

are finite, i.e., $x\in X$ and almost all $y\in Y$, the x-section f_x and y-section f^y of f are integrable.

Theorem 8.2.7 [**Fubini's Theorem**] Let $(X,\mathcal{A},\mu)$ and $(Y,\mathcal{B},\nu)$ be σ-finite measure spaces and let $f : X\times Y\to\overline{\mathbb{R}}$ an integrable function over $X\times Y$ with respect to the product measure $\mu\times\nu$. Then the extended real-valued functions g and h defined almost everywhere by

$$\begin{aligned}g(x) &= \int_Y f_x(y)\,d\nu(y),\\ h(y) &= \int_X f^y(x)\,d\mu(x)\end{aligned}$$

are integrable and

$$\int_X g\,d\mu = \int_{X\times Y} f\,d(\mu\times\nu) = \int_Y h\,d\nu.$$

[Proof] Since f is integrable over $\mu \times \nu$, f^+ and f^- are integrable. Then by Tonelli's Theorem (Theorem 8.2.6), the functions

$$X \ni x \longmapsto \int_Y f_x^{\pm}(y)\, d\nu(y)$$

are integrable and so finite almost everywhere, and

$$\int_X \int_Y f_x^{\pm}(y)\, d\nu(y)\, d\mu(x) = \int_{X\times Y} f^{\pm} d(\mu \times \nu).$$

Therefore, the function

$$\begin{aligned} X \ni x \longmapsto g(x) &= \int_Y f_x(y)\, d\nu(y) \\ &:= \int_Y f_x^+(y)\, d\nu(y) - \int_Y f_x^-(y)\, d\nu(y) \end{aligned}$$

is defined almost everywhere, and we obtain that

$$\begin{aligned} \int_{X\times Y} f\, d(\mu \times \nu) &= \int_{X\times Y} f^+\, d(\mu \times \nu) - \int_{X\times Y} f^-\, d(\mu \times \nu) \\ &= \int_X \int_Y f_x^+(y)\, d\nu(y)\, d\mu(x) - \int_X \int_Y f_x^-(y)\, d\nu(y)\, d\mu(x) \\ &= \int_X g\, d\mu, \end{aligned}$$

which gives the proof of the first part, and the proof of the second part is similar. □

Exercises

8.1 Let $(X, \mathcal{A})$ and $(Y, \mathcal{B})$ be measurable spaces. Let $\mathfrak{M}_0$ be the family of all finite unions of measurable rectangles.

(1) Give an example that $\mathcal{A} \times \mathcal{B} = \{A \times B \mid A \in \mathcal{A},\ \ B \in \mathcal{B}\}$ is not a σ-algebra.

(2) Prove that the smallest algebra containing all measurable rectangles of $X \times Y$ is $\mathfrak{M}_0$.

(3) Prove that every element in $\mathfrak{M}_0$ can be represented as a finite disjoint union of measurable rectangles.

8.2 Prove that every σ-algebra is a monotone class.

8.3 Prove that the intersection of any family of monotone classes is a monotone class.

8.4 Let X be a set and $\mathcal{C}$ a nonempty family of subsets of X. Prove that there exists the smallest monotone class containing $\mathcal{C}$.

8.5 Let X be a set and $\mathcal{S}$ a nonempty family of subsets of X. Prove that the σ-algebra $\sigma(\mathcal{S})$ contains the monotone class generated by $\mathcal{S}$.

8.6 Let $f : A = \mathbb{R} \times (0, 1] \to \mathbb{R}$ be a function defined by

$$f(x, y) = e^{-\frac{x^2}{2y^2}}, \qquad (x, y) \in A.$$

Compute the integral

$$\int_A f \, d(m \times m),$$

where $m \times m$ is the product measure of the Lebesgue measures m and m on $[0, 1]$ and $\mathbb{R}$, respectively.

8.7 Prove that the Tonelli's Theorem fails if we drop the hypothesis that μ and ν are σ-finite.

Chapter 9

Convergence of Measurable Functions

We study relations between various notions of convergence of measurable functions. Let $(X, \mathcal{M}, \mu)$ be a measure space.

9.1 Convergence in Measure

Definition 9.1.1 Let $\{f_n\}_{n=1}^{\infty}$ be a sequence of measurable ($\mathcal{M}$-measurable) functions and f be a measurable function. Then we say that

(i) $\{f_n\}_{n=1}^{\infty}$ **converges to** f **in measure** if for any $\epsilon > 0$,

$$\lim_{n \to \infty} \mu\left(\{x \in X \mid |f_n(x) - f(x)| > \epsilon\}\right) = 0,$$

(ii) $\{f_n\}_{n=1}^{\infty}$ is **Cauchy in measure** if for any $\epsilon > 0$,

$$\lim_{m,n \to \infty} \mu\left(\{x \in X \mid |f_n(x) - f_m(x)| > \epsilon\}\right) = 0.$$

Let $\{f_n\}_{n=1}^{\infty}$ be a sequence of measurable ($\mathcal{M}$-measurable) functions and f be a measurable function.

Theorem 9.1.2 If a sequence of measurable functions converges in measure, then the limit function is unique a.e.

[Proof] Let $\{f_n\}_{n=1}^{\infty}$ be a sequence of measurable functions and f, g be measurable functions. Suppose that $\{f_n\}_{n=1}^{\infty}$ converges to f and g in measure. Then for any $\epsilon > 0$, we obtain that

$$\mu(\{x \in X \,|\, |f(x) - g(x)| > \epsilon\}) \leq \mu(\{x \in X \,|\, |f(x) - f_n(x)| > \epsilon/2\}) \\ + \mu(\{x \in X \,|\, |f_n(x) - g(x)| > \epsilon/2\})$$

for any $n \in \mathbb{N}$. Therefore, by taking limit as $n \to \infty$, we see that

$$\mu(\{x \in X \,|\, |f(x) - g(x)| > \epsilon\}) = 0$$

for any $\epsilon > 0$. Hence by the continuity of measure, we see that $f = g$ a.e. □

Theorem 9.1.3 Let $\{f_n\}_{n=1}^{\infty}$ be a sequence of measurable functions on X. If $\{f_n\}_{n=1}^{\infty}$ is a Cauchy sequence in measure, then there exist a subsequence $\{f_{n_k}\}_{k=1}^{\infty}$ of $\{f_n\}_{n=1}^{\infty}$ and a measurable function f on X such that

(i) $\{f_{n_k}\}_{k=1}^{\infty}$ converges pointwise a.e. on X to f,

(ii) $\{f_{n_k}\}_{k=1}^{\infty}$ converges to f in measure.

(iii) $\{f_n\}_{n=1}^{\infty}$ converges to f in measure.

[Proof] Since $\{f_n\}_{n=1}^{\infty}$ is a Cauchy sequence in measure, we assign a monotone increasing sequence $\{n_k\}_{k=1}^{\infty}$ such that for each $k \in \mathbb{N}$ and for any $l, m \geq n_k$,

$$\mu\left(\left\{x \in X \,\middle|\, |f_l(x) - f_m(x)| \geq \frac{1}{2^k}\right\}\right) < \frac{1}{2^k}.$$

For each $k \in \mathbb{N}$, put

$$B_k = \left\{x \in X \,\middle|\, \left|f_{n_{k+1}}(x) - f_{n_k}(x)\right| \geq \frac{1}{2^k}\right\},$$

and

$$B = \limsup_{k\to\infty} B_k = \bigcap_{k=1}^{\infty} \bigcup_{m=k}^{\infty} B_m.$$

Then by the countable subadditivity of μ, we obtain that

$$\mu\left(\bigcup_{m=1}^{\infty} B_m\right) \leq \sum_{m=1}^{\infty} \mu(B_m) \leq \sum_{m=1}^{\infty} \frac{1}{2^m} = 1,$$

and so by the continuity of μ, we obtain that

$$\mu(B) = \lim_{k\to\infty} \mu\left(\bigcup_{m=k}^{\infty} B_m\right) \leq \lim_{k\to\infty} \sum_{m=k}^{\infty} \frac{1}{2^m} = \lim_{k\to\infty} \frac{1}{2^{k-1}} = 0,$$

which implies that $\mu(B) = 0$. On the other hand, for each

$$x \in X \setminus B = \bigcup_{k=1}^{\infty} \bigcap_{m=k}^{\infty} B_m^{\mathrm{c}},$$

there exists $k \in \mathbb{N}$ such that $x \in \bigcap_{m=k}^{\infty} B_m^{\mathrm{c}}$ and so for any $m \geq k$,

$$x \in B_m^{\mathrm{c}} = \left\{ x \in X \,\middle|\, \left| f_{n_{m+1}}(x) - f_{n_m}(x) \right| < \frac{1}{2^m} \right\}.$$

Then for any $m > l \geq N \geq k$, we obtain that

$$\begin{aligned} |f_{n_m}(x) - f_{n_l}(x)| &= \left| \sum_{i=l}^{m-1} \left(f_{n_{i+1}}(x) - f_{n_i}(x) \right) \right| \leq \sum_{i=l}^{m-1} \left| f_{n_{i+1}}(x) - f_{n_i}(x) \right| \\ &\leq \sum_{i=l}^{m-1} \frac{1}{2^i} < \frac{1}{2^{l-1}} \leq \frac{1}{2^{N-1}}, \end{aligned} \tag{9.1}$$

which implies that $\{f_{n_k}(x)\}_{k=1}^{\infty}$ is a Cauchy sequence in $\mathbb{R}$. Therefore, there exists a measurable function f such that $\{f_{n_k}(x)\}_{k=1}^{\infty}$ converges to $f(x)$ for any $x \in X \setminus B$, i.e., $\{f_{n_k}\}_{k=1}^{\infty}$ converges pointwise a.e. on X to f, which proved (i). Moreover, from (9.1), for any $x \in \bigcap_{m=k}^{\infty} B_m^{\mathrm{c}}$ and any $m > l \geq N \geq k$, we have

$$|f_{n_m}(x) - f_{n_l}(x)| \leq \frac{1}{2^{N-1}},$$

which implies that $\{f_{n_l}\}_{l=1}^{\infty}$ is uniformly Cauchy on $\bigcap_{m=k}^{\infty} B_m^{\mathrm{c}}$ for each $k \in \mathbb{N}$. Moreover, for any $x \in \bigcap_{m=k}^{\infty} B_m^{\mathrm{c}}$ and any $m \geq N \geq k$, we obtain that

$$|f_{n_m}(x) - f(x)| = \lim_{l \to \infty} |f_{n_m}(x) - f_{n_l}(x)| \leq \frac{1}{2^{N-1}},$$

which implies that $\{f_{n_l}\}_{l=1}^{\infty}$ converges uniformly to f on $E_k := \bigcap_{m=k}^{\infty} B_m^{\mathrm{c}}$ for each $k \in \mathbb{N}$. Let $\epsilon > 0$ be given. Then for any $k \in \mathbb{N}$, we obtain that

$$\begin{aligned} \lim_{l \to \infty} \mu\left(\{x \in X \mid |f_{n_l}(x) - f(x)| > \epsilon\}\right) &= \lim_{l \to \infty} \mu\left(\{x \in E_k \mid |f_{n_l}(x) - f(x)| > \epsilon\}\right) \\ &\quad + \lim_{l \to \infty} \mu\left(\{x \in E_k^{\mathrm{c}} \mid |f_{n_l}(x) - f(x)| > \epsilon\}\right) \\ &\leq 0 + \mu\left(\bigcup_{m=k}^{\infty} B_m \right) \leq \frac{1}{2^{k-1}}, \end{aligned}$$

which implies that $\{f_{n_l}\}_{l=1}^{\infty}$ converges to f in measure, which proved (ii). Also, we obtain that

$$\begin{aligned} \mu\left(\{x \in X \mid |f_n(x) - f(x)| > \epsilon\}\right) &\leq \mu\left(\{x \in X \mid |f_n(x) - f_{n_l}(x)| > \epsilon/2\}\right) \\ &\quad + \mu\left(\{x \in X \mid |f_{n_l}(x) - f(x)| > \epsilon/2\}\right), \end{aligned}$$

where the right hand side of the inequality converges to 0 as $n, l \to \infty$ and so the left hand side converges to 0 as $n \to \infty$. Hence $\{f_n\}_{n=1}^{\infty}$ converges to f in measure, which proved (iii). □

Corollary 9.1.4 Let $\{f_n\}_{n=1}^{\infty}$ be a sequence of measurable functions. If $\{f_n\}_{n=1}^{\infty}$ converges to a measurable function f in measure, there exists a subsequence $\{f_{n_k}\}_{k=1}^{\infty}$ of $\{f_n\}_{n=1}^{\infty}$ such that $\{f_{n_k}\}_{k=1}^{\infty}$ converges pointwise a.e. on X to f.

[Proof] Since $\{f_n\}_{n=1}^{\infty}$ converges to f in measure, $\{f_n\}_{n=1}^{\infty}$ is a Cauchy sequence in measure. Therefore, by (i) and (ii) in Theorem 9.1.3, there exist a subsequence $\{f_{n_k}\}_{k=1}^{\infty}$ of $\{f_n\}_{n=1}^{\infty}$ and a measurable function g such that $\{f_{n_k}\}_{k=1}^{\infty}$ converges pointwise a.e. on X to g and $\{f_{n_k}\}_{n=1}^{\infty}$ converges to g in measure. Then by Theorem 9.1.2, $g = f$ a.e. Hence $\{f_{n_k}\}_{k=1}^{\infty}$ converges pointwise a.e. on X to f. □

In general, the pointwise convergence does not imply the convergence in measure. For each $n \in \mathbb{N}$, we put

$$f_n = \chi_{[n,n+1]}.$$

Then $\{f_n\}_{n=1}^{\infty}$ converges pointwise on $\mathbb{R}$ to 0, but for any $1 > \epsilon > 0$, we have

$$m(\{x \in \mathbb{R} \mid |f_n(x)| > \epsilon\}) = 1,$$

where m is the Lebesgue measure, and so $\{f_n\}_{n=1}^{\infty}$ does not converge to 0 in measure. However, if μ is a finite measure, i.e., $\mu(X) < \infty$, then the pointwise convergence implies the convergence in measure.

Theorem 9.1.5 Let $\{f_n\}_{n=1}^{\infty}$ be a sequence of measurable functions. Suppose that $\mu(X) < \infty$. If $\{f_n\}_{n=1}^{\infty}$ converges pointwise a.e. on X to a measurable function f, then $\{f_n\}_{n=1}^{\infty}$ converges to f in measure.

[Proof] Let $\epsilon > 0$ be given. Then for any $\lambda > 0$, by the Egoroff's theorem (Theorem 4.1.38), there exists a measurable subset $E \in \mathcal{M}$ with $\mu(E) < \lambda$ such that $\{f_n\}_{n=1}^{\infty}$ converges uniformly on E^c to f. Therefore, we obtain that

$$\begin{aligned}\lim_{n\to\infty} \mu(\{x \in X \mid |f_n(x) - f(x)| > \epsilon\}) &= \lim_{n\to\infty} \mu(\{x \in E^c \mid |f_n(x) - f(x)| > \epsilon\}) \\ &\quad + \lim_{n\to\infty} \mu(\{x \in E \mid |f_n(x) - f(x)| > \epsilon\}) \\ &\leq \mu(E) < \lambda\end{aligned}$$

for arbitrary $\lambda > 0$, and so $\lim_{n\to\infty} \mu(\{x \in X \mid |f_n(x) - f(x)| > \epsilon\}) = 0$. Hence, $\{f_n\}_{n=1}^{\infty}$ converges to f in measure. □

Example 9.1.6 Consider the sequence $\{f_n\}_{n=1}^{\infty}$ of measurable functions f_n defined by $f_n = \chi_{[n,\infty)}$. Then $\{f_n\}_{n=1}^{\infty}$ converges pointwise to 0, but for $0 < \epsilon < 1$, we have

$$m(\{x \in \mathbb{R} \mid |f_n(x)| > \epsilon\}) = \infty$$

for all $n \in \mathbb{N}$, where m is the Lebesgue measure. Therefore, $\{f_n\}_{n=1}^{\infty}$ does not converge 0 in measure. Hence, in Theorem 9.1.5, if we drop the condition that $\mu(X) < \infty$, the statement is not true.

Theorem 9.1.7 Let $\{f_n\}_{n=1}^{\infty}$ be a sequence of measurable functions and $h : \mathbb{R} \to \mathbb{R}$ be continuous. Suppose that $\mu(X) < \infty$. If $\{f_n\}_{n=1}^{\infty}$ converges to a measurable function f in measure, then $\{h(f_n)\}_{n=1}^{\infty}$ converges to $h(f)$ in measure.

[Proof] Let $\epsilon > 0$ be given. Since $\{f_n\}_{n=1}^{\infty}$ converges to f in measure, f is finite a.e. Therefore, we have

$$\begin{aligned} 0 = \mu(\{x \in X \mid |f(x)| = \infty\}) &= \mu\left(\bigcap_{k=1}^{\infty} \{x \in X \mid |f(x)| > k\}\right) \\ &= \lim_{k\to\infty} \mu(\{x \in X \mid |f(x)| > k\}), \end{aligned}$$

and so for any $\lambda > 0$ there exists $k \in \mathbb{N}$ such that

$$\mu(\{x \in X \mid |f(x)| > k\}) < \lambda.$$

Since h is continuous, h is uniformly continuous on $[-k, k]$, and so there exists $\delta > 0$ such that $|h(x) - h(y)| < \epsilon$ if $|x - y| < \delta$ for any $x, y \in [-k, k]$. Therefore, we obtain that

$$\begin{aligned} &\{x \in X \mid |h(f_n(x)) - h(f(x))| \geq \epsilon\} \\ &\quad \subset \{x \in X \mid |h(f_n(x)) - h(f(x))| \geq \epsilon,\ |f(x)| \leq k\} \bigcup \{x \in X \mid |f(x)| > k\} \\ &\quad \subset \{x \in X \mid |f_n(x) - f(x)| \geq \delta,\ |f(x)| \leq k\} \bigcup \{x \in X \mid |f(x)| > k\} \\ &\quad \subset \{x \in X \mid |f_n(x) - f(x)| \geq \delta\} \bigcup \{x \in X \mid |f(x)| > k\}, \end{aligned}$$

and so we obtain that

$$\begin{aligned} \mu(\{x \in X \mid |h(f_n(x)) - h(f(x))| \geq \epsilon\}) &\leq \mu(\{x \in X \mid |f_n(x) - f(x)| \geq \delta\}) \\ &\quad + \mu(\{x \in X \mid |f(x)| > k\}) \\ &\leq \mu(\{x \in X \mid |f_n(x) - f(x)| \geq \delta\}) + \lambda, \end{aligned}$$

which implies that $\lim_{n\to\infty} \mu(\{x \in X \mid |h(f_n(x)) - h(f(x))| \geq \epsilon\}) \leq \lambda$ for arbitrary $\lambda > 0$. Hence we have $\lim_{n\to\infty} \mu(\{x \in X \mid |h(f_n(x)) - h(f(x))| \geq \epsilon\}) = 0$ and $\{h(f_n)\}_{n=1}^{\infty}$ converges to $h(f)$ in measure. □

9.2 Almost Uniform Convergence

Let $(X, \mathcal{M}, \mu)$ be a measure space.

Definition 9.2.1 Let $\{f_n\}_{n=1}^{\infty}$ be a sequence of measurable functions and f a measurable function. Then we say that $\{f_n\}_{n=1}^{\infty}$ converges to f **almost uniformly** if for any $\epsilon > 0$ there exists $E \in \mathcal{M}$ such that $\mu(E) < \epsilon$ and $\{f_n\}_{n=1}^{\infty}$ converges uniformly on E^c to f.

Example 9.2.2 Consider the measure space $([0,1], \mathfrak{M}([0,1]), m)$ and the sequence $\{f_n\}_{n=1}^{\infty}$ of functions f_n defined on $[0,1]$ by $f_n(x) = x^n$, where $\mathfrak{M}([0,1])$ is the σ-algebra of Lebesgue measurable subsets of $[0,1]$ and m is the Lebesgue measure. Then for given $0 < \epsilon < 1$ and any $x \in [0, 1-\epsilon]$, $x^n \leq (1-\epsilon)^n$ for all $n \in \mathbb{N}$. Therefore, $\{f_n\}_{n=1}^{\infty}$ converges to 0 (on $[0,1]$) almost uniformly.

More generally, we have the following theorem, for the proof, we refer to the Egoroff's Theorem (Theorem 4.1.38).

Theorem 9.2.3 (**Egoroff's Theorem**) Let $\{f_n\}_{n=1}^{\infty}$ be a sequence of measurable functions and f a measurable function. Suppose that $\mu(X) < \infty$. If $\{f_n\}_{n=1}^{\infty}$ converges pointwise a.e. on X to f, then $\{f_n\}_{n=1}^{\infty}$ converges to f almost uniformly.

Theorem 9.2.4 Let $\{f_n\}_{n=1}^{\infty}$ be a sequence of measurable functions on X. If $\{f_n\}_{n=1}^{\infty}$ is a Cauchy sequence in measure, then there exist a subsequence $\{f_{n_k}\}_{k=1}^{\infty}$ of $\{f_n\}_{n=1}^{\infty}$ and a measurable function f on X such that $\{f_{n_k}\}_{k=1}^{\infty}$ converges to f almost uniformly.

[Proof] Let $\{f_{n_k}\}_{k=1}^{\infty}$, f and $\{B_k\}_{k=1}^{\infty}$ be a subsequence of $\{f_n\}_{n=1}^{\infty}$, a measurable function and a sequence of measurable sets, respectively, constructed in the proof of Theorem 9.1.3. Then $\{f_{n_l}\}_{l=1}^{\infty}$ converges uniformly to f on $\bigcap_{m=k}^{\infty} B_m^{\mathrm{c}} = \left(\bigcup_{m=k}^{\infty} B_m\right)^{\mathrm{c}}$ for each $k \in \mathbb{N}$, and

$$\mu\left(\bigcup_{m=k}^{\infty} B_m\right) \leq \sum_{m=k}^{\infty} \frac{1}{2^m} = \frac{1}{2^{k-1}}.$$

Therefore, for given $\epsilon > 0$, there exists $k \in \mathbb{N}$ such that $2^{1-k} < \epsilon$ such that for $E = \bigcup_{m=k}^{\infty} B_m$, $\mu(E) < \epsilon$ and $\{f_{n_l}\}_{l=1}^{\infty}$ converges uniformly to f on E^{c}. Hence $\{f_{n_l}\}_{l=1}^{\infty}$ converges to f almost uniformly. □

Corollary 9.2.5 Let $\{f_n\}_{n=1}^{\infty}$ be a sequence of measurable functions. If $\{f_n\}_{n=1}^{\infty}$ converges to a measurable function f in measure, there exists a subsequence $\{f_{n_k}\}_{k=1}^{\infty}$ of $\{f_n\}_{n=1}^{\infty}$ such that $\{f_{n_k}\}_{k=1}^{\infty}$ converges to f almost uniformly.

[Proof] Since $\{f_n\}_{n=1}^{\infty}$ converges to f in measure, $\{f_n\}_{n=1}^{\infty}$ is a Cauchy sequence in measure (see Exercise **9.2**). Therefore, the proof is immediate from Theorem 9.2.4. □

Theorem 9.2.6 Let $\{f_n\}_{n=1}^{\infty}$ be a sequence of measurable functions and f a measurable function. If $\{f_n\}_{n=1}^{\infty}$ converges to f almost uniformly, then

(i) $\{f_n\}_{n=1}^{\infty}$ converges pointwise a.e. on X to f,

(ii) $\{f_n\}_{n=1}^{\infty}$ converges to f in measure.

[Proof] (i) For each $k \in \mathbb{N}$, there exists a measurable set $E_k \in \mathcal{M}$ such that $\mu(E_k) < 1/k$ and $\{f_n\}_{n=1}^{\infty}$ converges uniformly on E_k^c to f. Put

$$E = \bigcap_{k=1}^{\infty} E_k.$$

Then $\{f_n\}_{n=1}^{\infty}$ converges pointwise on E^c to f and $\mu(E_1) < 1$, and so by the continuity of measure, we have

$$\mu(E) = \lim_{k\to\infty} \mu(E_k) = 0.$$

Therefore, $\{f_n\}_{n=1}^{\infty}$ converges pointwise a.e on X to f.

(ii) Let $\epsilon > 0$ be given. For arbitrary $\lambda > 0$, there exists $E_\lambda \in \mathcal{M}$ such that $\mu(E_\lambda) < \lambda$ and $\{f_n\}_{n=1}^{\infty}$ converges uniformly on E_λ^c to f. Therefore, we obtain that

$$\begin{aligned}\lim_{n\to\infty} \mu\left(\{x \in X \,|\, |f_n(x) - f(x)| > \epsilon\}\right) &= \lim_{n\to\infty} \mu\left(\{x \in E_\lambda^c \,|\, |f_n(x) - f(x)| > \epsilon\}\right) \\ &\quad + \lim_{n\to\infty} \mu\left(\{x \in\in E_\lambda \,|\, |f_n(x) - f(x)| > \epsilon\}\right) \\ &\le 0 + \mu(E_\lambda) < \lambda\end{aligned}$$

for arbitrary $\lambda > 0$, which implies that

$$\lim_{n\to\infty} \mu\left(\{x \in X \,|\, |f_n(x) - f(x)| > \epsilon\}\right) = 0$$

and so $\{f_n\}_{n=1}^{\infty}$ converges to f in measure. □

9.3 Convergence in Mean

Let $(X, \mathcal{M}, \mu)$ be a measure space and let $1 \le p < \infty$.

Definition 9.3.1 Let $\{f_n\}_{n=1}^{\infty}$ be a sequence of measurable functions and f a measurable function. Then we say that $\{f_n\}_{n=1}^{\infty}$ converges to f **in the mean of order** p if

$$\lim_{n\to\infty} \|f_n - f\|_p = \lim_{n\to\infty} \left(\int_X |f_n - f|^p \, d\mu \right)^{1/p} = 0.$$

If $p = 1$, then $\{f_n\}_{n=1}^{\infty}$ is said to converge to f **in the mean**.

Example 9.3.2 Let $([0,1], \mathfrak{M}([0,1]), m)$ be a measure space with the Lebesgue measure m and let $f_n = n\chi_{[0,1/n]}$ for all $n \in \mathbb{N}$. Then it easy to see that $\{f_n\}_{n=1}^{\infty}$ converges pointwise on $[0,1]$ to 0 and for any given $\epsilon > 0$, we have

$$m\left(\{x \in [0,1] \,|\, |f_n(x)| > \epsilon\}\right) = \frac{1}{n} \to 0$$

as $n \to \infty$. Therefore, $\{f_n\}_{n=1}^{\infty}$ converges to 0 in measure (see Theorem 9.1.5). But, we obtain that

$$\|f_n\|_p^p = \int_0^1 \left(n\chi_{[0,1/n]}\right)^p dm = n^p \int_0^{1/n} 1dm = n^{p-1}.$$

Hence, $\{f_n\}_{n=1}^{\infty}$ does not converge to 0 in the mean of order p.

Theorem 9.3.3 Let $\{f_n\}_{n=1}^{\infty}$ be a sequence of measurable functions. If $\{f_n\}_{n=1}^{\infty}$ converges to a measurable function f in the mean of order p, then $\{f_n\}_{n=1}^{\infty}$ converges to f in measure.

[Proof] Let $\epsilon > 0$ be given. For each $n \in \mathbb{N}$, put

$$E_n = \{x \in X \,|\, |f_n(x) - f(x)| > \epsilon\}.$$

Then we obtain that

$$\begin{aligned}\mu(E_n) &= \int_{E_n} 1d\mu \le \int_{E_n} \frac{|f_n(x) - f(x)|^p}{\epsilon^p}\, d\mu \\ &\le \frac{1}{\epsilon^p}\|f_n - f\|_p^p \\ &\to 0\end{aligned}$$

as $n \to \infty$. Therefore, $\{f_n\}_{n=1}^{\infty}$ converges to f in measure. □

Corollary 9.3.4 Let $\{f_n\}_{n=1}^{\infty}$ be a sequence of measurable functions. If $\{f_n\}_{n=1}^{\infty}$ converges to a measurable function f in the mean of order p, then there exists a subsequence $\{f_{n_k}\}_{k=1}^{\infty}$ of $\{f_n\}_{n=1}^{\infty}$ such that $\{f_{n_k}\}_{k=1}^{\infty}$ converges pointwise a.e. on X to f.

[Proof] Suppose that $\{f_n\}_{n=1}^{\infty}$ converges to a measurable function f in the mean of order p. Then by Theorem 9.3.3, we see that $\{f_n\}_{n=1}^{\infty}$ converges to a measurable function f in measure. Therefore, by Corollary 9.1.4, there exists a subsequence $\{f_{n_k}\}_{k=1}^{\infty}$ of $\{f_n\}_{n=1}^{\infty}$ such that $\{f_{n_k}\}_{k=1}^{\infty}$ converges pointwise a.e. on X to f. □

Theorem 9.3.5 Let $\{f_n\}_{n=1}^{\infty}$ be a sequence of measurable functions. Suppose that there exists $g \in L^p(X, \mathcal{M}, \mu)$ such that $|f_n| \leq g$ for all $n \in \mathbb{N}$. If $\{f_n\}_{n=1}^{\infty}$ converges to a measurable function f in measure, then $\{f_n\}_{n=1}^{\infty}$ converges to f in the mean of order p.

[Proof] By assumption, we have

$$\int_X |f_n|^p \, d\mu \leq \int_X g^p \, d\mu < \infty,$$

and so $f_n \in L^p(X, \mathcal{M}, \mu)$ for all $n \in \mathbb{N}$. On the other hand, by Corollary 9.1.4, there exists a subsequence $\{f_{n_k}\}_{k=1}^{\infty}$ of $\{f_n\}_{n=1}^{\infty}$ such that $\{f_{n_k}\}_{k=1}^{\infty}$ converges pointwise a.e. on X to f. Therefore, we have

$$|f| = \lim_{k \to \infty} |f_{n_k}| \leq g,$$

which implies that $f \in L^p(X, \mathcal{M}, \mu)$. Suppose that $\{f_n\}_{n=1}^{\infty}$ does not converge to f in the mean of order p. Then there exist $\epsilon > 0$ and a subsequence $\{f_{n_k}\}_{n=1}^{\infty}$ of $\{f_n\}_{n=1}^{\infty}$ such that $\|f_{n_k} - f\|_p \geq \epsilon$ for all $k \in \mathbb{N}$. Since $\{f_{n_k}\}_{n=1}^{\infty}$ converges to f in measure, by Corollary 9.1.4, there exists a subsequence $\{f_{n_{k_j}}\}_{j=1}^{\infty}$ of $\{f_{n_k}\}_{k=1}^{\infty}$ such that $\{f_{n_{k_j}}\}_{j=1}^{\infty}$ converges pointwise a.e. on X to f. Therefore, by applying the dominated convergence theorem, we see that

$$\lim_{j \to \infty} \left\| f_{n_{k_j}} - f \right\|_p = 0,$$

which contradicts to that $\|f_{n_k} - f\|_p \geq \epsilon$ for all $k \in \mathbb{N}$. Hence $\{f_n\}_{n=1}^{\infty}$ converges to f in the mean of order p. □

Exercises

9.1 Let $\{f_n\}_{n=1}^{\infty}$ be a sequence of measurable functions and f a measurable function. Prove that $\{f_n\}_{n=1}^{\infty}$ converges to f in measure if and only if for any $\epsilon, \delta > 0$, there exists $N \in \mathbb{N}$ such that for any $n \geq N$,

$$\mu\left(\{x \in X \mid |f_n(x) - f(x)| > \epsilon\}\right) < \delta.$$

9.2 Let $\{f_n\}_{n=1}^{\infty}$ be a sequence of measurable functions. Prove that $\{f_n\}_{n=1}^{\infty}$ converges to a measurable function f in measure if and only if $\{f_n\}_{n=1}^{\infty}$ is a Cauchy sequence in measure.

9.3 Let $(X, \mathcal{M}, \mu)$ be a finite measure space and f be a measurable function which is finite a.e. Prove that for any $\epsilon > 0$ there exist $E \in \mathcal{M}$ and $M > 0$ such that $\mu(E) < \epsilon$ and $|f|$ is bounded by M on E^c.

9.4 Let $\{f_n\}_{n=1}^{\infty}$ and $\{g_n\}_{n=1}^{\infty}$ be sequences of measurable functions, and f, g be measurable functions. Suppose that $\{f_n\}_{n=1}^{\infty}$ and $\{g_n\}_{n=1}^{\infty}$ converge to f and g, respectively, in measure. Prove that

(1) $\{|f_n|\}_{n=1}^{\infty}$ converges to $|f|$ in measure,

(2) $\{f_n + g_n\}_{n=1}^{\infty}$ converges to $f + g$ in measure,

(3) $\{\alpha f_n\}_{n=1}^{\infty}$ converges to αf in measure, where α is a real number,

(4) if $\mu(X) < \infty$, then $\{f_n^2\}_{n=1}^{\infty}$ converges to f^2 in measure,

(5) if $\mu(X) < \infty$, then $\{f_n g_n\}_{n=1}^{\infty}$ converges to fg in measure.

9.5 Let $(X, \mathcal{M}, \mu)$ be a finite measure space and $L^0(X)$ be the space of all measurable function. The quotient space of the equivalent classes with respect to the equivalent relation $\sim$ on $L^0(X)$ defined by

$$f \sim g \quad \Leftrightarrow \quad f = g \text{ a.e.}$$

is denoted by $L^0(X)$ again and the equivalent class $[f]$ is denoted by f for simple notation. For $f, g \in L^0(X)$, define

$$\gamma(f, g) = \int_X \frac{|f-g|}{1+|f-g|} d\mu.$$

Let $\{f_n\}_{n=1}^{\infty}$ be a sequence of measurable functions. Prove that

(1) γ is a metric,

(2) $\{f_n\}_{n=1}^{\infty}$ converges to f in measure if and only if $\lim_{n\to\infty} \gamma(f_n, f) = 0$.

9.6 If a sequence of measurable functions converges almost uniformly, then the limit function is unique a.e.

9.7 Let $\{f_n\}_{n=1}^{\infty}$ be a sequence of measurable functions in $L^2(X, \mathcal{M}, \mu)$. Suppose that $\{f_n\}_{n=1}^{\infty}$ converges to a measurable function f in the mean of order 2. Prove that

$$\lim_{n\to\infty} \int_X f_n^2 \, d\mu = \int_X f^2 \, d\mu.$$

9.8 Let $(X, \mathcal{M}, \mu)$ be a finite measure space. Let $\{f_n\}_{n=1}^{\infty}$ be a sequence of measurable functions in $L^\infty(X, \mathcal{M}, \mu)$. Suppose that $\{f_n\}_{n=1}^{\infty}$ converges to a measurable function f in $L^\infty(X, \mathcal{M}, \mu)$. Prove that $\{f_n\}_{n=1}^{\infty}$ converges to f in $L^p(X, \mathcal{M}, \mu)$ for any $1 \le p < \infty$.

9.9 Let $(\mathbb{N}, \mathcal{P}(\mathbb{N}), c)$ be a measure space with the counting measure c, where $\mathcal{P}(\mathbb{N})$ is the power set of $\mathbb{N}$. Let $\{f_n\}_{n=1}^{\infty}$ be a sequence of functions on $\mathbb{N}$ and let f a function on $\mathbb{N}$. Prove that $\{f_n\}_{n=1}^{\infty}$ converges to f uniformly if and only if $\{f_n\}_{n=1}^{\infty}$ converges to f in measure.

9.10 Let $(X, \mathcal{M}, \mu)$ be a measure space. Let $\{f_n\}_{n=1}^{\infty}$ be a sequence of measurable functions and f a measurable function. Suppose that there exists an integrable function $g \in L^1(X, \mathcal{M}, \mu)$ such that $|f_n| \le g$ for all $n \in \mathbb{N}$. Prove that if $\{f_n\}_{n=1}^{\infty}$ converges pointwise a.e. on X to f, then $\{f_n\}_{n=1}^{\infty}$ converges to f almost uniformly.

Chapter 10

Further Topics in Functional Analysis

A **topological vector space** is a vector space equipped with a topology for which the fundamental algebraic operations of the vector space, i.e., the vector addition and the scalar multiplication are continuous. More precisely, we state the following definition.

Definition 10.0.1 [**Topological Vector Space**] Let V be a vector space over a scalar field $\mathbb{F}$ (usually $\mathbb{F} = \mathbb{R}$ or $\mathbb{F} = \mathbb{C}$) and let τ be a topology on V. Then the pair (V, τ) is a topological vector space if

(i) [Vector Addition] the map $+ : V \times V \ni (u, v) \mapsto u + v \in V$ is continuous,

(ii) [Scalar Multiplication] the map $\cdot : \mathbb{F} \times V \ni (\alpha, v) \mapsto \alpha v \in V$ is continuous.

Definition 10.0.2 Let V be a vector space and let $p : V \to [0, \infty)$ a functional. Then p is said to be

(i) **positively homogeneous** if for any $v \in V$ and any $\alpha > 0$,

$$p(\alpha v) = \alpha p(v),$$

(ii) **subadditive** if for any $u, v \in V$,

$$p(u + v) \leq p(u) + p(v).$$

Definition 10.0.3 Let V be a vector space (over $\mathbb{R}$ or $\mathbb{C}$) and $A \subset V$. Then A is said to be

(i) **absorbing** if for any $v \in V$ there is a $\lambda > 0$ such that $v \in \lambda A$,

(ii) **balanced** if for any scalar α with $|\alpha| < 1$, $\alpha A \subset A$,

(iii) **convex** if A is closed under convex combinations, i.e., for any $u, v \in A$ and $0 \leq \lambda \leq 1$, $\lambda u + (1 - \lambda)v \in A$.

Definition 10.0.4 [**Gauge Functional**] Let V be a vector space (over $\mathbb{R}$ or $\mathbb{C}$) and $A \subset V$ be an absorbing set. Then the **gauge functional** (or **Minkowski functional**) p_A of A is defined by

$$p_A(v) = \inf\{\lambda > 0 \,|\, v \in \lambda A\}, \qquad v \in V.$$

Theorem 10.0.5 Let $(V, \|\cdot\|)$ be a seminormed space and B be the unit open ball, i.e.,

$$B = \{v \in V \mid \|v\| < 1\}.$$

Then it holds that

(i) B is an absorbing, balanced, and convex subset of V,

(ii) $\|v\| = p_B(v)$ for all $v \in V$.

[Proof] (i) Let $v \in V$. Then for any $\lambda > 0$ with $\lambda\|v\| < 1$, $\|\lambda v\| = \lambda\|v\| < 1$ and so $\lambda v \in B$. Therefore, B is absorbing. Let α be a scalar with $|\alpha| < 1$. Then for any $v \in B$, we have $\|\alpha v\| = |\alpha|\|v\| < 1$ and so $\alpha B \subset B$. Therefore, B is balanced. For any $u, v \in B$ and $0 \leq \lambda \leq 1$, it holds that

$$\|\lambda u + (1 - \lambda)v\| \leq \lambda\|u\| + (1 - \lambda)\|v\| < \lambda + (1 - \lambda) = 1,$$

which implies that $\lambda u + (1 - \lambda)v \in B$ and B is convex.

(ii) For $v = 0$, it is obvious that $\|v\| = 0 = p_B(v)$. Suppose that $v \in V$ with $v \neq 0$. Let $\epsilon > 0$ be given. There exists a decreasing sequence $\{\lambda_n\}_{n=1}^{\infty}$ of positive numbers λ_n such that $\lambda_n^{-1}v \in B$ for all $n \in \mathbb{N}$ and $p_B(v) = \lim_{n\to\infty} \lambda_n$, equivalently, $\|v\| \leq \lambda_n$ for all $n \in \mathbb{N}$ and $p_B(v) = \lim_{n\to\infty} \lambda_n$.Therefore, $\|v\| \leq p_B(v)$. Suppose that $\|v\| < p_B(v)$. Then there exists $s > 0$ with $\|v\| < s < p_B(v)$ and then $\|s^{-1}v\| = s^{-1}\|v\| < 1$, which implies that $v \in sB$ and so $p_B(v) \le s$, which contradicts that $s < p_B(v)$. Hence $\|v\| \ge p_B(v)$ and so $\|v\| = p_B(v)$. □

Theorem 10.0.6 Let V be a vector space (over $\mathbb{R}$ or $\mathbb{C}$) and $C \subset V$ be a convex subset containing 0 as an interior point. Let p_C be the gauge functional of C. Then p_C is subadditive and positively homogeneous.

[Proof] Let $u, v \in V$ and let $\alpha, \beta > 0$ such that $u \in \alpha C$ and $v \in \beta C$. Then we obtain that

$$\frac{1}{\alpha+\beta}(u+v) = \frac{\alpha}{\alpha+\beta}\frac{u}{\alpha} + \frac{\beta}{\alpha+\beta}\frac{v}{\beta} \in C,$$

which implies that

$$p_C(u+v) \le \alpha + \beta,$$

and then by taking infima over all such α and β, respectively, we see that $p_C(u+v) \le p_C(u) + p_C(v)$. Let $v \in V$ and $\lambda > 0$ be given. Then we obtain that

$$p_C(\lambda v) = \inf\{\alpha > 0 \,|\, \lambda v \in \alpha C\} = \lambda \inf\{\gamma > 0 \,|\, v \in \gamma C\} = \lambda p_C(v),$$

which says that p_C is positively homogeneous. □

10.1 Hahn-Banach Theorem

Lemma 10.1.1 Let V be a real vector space and p be a subadditive, positively homogeneous functional on V. Let V_0 be a linear subspace of V and let L_0 be a linear functional on V_0 such that

$$L_0 v \le p(v), \qquad v \in V_0.$$

Suppose that $z \in V \setminus V_0$. Then there exists a linear functional $F : V_0 + \mathbb{R}z \to \mathbb{R}$ such that $Fv = L_0 v$ for all $v \in V_0$ and $Fw \le p(w)$ for all $w \in V_0 + \mathbb{R}z$.

[Proof] We observe that for any $v_1, v_2 \in V_0$,

$$L_0 v_1 + L_0 v_2 = L_0(v_1 + v_2) \le p(v_1 + v_2) \le p(v_1 - z) + p(v_2 + z),$$

equivalently,

$$L_0 v_1 - p(v_1 - z) \le -L_0 v_2 + p(v_2 + z)$$

for all $v_1, v_2 \in V_0$, which implies that

$$\sup_{v \in V_0}\{L_0 v - p(v - z)\} \le \inf_{v \in V_0}\{-L_0 v + p(v + z)\}.$$

Take a constant $\gamma \in \mathbb{R}$ with

$$\sup_{v \in V_0}\{L_0 v - p(v - z)\} \le \gamma \le \inf_{v \in V_0}\{-L_0 v + p(v + z)\},$$

and define a functional $F : V_0 + \mathbb{R}z \to \mathbb{R}$ by

$$F(v + \alpha z) = L_0 v + \alpha\gamma, \qquad v \in V_0.$$

Then we can easily see that F is linear and $F = L_0$ on V_0, and so $F(v) = L_0 v \le p(v)$ for all $v \in V_0$. Let $v \in V_0$ and $\alpha \in \mathbb{R}$ with $\alpha \neq 0$. If $\alpha > 0$, then

$$\begin{aligned} F(v + \alpha z) &= L_0 v + \alpha\gamma \le L_0 v + \alpha\left(-L_0 \frac{v}{\alpha} + p\left(\frac{v}{\alpha} + z\right)\right) \\ &= p(v + \alpha z). \end{aligned}$$

If $\alpha < 0$ and $\alpha = -\beta$ for $\beta > 0$, then we have

$$\begin{aligned} F(v + \alpha z) &= L_0 v - \beta\gamma \le L_0 v - \beta\left(L_0 \frac{v}{\beta} - p\left(\frac{v}{\beta} - z\right)\right) \\ &= p(v + \alpha z). \end{aligned}$$

Hence $Fw \le p(w)$ for all $w \in V_0 + \mathbb{R}z$. □

Theorem 10.1.2 **[Hahn-Banach Theorem]** Let V be a real vector space and p be a subadditive, positively homogeneous functional on V. Let V_0 be a linear subspace of V and let L_0 be a linear functional on V_0 such that

$$L_0 v \le p(v), \qquad v \in V_0. \tag{10.1}$$

Then there exists an extension L of L_0 to the whole space V, i.e., L is a linear functional on V such that $Lv = L_0 v$ for all $v \in V_0$ and

$$Lw \le p(w), \qquad w \in V. \tag{10.2}$$

[Proof] Let $\mathfrak{X}$ be the family of all pairs of the form (U, F), where

- U is a subspace of V such that $V_0 \subset U \subset V$,
- F is a linear functional on U such that $F = L_0$ on V_0 and $Fw \le p(w)$ for all $w \in U$.

Then it is obvious that $(V_0, L_0) \in \mathfrak{X}$. For each $(U, F), (U', F') \in \mathfrak{X}$, we say that

$$(U, F) \preceq (U', F') \quad \text{if } U \subset U' \text{ and } F = F' \text{ on } U.$$

Then $(\mathfrak{X}, \preceq)$ is a partially ordered set. Let $\{(U_\lambda, F_\lambda)\}_{\lambda \in \Lambda}$ be a linearly (totally) ordered subset of $\mathfrak{X}$. We put

$$U = \bigcup_{\lambda \in \Lambda} U_\lambda.$$

and define a linear functional $F : U \to \mathbb{R}$ as following: for any $v \in U$, there exists $\lambda \in \Lambda$ such that $v \in U_\lambda$ and then $Fv = F_\lambda v$. Then we can easily see that $(U, F) \in \mathfrak{X}$ and (U, F) is an upper bound of $\{(U_\lambda, F_\lambda)\}_{\lambda \in \Lambda}$. Therefore, by Zorn's Lemma, there exists a maximal element $(W, G) \in \mathfrak{X}$. Then $W = V$. If there exists $z \in V \setminus W$, then by applying Lemma 10.1.1, we can easily see that there exists an extension G' of G to $W + \mathbb{R}z$ such that $(W + \mathbb{R}z, G') \in \mathfrak{X}$, which contradicts the maximality of $(W, G) \in \mathfrak{X}$. Hence by taking $L = G$, we complete the proof. □

Theorem 10.1.3 **[Hahn-Banach Theorem]** Let V be a complex vector space and p be a seminorm on V. Let V_0 be a linear subspace of V and let L_0 be a complex linear functional on V_0 such that

$$|L_0 v| \leq p(v), \qquad v \in V_0. \tag{10.3}$$

Then there exists an extension L of L_0 to the whole space V, i.e., L is a linear functional on V such that $Lv = L_0 v$ for all $v \in V_0$ and

$$|Lw| \leq p(w), \qquad w \in V. \tag{10.4}$$

[Proof] Let $L_1 = \Re(L_0)$. Then by Theorem 7.2.8, L_1 is real linear on V_0 and from (10.3), we have

$$L_1 v \leq |L_0 v| \leq p(v), \qquad v \in V_0.$$

Then by applying Theorem 10.1.2, we see that there exists an extension L_2 of L_1 to V such that $L_2 = L_1$ on V_0 and

$$L_2 w \leq p(w), \qquad w \in V.$$

Then by taking $-w$ instead of w, we see that

$$-L_2 w \leq p(w), \qquad w \in V.$$

Hence, we have

$$|L_2 w| \leq p(w), \qquad w \in V.$$

Then by applying Theorem 7.2.8, we define a complex linear functional L on V by

$$Lw = L_2(w) - iL_2(iw), \qquad w \in V.$$

Then for any $v \in V_0$, we have

$$Lv = L_2(v) - iL_2(iv) = L_1(v) - iL_1(iv) = L_0 v$$

and for any $w \in V$ with $Lw \neq 0$, we also obtain that

$$\mathbb{R} \ni |Lw| = \frac{|Lw|}{Lw} Lw = L\left(\frac{|Lw|}{Lw} w\right),$$

which implies that

$$\begin{aligned} |Lw| &= L\left(\frac{|Lw|}{Lw}w\right) = L_2\left(\frac{|Lw|}{Lw}w\right) \\ &\le p\left(\frac{|Lw|}{Lw}w\right) = \left|\frac{|Lw|}{Lw}\right| p(w) \\ &= p(w), \end{aligned}$$

which proves (10.4). □

Theorem 10.1.4 **[Hahn-Banach Theorem]** Let $(V, \|\cdot\|)$ be a (complex) normed space and V_0 be a linear subspace of V. Let L_0 be a bounded linear functional on V_0. Then there exists a bounded linear functional L on V such that $Lv = L_0 v$ for all $v \in V_0$ and

$$\|L\| = \|L_0\|. \tag{10.5}$$

[Proof] Since L_0 is a bounded linear functional on V_0,

$$|L_0 v| \le \|L_0\|\|v\|, \qquad v \in V_0.$$

For each $v \in V$, we put

$$p(v) = \|L_0\|\|v\|.$$

Then p is a seminorm on V and $|L_0 v| \le p(v)$ for all $v \in V_0$. Therefore, by Theorem 10.1.3, there exists a linear functional L on V such that $Lv = L_0 v$ for all $v \in V_0$ and $|Lw| \le p(w)$ for all $w \in V$, i.e.,

$$|Lw| \le \|L_0\|\|w\|, \qquad w \in V,$$

which implies that $\|L\| \le \|L_0\|$. On the other hand, we have

$$\|L\| \ge \sup_{v \in V_0; \|v\| \le 1} |L_0 v| = \|L_0\|.$$

Hence, $\|L\| = \|L_0\|$. □

Theorem 10.1.5 Let $(V, \|\cdot\|)$ be a (complex) normed space, K a closed subspace of V and $w \in V \setminus K$ be fixed. Then there exists a bounded linear functional $L \in V^*$ such that $L(w) \neq 0$ and $L = 0$ on K. In particular, if $\lambda = \inf_{u \in K}\{\|w - u\|\}$, then L can be chosen as $\|L\| = 1$ and $L(w) = \lambda$.

[Proof] Consider the bounded linear functional $L_0 : K + \mathbb{C}w \to \mathbb{C}$ defined by

$$L_0(u + \alpha w) = \alpha\lambda,$$

where $\lambda > 0$ since K is closed. Then by the construction, we have $L_0(w) = \lambda$, and for $\alpha \in \mathbb{C}$ with $\alpha \neq 0$, we have

$$|L_0(u + \alpha w)| = |\alpha|\lambda \leq |\alpha| \left\|w + \alpha^{-1}u\right\| = \|u + \alpha w\|,$$

which implies that $\|L_0\| \leq 1$. On the other hand, by the definition of λ, there exists a sequence $\{u_n\}_{n=1}^{\infty} \subset K$ such that

$$\lambda = \lim_{n\to\infty} \|w - u_n\|,$$

and then we obtain that

$$\lambda = \lim_{n\to\infty} L_0(w - u_n) \leq \lim_{n\to\infty} \|L_0\|\|w - u_n\| = \|L_0\|\lambda,$$

which implies that $\|L_0\| \geq 1$ and so $\|L_0\| = 1$. Therefore, by Hahn-Banach Theorem (Theorem 10.1.4), there exists a bounded linear functional L on V such that $Lv = L_0v$ for all $v \in K + \mathbb{C}w$ and so $L = 0$ on K, $Lw = L_0w = \lambda$, and $\|L\| = \|L_0\| = 1$. □

10.2 Baire Category Theorem

Definition 10.2.1 A subset A of a topological space X is said to be **nowhere dense** if

$$\overset{\circ}{\overline{A}} = \emptyset,$$

where $\overline{A}$ is the closure of A and $\overset{\circ}{B}$ is the interior of B.

Let A be a subset of a topological space X. Then A is nowhere dense if and only if $\left(\overline{A}\right)^{c}$ is dense in X (see Exercise **10.6**).

Definition 10.2.2 Let E be a subset of a topological space X. Then E is said to be of

(i) **the first category** if it can be expressed as the countable union of nowhere dense subsets,

(ii) **the second category** if it is not of the first category.

It is obvious that a subset E of a topological space is of the first category if and only if E is the countable union of nowhere dense closed subsets. Therefore, it is obvious that if a set E is the second category and E can be expressed by a union of a countable family $\{F_n\}_{n=1}^{\infty}$ of closed subsets, then there exists $k \in \mathbb{N}$ such that $\overset{\circ}{F_k} \neq \emptyset$.

Proposition 10.2.3 Let (X, τ) be a topological space. Then X is of the second category if and only if for any countable family $\{O_n\}_{n=1}^{\infty}$ of open dense subsets of X,

$$\bigcap_{n=1}^{\infty} O_n \neq \emptyset.$$

[Proof] The proof is straightforward. □

Definition 10.2.4 A topological space X is a **Baire space** if for any countable family $\{O_n\}_{n=1}^{\infty}$ of open dense subsets O_n in X,

$$\bigcap_{n=1}^{\infty} O_n$$

is dense in X.

It is obvious that every Baire space is of the second category.

Theorem 10.2.5 **[Baire Category Theorem]** If (X, d) is a complete metric space, then X is a Baire space.

[Proof] Let $\{O_n\}_{n=1}^{\infty}$ be a countable family of open dense subsets O_n of X. We have to prove that $\cap_{n=1}^{\infty} O_n$ is dense in X, i.e., for any given nonempty open subset $U \subset X$,

$$U \cap \left(\bigcap_{n=1}^{\infty} O_n \right) \neq \emptyset.$$

We start with $x_0 \in U$ and $r_0 > 0$ such that

$$B(x_0, r_0) = \{x \in X \,|\, d(x, x_0) < r_0\} \subset U.$$

By assumption, for any $n \in \mathbb{N}$, O_n is dense in X and so it holds that

$$B\left(x_0, \frac{r_0}{2}\right) \bigcap O_1 \neq \emptyset.$$

Then take $x_1 \in B\left(x_0, r_0/2\right) \cap O_1$ and $r_1 > 0$ with $r_1 < r_0/2$ such that

$$B(x_1, r_1) \subset B\left(x_0, \frac{r_0}{2}\right) \cap O_1.$$

Inductively, we take $x_n \in B\left(x_{n-1}, \frac{r_{n-1}}{2}\right) \cap O_n$ and $r_n > 0$ with $r_n < \frac{r_{n-1}}{2}$ such that

$$B(x_n, r_n) \subset B\left(x_{n-1}, \frac{r_{n-1}}{2}\right) \cap O_n.$$

Then for any $n > m$ we obtain that

$$\begin{aligned} d(x_m, x_n) &\leq d(x_m, x_{m+1}) + d(x_{m+1}, x_n) \leq d(x_m, x_{m+1}) + \cdots + d(x_{n-1}, x_n) \\ &\leq \frac{r_m}{2} + \frac{r_{m+1}}{2} + \cdots + \frac{r_{n-1}}{2} \leq \frac{r_0}{2^m}\left(1 - \frac{1}{2^{n-m+1}}\right) \\ &\leq \frac{r_0}{2^m}, \end{aligned} \tag{10.6}$$

which implies that $\{x_n\}_{n=1}^{\infty}$ is a Cauchy sequence in X. Since X is complete, there exists $x \in X$ such that x_n converges to x in X. For any $m \geq 1$, from (10.6) we have

$$d(x_m, x) \leq d(x_m, x_n) + d(x_n, x) \leq \frac{r_m}{2} + d(x_n, x)$$

for any $n \geq m$ and so, since $\lim_{n\to\infty} d(x_n, x) = 0$, we have

$$d(x_m, x) \leq \frac{r_m}{2},$$

which implies that $x \in B(x_m, r_m)$ for any $m \geq 1$. On the other and, by construction

$$B(x_n, r_n) \subset B\left(x_{n-1}, \frac{r_{n-1}}{2}\right) \subset U, \qquad B(x_n, r_n) \subset O_n$$

for any $n \in \mathbb{N}$. Therefore, for any $n \in \mathbb{N}$,

$$x \in B(x_n, r_n) \subset U, \qquad x \in B(x_n, r_n) \subset O_n,$$

which implies that

$$x \in U \cap \left(\bigcap_{n=1}^{\infty} O_n\right) \neq \emptyset.$$

Hence $\cap_{n=1}^{\infty} O_n$ is dense in X and X is a Baire space. □

Proposition 10.2.6 Every (nonempty) open subset of a Baire space is a Baire space.

[Proof] Let X be a Baire space and U be a nonempty open subset in X. Let $\{O_n\}_{n=1}^{\infty}$ be a family of open dense subsets in U. Put

$$V = \left(\overline{U}\right)^{\mathrm{c}}, \qquad G = \bigcap_{n=1}^{\infty} O_n.$$

Then for any $n \in \mathbb{N}$, $O_n \cup V$ is open in X. Moreover, for any $n \in \mathbb{N}$, we can see that $O_n \cup V$ is dense in X. In fact, for any nonempty open subset O in X, consider

$$O \cap (O_n \cup V) = (O \cap O_n) \cup (O \cap V).$$

If $O \cap V \neq \emptyset$, then $O \cap (O_n \cup V) \neq \emptyset$ and so $O_n \cup V$ is dense in X. If $O \cap V = \emptyset$, then $O \subset V^{\mathrm{c}} = \overline{U}$ and so $O \cap U$ is a nonempty open subset of U. Since O_n is dense in U for any $n \in \mathbb{N}$, $(O \cap U) \cap O_n \neq \emptyset$, which implies that $O \cap O_n \neq \emptyset$ and so $O \cap (O_n \cup V) \neq \emptyset$. Therefore, $O_n \cup V$ is dense in X. Then $\{O_n \cup V\}_{n=1}^{\infty}$ is a countable family of dense open subset of X. Since X is a Baire space,

$$G \cup V = \left(\bigcap_{n=1}^{\infty} O_n\right) \cup V = \bigcap_{n=1}^{\infty} (O_n \cup V) \quad \text{is dense in} \quad X.$$

Then for any nonempty open subset S of U, we obtain that

$$S \cap V = S \cap \left(\overline{U}\right)^{\mathrm{c}} \subset U \cap \left(\overline{U}\right)^{\mathrm{c}} \subset \overline{U} \cap \left(\overline{U}\right)^{\mathrm{c}} = \emptyset,$$

and

$$\emptyset \neq S \cap (G \cup V) = (S \cap G) \cup (S \cap V) = S \cap G,$$

which implies that G is dense in U and so U is a Baire space. □

Theorem 10.2.7 Let X be a (nonempty) topological space. Then the following are equivalent:

(1) X is a Baire space,

(2) every nonempty open subset of X is of the second category.

[Proof] $\Rightarrow$) Suppose that a nonempty open set U is not the second category, i.e., U is the first category. Then U can be expresses as a union of countable family $\{F_n\}_{n=1}^{\infty}$ of nowhere dense closed subsets F_n, i.e.,

$$U = \bigcup_{n=1}^{\infty} F_n,$$

where $\overset{\circ}{F_n} = \emptyset$ for any $n \in \mathbb{N}$. Then for any $n \in \mathbb{N}$, F_n^{c} is dense in X (see Exercise **10.6**), and so $G_n = F_n^{\mathrm{c}} \cap U$ is a dense open subset in U (see Exercise **10.7**). By assumption

$$\emptyset = U \setminus U = U \cap U^{\mathrm{c}} = \bigcap_{n=1}^{\infty} (U \cap F_n^{\mathrm{c}}),$$

On the other hand, by Proposition 10.2.6, U is Baire space. Therefore, $\bigcap_{n=1}^{\infty}(F_n^c \cap U) = \emptyset$ is dense in $U \neq \emptyset$, which contradicts.

$\Leftarrow$) Suppose that X is not a Baire space. There there exists a countable family $\{O_n\}_{n=1}^{\infty}$ of open dense subsets O_n in X such that $\bigcap_{n=1}^{\infty} O_n$ is not dense in X. Therefore, there exists a nonempty open subset V in X such that

$$V \bigcap \left(\bigcap_{n=1}^{\infty} O_n \right) = \emptyset.$$

Then we have

$$\emptyset \neq V \subset \bigcup_{n=1}^{\infty} O_n^c, \qquad V = \bigcup_{n=1}^{\infty} \left(V \bigcap O_n^c \right).$$

On the other hand, for any $n \in \mathbb{N}$, O_n^c is nowhere dense (see Exercise **10.8**) and so $V \bigcap O_n^c$ is nowhere dense. Therefore, V is of the first category which contradicts the assumption. □

Corollary 10.2.8 Every complete metric space is of the second category.

[Proof] Since every Baire space is of the second category (see also Theorem 10.2.7). The proof is immediate from Baire Category Theorem (Theorem 10.2.5). □

10.3 Uniform Boundedness Principle

Theorem 10.3.1 **[Uniform Boundedness Principle]** Let $\{T_\lambda\}_{\lambda \in \Lambda}$ be a family of bounded linear operators T_λ from a Banach space V into a normed space W. Suppose that for any $v \in V$ there exists a non-negative constant $C_v \geq 0$ such that

$$\sup_{\lambda \in \Lambda} \| T_\lambda v \| \leq C_v. \tag{10.7}$$

Then there exists a non-negative constant $C \geq 0$ such that

$$\sup_{\lambda \in \Lambda} \| T_\lambda \| \leq C.$$

[Proof] For each $k \in \mathbf{N}$, put

$$A_k = \left\{ v \in V \,\middle|\, \sup_{\lambda \in \Lambda} \| T_\lambda v \| \leq k \right\} = \bigcap_{\lambda \in \Lambda} \{ v \in V \mid \| T_\lambda v \| \leq k \}.$$

Then A_k is closed since T_λ is continuous for all $\lambda \in \Lambda$. By (10.7), we see that for any $v \in V$ there exists $k \in \mathbb{N}$ such that $C_v \le k$ and so $v \in A_k$, which implies that

$$V = \bigcup_{k \ge 1} A_k.$$

Then, since V is complete, by the Baire Category Theorem (see Corollary 10.2.8), V is of the second category. Therefore, there exists $k_0 \in \mathbb{N}$ such that $\overset{\circ}{A}_{k_0} \neq \emptyset$ and so A_{k_0} contains a nonempty open ball $B(v_0, \epsilon)$, i.e.,

$$B(v_0, \epsilon) = \{v \in V \mid \| v - v_0 \| < \epsilon\} \subset A_{k_0}.$$

Then for any

$$0 \neq v \in V, \qquad z = v_0 + \frac{\epsilon}{2 \| v \|} v,$$

we have that $\| z - v_0 \| = \epsilon/2 < \epsilon$ and $z \in B(v_0, \epsilon) \subset A_{k_0}$, and so we have

$$v = \frac{2 \| v \|}{\epsilon} (z - v_0).$$

Therefore, we have

$$\begin{aligned}
\sup_{\lambda \in \Lambda} \| T_\lambda v \| &= \sup_{\lambda \in \Lambda} \left\| T_\lambda \left(\frac{2 \| v \|}{\epsilon} (z - v_0) \right) \right\| = \sup_{\lambda \in \Lambda} \frac{2 \| v \|}{\epsilon} \| T_\lambda (z - v_0) \| \\
&\le \sup_{\lambda \in \Lambda} \frac{2 \| v \|}{\epsilon} (\| T_\lambda z \| + \| T_\lambda v_0 \|) \\
&= \frac{4k_0}{\epsilon} \| v \|,
\end{aligned}$$

which implies that

$$\sup_{n \in \mathbb{N}} \| T_n \| \le \frac{4k_0}{\epsilon}.$$

By taking $C = 4k_0/\epsilon$, the proof is completed. □

The uniform boundedness principle (Theorem 10.3.1) is also referred to as **Banach-Steinhaus Theorem**.

10.4 Open Mapping Theorem

Let X and Y be topological spaces. A map $f : X \to Y$ is said to be **open** if for any open subset U of X, $f(U)$ is open in Y.

Lemma 10.4.1 Let T be a bounded linear operator from a Banach space X onto another Banach space Y. Then the image $T(B_{1,X}(0))$ of the unit open ball

$$B_{1,X}(0) = \{v \in X \mid \| v \| < 1\} \subset X$$

contains an open ball $B_{r,Y}(0) \subset Y$ for some $r > 0$.

[Proof] Since $X = \bigcup_{n=1}^{\infty} B_{n,X}(0)$ and T is surjective, we have

$$Y = T(X) = T\left(\bigcup_{n=1}^{\infty} B_{n,X}(0)\right) = \bigcup_{n=1}^{\infty} T\left(B_{n,X}(0)\right) = \bigcup_{n=1}^{\infty} \overline{T\left(B_{n,X}(0)\right)}.$$

Since Y is a Banach space which is complete, by Baire category theorem (Theorem 10.2.5), there exists $k \geq 1$ such that $\overline{T\left(B_{k,X}(0)\right)}$ contains an open ball $B_{\epsilon,Y}(y_0)$, i.e., $B_{\epsilon,Y}(y_0) \subset \overline{T\left(B_{k,X}(0)\right)}$ and so

$$B_{\epsilon,Y}(0) = B_{\epsilon,Y}(y_0) - y_0 \subset \overline{T\left(B_{k,X}(0)\right)} - y_0 = \overline{T\left(B_{k,X}(0)\right)} - T(v_0),$$

where $T(v_0) = y_0$ for some $v_0 \in X$ since T is surjective. Since T is continuous and linear, we can choose a sufficiently large $k_1 \geq 1$ such that

$$\overline{T\left(B_{k,X}(0)\right)} - T(v_0) \subset T\left(B_{k_1,X}(0)\right),$$

and so, by using the fact that $B_{n,X}(0) = nB_{1,X}(0)$ for any $n > 0$, we have

$$B_{\epsilon,Y}(0) \subset T\left(B_{k_1,X}(0)\right) = k_1 T\left(B_{1,X}(0)\right),$$

which implies that $B_{\frac{\epsilon}{k_1},Y}(0) \subset T\left(B_{1,X}(0)\right)$, which gives the proof. □

> **Theorem 10.4.2** **[Open Mapping Theorem]** A bounded linear operator T from a Banach space onto another Banach space is an open mapping.

[Proof] Let U be an open subset in X. Then we claim that $T(U)$ is open in Y. For any $y \in T(U)$, since T is surjective, there exists $v_0 \in U$ such that $T(v_0) = y \in T(U)$. Since U is an open set and so it contains an open ball $B_{\epsilon,X}(v_0)$ centered by $v_0 \in U$ and then

$$U - v_0 \supset B_{\epsilon,X}(0), \quad \text{which implies that} \quad \frac{1}{\epsilon}(U - v_0) \supset B_{1,X}(0).$$

Therefore, we have

$$T(B_{1,X}(0)) \subset \frac{1}{\epsilon} T\left(U - v_0\right) = \frac{1}{\epsilon}\left(T(U) - T(v_0)\right),$$

and by Lemma 10.4.1 there exists $\delta > 0$ such that

$$B_{\delta,Y}(0) \subset \frac{1}{\epsilon}\left(T(U) - T(v_0)\right),$$

equivalently, $B_{\epsilon\delta,Y}(y) \subset T(U)$, which implies that $T(U)$ is open. Therefore, T is an open mapping. □

> **Theorem 10.4.3** **[Bounded Inverse Theorem]** Let T be a bijective linear mapping from a Banach space V onto another Banach space W. If T is bounded, then T^{-1} is bounded.

[Proof] Since T is a bounded linear operator from a Banach space V onto another Banach space W , by Theorem 10.4.2, T is an open mapping. Therefore, for any open subset O in V, $\left(T^{-1}\right)^{-1}(O) = T(O)$ is open in W, which implies that T^{-1} is bounded. □

10.5 Closed Graph Theorem

Let V and W be two normed spaces. Then the two algebraic operations of the vector space $V \times W$ are defined as usual:

$$(v_1, y_1) + (v_2, y_2) = (v_1 + v_2, y_1 + y_2),$$
$$\alpha(v_1, y_1) = (\alpha v_1, \alpha v_2), \qquad v_i \in V, \quad y_i \in W, \quad i = 1, 2, \quad \alpha : \text{scalar},$$

and then $V \times W$ becomes a normed space with the norm defined by

$$\|(v, y)\| = \|v\| + \|y\|, \qquad v \in V, \quad y \in W. \tag{10.8}$$

Proposition 10.5.1 Let V and W be two Banach spaces. Then $V \times W$ is a Banach space with the norm given as in (10.8).

[Proof] Straightforward. □

Let T be a linear operator from a normed space V into another normed space W. Then

$$\mathcal{G}(T) = \{(v, Tv) \in V \times W \,|\, v \in V\}$$

is called the **graph** of T.

Definition 10.5.2 Let V and W be two normed spaces. Then a linear operator $T : V \to W$ is said to be **closed** if $\mathcal{G}(T)$ is closed subset of $V \times W$.

Note that a linear operator T from a normed space V into another normed space W is closed if and only if the following property holds: if $\{v_n\}_{n=1}^{\infty}$ converges to v in V and $\{Tv_n\}_{n=1}^{\infty}$ converges to w in W, then $w = Tv$, i.e., $(v, w) \in \mathcal{G}(T)$.

Theorem 10.5.3 **[Closed Graph Theorem]** Let V and W be Banach spaces. If a linear operator $T : V \to W$ is closed, then T is bounded.

[Proof] Since $\mathcal{G}(T)$ is a closed subspace of $V \times W$, $\mathcal{G}(T)$ is a Banach space. Consider the operator P defined by

$$P : \mathcal{G}(T) \ni (v, Tv) \longmapsto v \in V,$$

and then P is a bounded linear bijection map. In fact, for any $v \in V$, we have

$$\| P(v, Tv) \| = \| v \| \leq \| v \| + \| Tv \| = \| (v, Tv) \| .$$

Similarly, the linear operator Q defined by

$$Q : \mathcal{G}(T) \ni (v, Tv) \longmapsto Tv \in W$$

is bounded. On the other hand, by Theorem 10.4.3, P^{-1} is bounded. Therefore, we have that $T = Q \circ P^{-1}$ and so T is bounded. $\square$

10.6 Weak Topology

Consider functions f_n ($n \in \mathbb{N}$) defined by

$$f_n(x) = \begin{cases} 0, & 1/n \leq |x| \leq 1, \\ n/2, & |x| \leq 1/n. \end{cases}$$

Then for each $n \in \mathbb{N}$, f_n is a nonnegative function and one can easily see that $f_n \in L^1([-1,1])$ with $\|f_n\|_1 = 1$. Also, for any $m, n \in \mathbb{N}$, one can see that

$$\|f_n - f_m\|_1 = |n-m|\frac{1}{\max\{n,m\}} + \left|\frac{1}{n} - \frac{1}{m}\right| \min\{n,m\},$$

which implies that $\{f_n\}_{n=1}^{\infty}$ is not a Cauchy sequence in $L^1([-1,1])$ and so $\{f_n\}_{n=1}^{\infty}$ does not converge in $L^1([-1,1])$. Nevertheless, for any continuous function g on $[-1,1]$, we see that

$$\begin{aligned} \lim_{n\to\infty} \int_{-1}^{1} f_n g \, dm &= \lim_{n\to\infty} \frac{n}{2} \int_{-\frac{1}{n}}^{\frac{1}{n}} g(x) dm(x) \\ &= \frac{1}{2}\left[\lim_{n\to\infty} \frac{1}{\frac{1}{n}} \left(\int_{-\frac{1}{n}}^{0} g(x)m(x) + \int_{0}^{\frac{1}{n}} g(x)m(x) \right)\right] \\ &= g(0). \end{aligned}$$

This observation gives a motivation of the notion of **weak convergence**.

Definition 10.6.1 Let V be a normed space and $\{v_n\}_{n=1}^{\infty}$ be a sequence in V. Then we say that $\{v_n\}_{n=1}^{\infty}$ **converges weakly** to v in V if for any $f \in V^*$,

$$\lim_{n\to\infty} f(v_n) = f(v).$$

In this case, we write $v_n \to v$ weakly in V.

Remark 10.6.2 We recall that a sequence $\{v_n\}_{n=1}^{\infty}$ in a normed space V converges to v in V means that $\lim_{n\to\infty} \|v_n - v\| = 0$. Therefore, if a sequence $\{v_n\}_{n=1}^{\infty} \subset V$ converges to v in V, then $\{v_n\}_{n=1}^{\infty} \subset V$ converges weakly to v in V. In fact, we have that for any $f \in V^*$,

$$|f(v_n) - f(v)| = |f(v_n - v)| \leq \|f\|\|v_n - v\|.$$

But, the converse is false in general.

Proposition 10.6.3 Let $(V, \|\cdot\|)$ be a normed space and $\{v_n\}_{n=1}^{\infty}$ be a sequence in V. If $\{v_n\}_{n=1}^{\infty}$ converges weakly to v in V, then v is unique.

[Proof] Suppose that $\{v_n\}_{n=1}^{\infty}$ converges weakly to w and $v \neq w$. Then by Hahn-Banach Theorem (Theorem 10.1.4), there exists a bounded linear functional $f \in V^*$ such that $f(v-w) = \|v-w\| \neq 0$ and $\|f\| = 1$ (see Theorem 10.1.5). But for any $f \in V^*$, it holds that

$$f(v) = \lim_{n\to\infty} f(v_n) = f(w),$$

which implies that $f(v-w) = 0$ for all $f \in V^*$, which contradicts. □

Remark 10.6.4 Let $(X, \mathcal{M}, \mu)$ be a σ-finite measure space. Let $1 \leq p < \infty$, $\{f_n\}_{n=1}^{\infty}$ a sequence in $L^p(X)$ and $f \in L^p(X)$. Suppose that $f_n \to f$ weakly in $L^p(X)$ and $f_n \to g$ weakly in $L^p(X)$. Then for the conjugate function $(f-g)^*$ of $f-g$, we obtain that

$$\int_X (f-g)^* f \, d\mu = \lim_{n\to\infty} \int_X (f-g)^* f_n \, d\mu = \int_X (f-g)^* g \, d\mu,$$

which implies that $\|f-g\|_p = \int_X (f-g)^*(f-g)\, d\mu = 0$, and so $f = g$.

Proposition 10.6.5 Let $(X, \mathcal{M}, \mu)$ be a σ-finite measure space and $1 \leq p < \infty$. Let $\{f_n\}_{n=1}^{\infty}$ be a sequence in $L^p(X)$ and $f \in L^p(X)$. Then $f_n \to f$ weakly in $L^p(X)$ if and only if for any $g \in L^q(X)$ (q is the conjugate number of p),

$$\lim_{n\to\infty} \int_X g \cdot f_n \, d\mu = \int_X g \cdot f \, d\mu.$$

[Proof] By the definition, $f_n \to f$ weakly in $L^p(X)$ if and only if for any $T \in (L^p(X))^*$,

$$\lim_{n\to\infty} T(f_n) = T(f)$$

if and only if, by Riesz Representation Theorem (Theorem 7.6.5), for any $g \in L^q(X)$ (q is the conjugate number of p),

$$\lim_{n\to\infty} \int_X g \cdot f_n \, d\mu = \int_X g \cdot f \, d\mu.$$

Thus we prove the desired assertion. □

Let $(V, \|\cdot\|)$ be a normed space. For each $v \in V$, we define a linear functional v^{**} on V^* by

$$v^{**}(f) = f(v), \qquad f \in V^{**} = (V^*)^*.$$

Here V^{**} is called the **bidual space** (or **second conjugate space**) of V. Then we obtain that

$$|v^{**}(f)| = |f(v)| \leq \|v\|\|f\|,$$

which implies that v^{**} is bounded and $\|v^{**}\| \leq \|v\|$. Therefore, $v^{**} \in V^{**}$ for each $v \in V$. Moreover, for each $0 \neq v \in V$, by Hahn-Banach Theorem (Theorem 10.1.4), there exists a bounded linear functional $f_0 \in V^*$ such that $f_0(v) = \|v\| \neq 0$ and $\|f_0\| = 1$ (see Theorem 10.1.5). Therefore, we see that

$$\|v^{**}\| \geq |v^{**}(f_0)| = |f_0(v)| = \|v\|.$$

Hence, we see that $\|v^{**}\| = \|v\|$ for all $v \in V$.

Theorem 10.6.6 Every normed space can be considered as a dense subspace of a Banach space.

[Proof] Let $(V, \|\cdot\|)$ be a normed space. Define a map J_V from V into V^{**} by

$$J_V(v) = v^{**}, \qquad v \in V.$$

From the previous observation, J_V is an isometric isomorphism from V onto $J_V(V)$, which means that we can identify V and $J_V(V)$. Since the closure $\overline{J_V(V)}$ of $J_V(V)$ is a closed subspace of the Banach space V^{**} (see Corollary 7.2.16), $\overline{J_V(V)}$ is a Banach space. Hence V can be considered as a dense subspace of the Banach space $\overline{J_V(V)}$. □

Theorem 10.6.7 Let $(V, \|\cdot\|)$ be a normed space and let $\{v_n\}_{n=1}^{\infty}$ be a sequence in V. If $\{v_n\}_{n=1}^{\infty}$ converges weakly to v in V, then $\{v_n\}_{n=1}^{\infty}$ is bounded in V.

[Proof] Suppose that $\{v_n\}_{n=1}^{\infty}$ converges weakly to v in V. Then for each $f \in V^*$, $\{f(v_n)\}_{n=1}^{\infty}$ converges to $f(v)$ and so $\{f(v_n)\}_{n=1}^{\infty}$ is bounded. Hence, for each $f \in V^*$,

$$\sup_{n\in\mathbb{N}}\{|v_n^{**}(f)|\} = \sup_{n\in\mathbb{N}}\{|f(v_n)|\}$$

is bounded. Therefore, by Uniform Boundedness Principle (Theorem 10.3.1), $\{v_n^{**}\}_{n=1}^{\infty}$ is bounded. Since $\|v^{**}\| = \|v\|$ for all $v \in V$, $\{v_n\}_{n=1}^{\infty}$ is bounded. □

Theorem 10.6.8 Let $(V, \|\cdot\|)$ be a normed space and let $\{v_n\}_{n=1}^{\infty}$ be a sequence in V. If $\{v_n\}_{n=1}^{\infty}$ converges weakly to v in V, then it holds that

$$\|v\| \le \liminf_{n\to\infty} \|v_n\|.$$

[Proof] If $v = 0$, then the assertion is obvious. Suppose that $v \neq 0$. Then by applying Hahn-Banach Theorem (Theorem 10.1.4), we see that there exists a bounded linear functional $f_0 \in V^*$ such that $f_0(v) = \|v\| \neq 0$ and $\|f_0\| = 1$ (see Theorem 10.1.5). Therefore, we obtain that

$$\|v\| = f_0(v) = \lim_{n\to\infty} f_0(v_n) \le \liminf_{n\to\infty} \|f_0\|\|v_n\| = \liminf_{n\to\infty} \|v_n\|,$$

which proves the desired assertion. □

Corollary 10.6.9 Let $(V, \|\cdot\|)$ be a normed space. Suppose that $\{v_n\}_{n=1}^{\infty}$ converges weakly to v in V and $\{f_n\}_{n=1}^{\infty}$ converges (strongly) to f in V^*. Then it holds that

$$\lim_{n\to\infty} f_n(v_n) = f(v).$$

[Proof] We obtain that

$$\begin{aligned} |f_n(v_n) - f(v)| &\le |f_n(v_n) - f(v_n)| + |f(v_n) - f(v)| \\ &\le \|f_n - f\| \, \|v_n\| + |f(v_n) - f(v)| \,. \end{aligned}$$

On the other hand, by Theorem 10.6.7 $\{\|v_n\|\}_{n=1}^{\infty}$ is bounded. Therefore,

$$\lim_{n\to\infty} |f_n(v_n) - f(v)| = 0,$$

which proves the desired assertion. □

Corollary 10.6.10 Let $(V, \|\cdot\|)$ be a normed space. Suppose that $\{v_n\}_{n=1}^{\infty}$ converges to v in V and $\{f_n\}_{n=1}^{\infty}$ converges weakly to f in V^*. Then it holds that

$$\lim_{n\to\infty} f_n(v_n) = f(v).$$

[Proof] We obtain that

$$\begin{aligned}|f_n(v_n) - f(v)| &\le |f_n(v_n) - f_n(v)| + |f_n(v) - f(v)| \\ &\le \|f_n\| \, \|v_n - v\| + |v^{**}(f_n) - v^{**}(f)| \,.\end{aligned}$$

On the other hand, by Theorem 10.6.7 $\{\|f_n\|\}_{n=1}^{\infty}$ is bounded. Therefore,

$$\lim_{n\to\infty} |f_n(v_n) - f(v)| = 0,$$

which proves the desired assertion. □

Corollary 10.6.11 Let $(V, \|\cdot\|)$ be a normed space. Let S be a dense subset of V^*. Suppose that $\{v_n\}_{n=1}^{\infty} \subset V$ is bounded and $v \in V$. Then $\{v_n\}_{n=1}^{\infty}$ converges weakly to v in V if and only if for any $g \in S$,

$$\lim_{n\to\infty} g(v_n) = g(v).$$

[Proof] $\Rightarrow$) It is obvious by definition.

$\Leftarrow$) Put

$$M = \max\{1, \|v\|, \|v_n\| \,|\, n \in \mathbb{N}\} > 0.$$

Let $\epsilon > 0$ be given. Since S is dense in V^*, for any $h \in V^*$, there exists $g \in S$ such that

$$\|h - g\| < \epsilon/(3M).$$

Also, for $g \in S$, by assumption, there exists a $N \in \mathbb{N}$ such that for any $n > N$, $|g(v_n - v)| < \epsilon/3$. Therefore, there exists a $N \in \mathbb{N}$ such that for any $n > N$,

$$\begin{aligned}|h(v_n) - h(v)| &\le |h(v_n) - g(v_n)| + |g(v_n) - g(v)| + |g(v) - h(v)| \\ &< \frac{\epsilon}{3M}M + \frac{\epsilon}{3} + \frac{\epsilon}{3M}M = \epsilon,\end{aligned}$$

which gives the proof. □

Let X be a vector space and $\emptyset \ne S \subset X$. Put

$$\begin{aligned}[S] &:= \mathrm{LS}(S) \\ &:= \{\alpha_1 v_1 + \cdots + \alpha_n v_n \,|\, n \in \mathbb{N},\ \ v_i \in S,\ \ \alpha_i \in \mathbb{C}\ (\text{scalars})\ ,\ \ i = 1, 2, \cdots, n\},\end{aligned}$$

which is called the **linear span** of S.

Proposition 10.6.12 Let $(V, \|\cdot\|)$ be a normed space. Let S be a nonempty subset of V^*. Suppose that the linear span $[S]$ of S is dense in V^* and $\{v_n\}_{n=1}^{\infty} \subset V$ is bounded and $v \in V$. Then $\{v_n\}_{n=1}^{\infty}$ converges weakly to v in V if and only if for any $g \in S$,

$$\lim_{n\to\infty} g(v_n) = g(v). \tag{10.9}$$

[Proof] By Corollary 10.6.11 $\{v_n\}_{n=1}^{\infty}$ converges weakly to v in V if and only if for any $g \in [S]$, (10.9) holds if and only if for any $g \in S$, (10.9) holds. In fact, it is obvious that if for any $g \in [S]$, (10.9) holds, then for any $g \in S$, (10.9) holds. Suppose that for any $g \in S$, (10.9) holds. Then for any $g \in [S]$, there exist $m \in \mathbb{N}$ and $g_i \in S$, $\alpha_i \in \mathbb{C}$ for $i = 1, 2, \cdots, m$ such that

$$g = \sum_{i=1}^{m} \alpha_i g_i.$$

Therefore, we have

$$\begin{aligned}\lim_{n\to\infty} g(v_n) &= \sum_{i=1}^{m} \alpha_i \left(\lim_{n\to\infty} g_i(v_n)\right)\\ &= \sum_{i=1}^{m} \alpha_i \left(g_i(v)\right)\\ &= g(v).\end{aligned}$$

which implies that for any $g \in [S]$, (10.9) holds. □

Corollary 10.6.13 Let $(X, \mathcal{M}, \mu)$ be a σ-finite measure space. Let $1 \leq p < \infty$ and q the conjugate number of p. Let S be a nonempty subset of $L^q(X)$. Suppose that the linear span $[S]$ of S is dense in $L^q(X)$ and $\{f_n\}_{n=1}^{\infty} \subset L^p(X)$ is bounded and $f \in L^p(X)$. Then $\{f_n\}_{n=1}^{\infty}$ converges weakly to f in $L^p(X)$ if and only if for any $g \in S$,

$$\lim_{n\to\infty} \int_X g \cdot f_n \, d\mu = \int_X g \cdot f \, d\mu. \tag{10.10}$$

[Proof] By Riesz Representation (Theorem 7.6.5), we see that $(L^p(X))^* \cong L^q(X)$. Therefore, the proof is immediate from Proposition 10.6.12. □

Theorem 10.6.14 Let $(X, \mathcal{M}, \mu)$ be a σ-finite measure space and let $1 \leq p < \infty$. Suppose that $\{f_n\}_{n=1}^{\infty} \subset L^p(X)$ is bounded and $f \in L^p(X)$. Then $\{f_n\}_{n=1}^{\infty}$ converges weakly to f in $L^p(X)$ if and only if

(1) [case of $p = 1$] for any measurable subset $A \subset X$,

(2) [case of $p > 1$] for any measurable subset $A \subset X$ of finite measure,

$$\lim_{n\to\infty} \int_A f_n \, d\mu = \int_A f \, d\mu. \tag{10.11}$$

[Proof] By Proposition 7.7.1, The subspace LS$\{\chi_A \,|A$ measurable subset of $X\}$ of simple functions in $L^q(X)$ $(1 \le q \le \infty)$ is dense in $L^q(X)$. Therefore, the assertion is immediate from Corollary 10.6.13. Moreover, if $p > 1$ and the characteristic (indicator) function χ_A belongs to $L^q(X)$ $(q < \infty)$, then the measure $\mu(A)$ of $A \subset X$ should be finite. Therefore, it is necessary to consider sets A of finite measure. □

Theorem 10.6.15 Let $[a, b]$ be a closed and bounded interval and let $1 < p < \infty$. Suppose that $\{f_n\}_{n=1}^{\infty} \subset L^p([a, b])$ is bounded and $f \in L^p([a, b])$. Then $\{f_n\}_{n=1}^{\infty}$ converges weakly to f in $L^p([a, b])$ if and only if for all $x \in [a, b]$,

$$\lim_{n\to\infty} \int_a^x f_n \, dm = \int_a^x f \, dm. \tag{10.12}$$

[Proof] By Proposition 7.7.2, if $1 \le q < \infty$ and $[a, b]$ is a closed and bounded interval, then the subspace of all step functions on $[a, b]$ is dense in $L^q([a, b])$. Therefore, by Corollary 10.6.13, $\{f_n\}_{n=1}^{\infty}$ converges weakly to f in $L^p([a, b])$ if and only if for all $x, y \in [a, b]$ with $x < y$,

$$\lim_{n\to\infty} \int_x^y f_n \, dm = \int_x^y f \, dm$$

if and only if for all $x \in [a, b]$,

$$\lim_{n\to\infty} \int_a^x f_n \, dm = \int_a^x f \, dm,$$

which gives the proof. □

Remark 10.6.16 Note that the set of all step functions are not dense in $L^\infty([a, b])$. Therefore, if $p = 1$, then Theorem 10.6.15 is false.

Example 10.6.17 [**Riemann-Lebesgue Lemma**] Let $1 < p < \infty$. For each $n \in \mathbb{N}$, put

$$f_n(x) = \cos nx, \qquad x \in [-\pi, \pi].$$

Then we have

$$\|f_n\|_p = \left(\int_{-\pi}^{\pi} |\cos nx|^p dx \right)^{1/p} \le (2\pi)^{1/p},$$

which implies that $\{f_n\}_{n=1}^{\infty} \subset L^p([-\pi, \pi])$ is bounded. Also, for all $x \in [-\pi, \pi]$, we have

$$\begin{aligned}
\lim_{n\to\infty} \int_{-\pi}^{x} \cos nt \, dt &= \lim_{n\to\infty} \left[\frac{1}{n} \sin nt \right]_{-\pi}^{x} \\
&= \lim_{n\to\infty} \frac{1}{n} (\sin nx + \sin n\pi) \\
&= 0.
\end{aligned}$$

Therefore, by Theorem 10.6.15, $\{f_n\}$ converges weakly to $f = 0$ in $L^p([-\pi, \pi])$.

On the other hand, we have

$$\begin{aligned}\|f_n\|_2 &= \left(\int_{-\pi}^{\pi} |\cos nx|^2 dx\right)^{1/2} \\ &= \left(\int_{-\pi}^{\pi} \cos^2 nx\, dx\right)^{1/2} \\ &= \left(\int_{-\pi}^{\pi} \frac{1+\cos 2nx}{2}\, dx\right)^{1/2} \\ &= \sqrt{\pi},\end{aligned}$$

which implies that there is no subsequence of $\{f_n\}$ which converges (strongly) to $f = 0$ in $L^2([-\pi, \pi])$.

Suppose that $\{f_{n_k}\}_{k=1}^{\infty}$ is a subsequence of $\{f_n\}_{n=1}^{\infty}$ and converges to $f = 0$ pointwise almost everywhere on $[-\pi, \pi]$. Then by the Lebesgue Dominated Convergence Theorem (Theorem 5.1.26), it holds that

$$0 = \|f\|_2^2 = \lim_{k\to\infty} \int_{[-\pi,\pi]} |f_{n_k}|^2\, dm = \pi,$$

which contradicts. Therefore, there is no subsequence of $\{f_n\}$ which converges to $f = 0$ pointwise almost everywhere on $[-\pi, \pi]$.

Theorem 10.6.18 Let $(X, \mathcal{M}, \mu)$ be a σ-finite measure space and let $1 < p < \infty$. Let $\{f_n\}_{n=1}^{\infty}$ be a bounded sequence in $L^p(X)$ which converges pointwise almost everywhere on X to f. Then $\{f_n\}_{n=1}^{\infty}$ converges weakly to f in $L^p(X)$.

[Proof] Since $\{f_n\}_{n=1}^{\infty}$ is bounded in $L^p(X)$, there exists a $M \geq 0$ such that $\|f_n\|_p \leq M$ for all $n \in \mathbb{N}$. Therefore, by Corollary 7.4.10, $\{f_n\}_{n=1}^{\infty}$ is uniformly integrable on X, and, also, by applying Fatou's Lemma we have

$$\int_X |f|^p\, d\mu \leq \liminf_{n\to\infty} \int_X |f_n|^p\, d\mu \leq M^p,$$

which implies that $f \in L^p(X)$. Let A be a measurable subset of X of finite measure. Then $\{f_n\}_{n=1}^{\infty}$ is tight on A. Hence by Vitali Convergence Theorem (Theorem 6.1.9),

$$\lim_{n\to\infty} \int_A f_n\, d\mu = \int_A f\, d\mu.$$

Therefore, by Theorem 10.6.14, the assertion is proved. □

Exercises

10.1 Let U be a linear subspace of a normed space $(V, \|\cdot\|)$ and let $w \in V$ be fixed. Suppose that

$$d(w, U) = \inf\{\|w - u\| \,|\, u \in U\} > 0.$$

Prove that for any $\delta > 0$, there exists a bounded linear functional $L \in V^*$ such that $L(w) = \delta$, $L = 0$ on U and $\|L\| = \delta/d(w, U)$.

10.2 Let U be a linear subspace of a normed space $(V, \|\cdot\|)$ and let $w \in V$ be fixed. Prove that $w \in \overline{U}$ if and only if for bounded linear functional $L \in V^*$ vanishing on U, $L(w) = 0$.

10.3 Let U be a linear subspace of a normed space $(V, \|\cdot\|)$. Prove that U is dense in V if and only if for bounded linear functional $L \in V^*$ vanishing on U, $L = 0$ on V.

10.4 Let $(V, \|\cdot\|)$ be a normed space and let $0 \neq w \in V$ be fixed. Prove that for any $\delta > 0$, there exists a bounded linear functional $L \in V^*$ such that $L(w) = \delta$ and $\|L\| = \delta/\|w\|$.

10.5 Let $(V, \|\cdot\|)$ be a normed space and let $z, w \in V$ such that $z \neq w$. Prove that there exists a bounded linear functional $L \in V^*$ such that $L(z) \neq L(w)$.

10.6 Let (X, τ) be a topological space and A a subset of X. Prove that $\mathring{A} = \emptyset$ if and only if A^c is dense in X, where $\mathring{A}$ is the interior of A.

10.7 Let (X, τ) be a topological space and E a subset of X. Suppose that E is dense in X. Prove that for any (nonempty) open subset U of X, $E \cap U$ is dense in U.

10.8 Let (X, τ) be a topological space. Prove that for any open dense subset O of X, O^c is nowhere dense.

10.9 A Hausdorff topological space (X, τ) is called a locally compact Hausdorff space if every point in X has a compact neighborhood. Prove that every locally compact Hausdorff space is a Baire space.

10.10 Let $(V, \|\cdot\|)$ be a normed space and $v \in V$. Define a linear functional $J_v : V^* \to \mathbb{C}$ by

$$J_v(F) = F(v), \qquad F \in V^*.$$

Prove that $\|J_v\| = \|v\|$, where $\|J_v\|$ is the functional norm of J_v.

10.11 Let V and W be normed spaces and $T : V \to W$ a linear operator. Prove that if $F \circ T \in V^*$ for any $F \in W^*$, then T is bounded.

10.12 Let $\{T_\lambda\}_{\lambda \in \Lambda}$ be a family of bounded linear operators from a Banach space V into a normed space W. Suppose that $\sup_{\lambda \in \Lambda}\{|F(T_\lambda v)|\} < \infty$ for any $F \in W^*$ and $v \in V$. Prove that

$$\sup_{\lambda \in \Lambda}\{\|T_\lambda\|\} < \infty.$$

10.13 Let V be a vector space and let $\|\cdot\|_1, \|\cdot\|_2$ be norms on V. Suppose that there exists a constant $C > 0$ such that

$$\|v\|_2 \leq C\|v\|_1, \qquad v \in V.$$

Prove that the norms $\|\cdot\|_1$ and $\|\cdot\|_2$ are equivalent.

Solutions

1.1 It is obvious that $\mathbb{N}$ is an inductive set. Suppose that $\mathbf{M}$ is an inductive subset of $\mathbb{R}$. Then for any $n \in \mathbb{N}$ we have

$$\begin{aligned} 2 &= 1+1 \in \mathbf{M}, \\ 3 &= 2+1 \in \mathbf{M}, \\ &\cdots \\ n &= (n-1)+1 \in \mathbf{M}, \end{aligned}$$

which implies that $\mathbb{N} \subset \mathbf{M}$.

1.2 The set $\mathbb{Q} \setminus \{0\}$ can be considered as a subset of the countable set $\bigcup_{i=1}^{2} \mathbb{N} \times \mathbb{N}$.

1.4 The proof is a simple modification of the proof of Theorem 1.4.18. Let K be a compact subset of the Hausdorff space (X, τ). If $K = X$ or $K = \emptyset$, then K is closed. Suppose that $K \neq X$ and $K \neq \emptyset$. Let $x \in K^c$. Then for each $y \in K$ there exist $U_y, V_y \in \tau$ such that

$$y \in U_y, \quad x \in V_y, \quad U_y \cap V_y = \emptyset.$$

Then $\{U_y | y \in K\}$ is an open cover of K. Since K is compact, there exists a finite set $\{y_1, \cdots, y_n\} \subset K$ such that $\{U_{y_1}, \cdots, U_{y_n}\}$ is a finite cover of K, i.e.,

$$K \subset \bigcup_{i=1}^{n} U_{y_i} \iff \left(\bigcup_{i=1}^{n} U_{y_i} \right)^{c} \subset K^c.$$

Put

$$V = \bigcap_{i=1}^{n} V_{y_i}$$

and then V is an open set and

$$x \in V \subset \left(\bigcup_{i=1}^{n} U_{y_i} \right)^{c} \subset K^c$$

and so K^c is open. Therefore, K is closed.

1.5 The proof is a simple modification of the Nested Set Theorem (Theorem 1.4.19). Suppose that the intersection is empty, i.e.,

$$\bigcap_{n=1}^{\infty} F_n = \emptyset \quad \Longrightarrow \quad F_1 \subset X = \left(\bigcap_{n=1}^{\infty} F_n\right)^{\mathrm{c}} = \bigcup_{n=1}^{\infty} F_n^{\mathrm{c}},$$

i.e., $\{F_n^{\mathrm{c}}\}_{n=1}^{\infty}$ is an open cover of the compact set F_1, then by the definition of compact set, there exists a finite family $\{F_{\lambda_1}^{\mathrm{c}}, \cdots, F_{\lambda_k}^{\mathrm{c}}\}$ covering F_1. Suppose that $\{F_{\lambda_1}^{\mathrm{c}}, \cdots, F_{\lambda_k}^{\mathrm{c}}\}$ is ascending and so

$$F_1 \subset \bigcup_{\ell=1}^{k} F_{\lambda_\ell}^{\mathrm{c}} = F_{\lambda_k}^{\mathrm{c}}$$

and so for any $n \geq \lambda_k$, since $F_{\lambda_k} \supset F_n$, $F_1 \subset F_{\lambda_k}^{\mathrm{c}} \subset F_n^{\mathrm{c}}$, which implies that

$$F_n = F_n \cap F_1 \subset F_n \cap F_n^{\mathrm{c}} = \emptyset,$$

which is a contradiction.

1.6 Suppose that (a, b) is countably infinite. Then there is a one-to-one and onto function $f : \mathbb{N} \to (a, b)$. Let $\{[x_n, y_n]\}_{n=1}^{\infty}$ be a descending countable family of nonempty closed intervals such that for any $n \in \mathbb{N}$,

$$[x_n, y_n] \subset (a, b) \quad \text{and} \quad f(n) \notin [x_n, y_n].$$

Then by the Nested Set Theorem (Theorem 1.4.19), we have

$$\bigcap_{n=1}^{\infty} [x_n, y_n] \neq \emptyset,$$

and so $x \in \cap_{n=1}^{\infty}[x_n, y_n] \subset (a, b)$ and there is no $\ell \in \mathbb{N}$ such that $f(\ell) = x$, which is a contradiction to that f is onto. Hence (a, b) is uncountable.

1.8 For any closed subset F of $\mathbb{R}$ with $E \subset F$, we have $\overline{E} \subset \overline{F} = F$ and

$$\overline{E} \subseteq \bigcap_{F:\text{closed};E\subset F} F,$$

$$\overline{E} = \bigcap_{F:\text{closed};\overline{E}\subset F} F \supseteq \bigcap_{F:\text{closed};E\subset F} F,$$

where for the last identity we used the fact that $\overline{E}$ is a closed subset containing E.

1.9 Let $\{a_n\}_{n=1}^{\infty}$ be a sequence of real numbers which converges to a. Then for the given $\epsilon = 1$, there exists a $N \in \mathbb{N}$ such that for any $n \geq N$,

$$|a_n - a| < 1,$$

which implies that $a - 1 < a_n < a + 1$ for any $n \geq N$, and so $|a_n| < |a| + 1$ for any $n \geq N$. Put

$$M = \{|a_1|, \cdots, |a_{N-1}|, |a| + 1\}.$$

Then for any $n \in \mathbb{N}$, $|a_n| \leq M$ and so $\{a_n\}_{n=1}^{\infty}$ is bounded.

1.11 Let K be a subset of $\mathbb{R}$.

$\Leftarrow$) Suppose that K is closed and bounded and $\{a_n\}_{n=1}^{\infty} \subset K$ is a sequence. Then since K is bounded, $\{a_n\}_{n=1}^{\infty}$ is bounded. Therefore, by the Bolzano-Weierstrass Theorem (Theorem 1.5.7), $\{a_n\}_{n=1}^{\infty}$ has a convergent subsequence.

$\Rightarrow$) **Closedness.** Let K be a sequentially compact subset. Suppose that K is not closed. Then there exists a closure point $a \in \overline{K}$ with $a \in K$. Since $a \in \overline{K}$, there exists a sequence $\{a_n\}_{n=1}^{\infty} \subset K$ such that $\lim_{n\to\infty} a_n = a$. Since K is sequentially compact, there exists a subsequence $\{a_{n_k}\}_{k=1}^{\infty}$ of $\{a_n\}_{n=1}^{\infty}$ such that $\{a_{n_k}\}_{k=1}^{\infty}$ converges to a point $c \in K$. On the other hand, since $\lim_{n\to\infty} a_n = a$, every subsequence of $\{a_n\}_{n=1}^{\infty}$ converges to a. Therefore, the subsequence $\{a_{n_k}\}_{k=1}^{\infty}$ also converges to a. Hence the subsequence $\{a_{n_k}\}_{k=1}^{\infty}$ also converges to a and c, and so $a = c$. Thus $a = c \in K$, which is contradicts.

Boundness. Suppose that K is not bounded. Then there exists a sequence $\{a_n\}_{n=1}^{\infty}$ with $|a_n| \geq n$ for any $n \in \mathbb{N}$, and then every subsequence of $\{a_n\}_{n=1}^{\infty}$ diverges, which contradicts.

2.3 Since $\mathcal{A}_1$ and $\mathcal{A}_2$ are nonempty families, $\mathcal{A}_1 \cup \mathcal{A}_2$ is a nonempty family. For any $A \in \mathcal{A}_1 \cup \mathcal{A}_2$, there exist $A_1 \in \mathcal{A}_1$ and $A_2 \in \mathcal{A}_2$ such that $A = A_1 \cup A_2$ and so

$$A^c = X \setminus A = (X_1 \setminus A_1) \cup (X_2 \setminus A_2) \in \mathcal{A}_1 \cup \mathcal{A}_2.$$

For any finite family $\{A_i \cup B_i\}_{i=1}^{n}$ with $A_i \in \mathcal{A}_1$ and $B_i \in \mathcal{A}_2$, we have

$$\bigcup_{i=1}^{n}(A_i \cup B_i) = \left(\bigcup_{i=1}^{n} A_i\right) \cup \left(\bigcup_{i=1}^{n} B_i\right) \in \mathcal{A}_1 \cup \mathcal{A}_2.$$

2.4 Since $\mathcal{A}$ is nonempty, $X_0 \cap \mathcal{A}$ is nonempty. For any $A \in \mathcal{A}$,

$$(X_0 \cap A)^c = X_0 \setminus (X_0 \cap A) = X_0 \cap A^c \in X_0 \cap \mathcal{A}$$

and for any finite family $\{X_0 \cap A_i\}_{i=1}^{n} \subset X_0 \cap \mathcal{A}$,

$$\bigcup_{i=1}^{n}(X_0 \cap A_i) = X_0 \cap \left(\bigcup_{i=1}^{n} A_i\right) \in X_0 \cap \mathcal{A}.$$

2.5 It is obvious that $\sigma(\mathcal{A} \cap X_0) \subset \sigma(\mathcal{A})|_{X_0}$. Put

$$\mathcal{S} = \{E \in \sigma(\mathcal{A}) \mid E \cap X_0 \in \sigma(\mathcal{A} \cap X_0)\}.$$

Then $\mathcal{S}$ is a σ-algebra. In fact, we have

(i) $\emptyset, X \in \sigma(\mathcal{A})$ and $\emptyset \cap X_0 = \emptyset, X \cap X_0 = X_0 \in \sigma(\mathcal{A} \cap X_0)$ and so $\emptyset, X \in \mathcal{S}$;

(ii) if $E \in \mathcal{S}$, i.e., $E \in \sigma(\mathcal{A})$ and $E \cap X_0 \in \sigma(\mathcal{A} \cap X_0)$, then $E^{cX} \in \sigma(\mathcal{A})$ and $E^{cX} \cap X_0 = (E \cap X_0)^{cX_0} \in \sigma(\mathcal{A} \cap X_0)$, where E^{cX} is the complement of E with respect to X and $(E \cap X_0)^{cX_0}$ is the complement of $E \cap X_0$ with respect to X_0, and so $E^{cX} \in \mathcal{S}$;

(iii) if $\{E_i\}_{i=1}^{\infty} \subset \mathcal{S}$, then for each $i \in \mathbb{N}$, $E_i \in \sigma(\mathcal{A})$ and $E_i \cap X_0 \in \sigma(\mathcal{A} \cap X_0)$ and so $\cup_{i=1}^{\infty} E_i \in \sigma(\mathcal{A})$ and

$$\left(\bigcup_{i=1}^{\infty} E_i\right) \cap X_0 = \bigcup_{i=1}^{\infty} (E_i \cap X_0) \in \sigma(\mathcal{A} \cap X_0),$$

which implies that $\cup_{i=1}^{\infty} E_i \in \mathcal{A}$.

Therefore, $\mathcal{S}$ is a σ-algebra. On the other hand, for any $A \in \mathcal{A}$, $A \in \sigma(\mathcal{A})$ and $A \cap X_0 \in \mathcal{A} \cap X_0 \subset \sigma(\mathcal{A} \cap X_0)$, and so $\mathcal{A} \subset \mathcal{S}$. Therefore, $\sigma(\mathcal{A}) \subset \mathcal{S}$ and then $\sigma(\mathcal{A}) = \mathcal{S}$. Hence for any $E \in \sigma(\mathcal{A}) = \mathcal{S}$, $E \cap X_0 \in \sigma(\mathcal{A} \cap X_0)$, i.e., $\sigma(\mathcal{A})|_{X_0} \subset \sigma(\mathcal{A} \cap X_0)$.

[**Another solution:**] Put $\mathcal{B} = \sigma(\mathcal{A})$. If $\mathcal{C}$ is a σ-algebra of subsets of X_0 containing $X_0 \cap \mathcal{A}$, then $\mathcal{C} \cup (X_0^c \cap \mathcal{B})$ is a σ-algebra of subsets of X containing $\mathcal{A}$ and so, since $\mathcal{B}$ is the smallest σ-algebra containing $\mathcal{A}$,

$$\mathcal{B} \subset \mathcal{C} \cup (X_0^c \cap \mathcal{B}),$$

and

$$\mathcal{C} = X_0 \cap (\mathcal{C} \cup (X_0^c \cap \mathcal{B})) \supset X_0 \cap \mathcal{B}.$$

2.6 Put

$$\mathcal{H} = \bigcap_{\mathcal{F}:\sigma\text{-algebra containing } \mathcal{C}} \mathcal{F}.$$

Then $\mathcal{H}$ is a σ-algebra containing $\mathcal{C}$ and so, $\sigma(\mathcal{C}) \subset \mathcal{H}$ since $\sigma(\mathcal{C})$ is a smallest σ-algebra containing $\mathcal{C}$. Also, since $\sigma(\mathcal{C})$ is a σ-algebra containing $\mathcal{C}$, $\mathcal{H} \subset \sigma(\mathcal{C})$.

2.8 For each $\lambda \in \Lambda$ and $F_\lambda \in \mathcal{F}_\lambda$,

$$\mathrm{Pr}_\lambda^{-1}(F_\lambda) = \prod_{\alpha \in \Lambda} E_\alpha, \qquad E_\alpha = \begin{cases} F_\lambda, & \alpha = \lambda, \\ X_\alpha, & \alpha \neq \lambda. \end{cases}$$

Therefore, we have

$$\left\{\mathrm{Pr}_\lambda^{-1}(F_\lambda) \mid F_\lambda \in \mathcal{F}_\lambda, \ \ \lambda \in \Lambda\right\} \subset \left\{\prod_{\lambda \in \Lambda} F_\lambda \,\middle|\, F_\lambda \in \mathcal{F}_\lambda\right\},$$

which implies that

$$\bigotimes_{\lambda \in \Lambda} \mathcal{F}_\lambda = \sigma\left(\left\{\mathrm{Pr}_\lambda^{-1}(F_\lambda) \mid F_\lambda \in \mathcal{F}_\lambda, \ \ \lambda \in \Lambda\right\}\right) \subset \sigma\left(\left\{\prod_{\lambda \in \Lambda} F_\lambda \,\middle|\, F_\lambda \in \mathcal{F}_\lambda\right\}\right).$$

Also, since Λ is countable,

$$\prod_{\lambda \in \Lambda} F_\lambda = \bigcap_{\lambda \in \Lambda} \mathrm{Pr}_\lambda^{-1}(F_\lambda)$$

is a countable intersection, and so

$$\prod_{\lambda\in\Lambda} F_\lambda \in \sigma\left(\{\mathrm{Pr}_\lambda^{-1}(F_\lambda) \,|\, F_\lambda \in \mathcal{F}_\lambda,\ \ \lambda\in\Lambda\}\right) = \bigotimes_{\lambda\in\Lambda}\mathcal{F}_\lambda.$$

Hence, we have

$$\left\{\prod_{\lambda\in\Lambda} F_\lambda \,\middle|\, F_\lambda \in \mathcal{F}_\lambda\right\} \subset \bigotimes_{\lambda\in\Lambda}\mathcal{F}_\lambda,$$

which implies that

$$\sigma\left(\{\mathrm{Pr}_\lambda^{-1}(F_\lambda) \,|\, F_\lambda \in \mathcal{F}_\lambda,\ \ \lambda\in\Lambda\}\right) \subset \bigotimes_{\lambda\in\Lambda}\mathcal{F}_\lambda.$$

2.14 Let $(X, \mathcal{M}, \mu)$ be a σ-finite measure space and $E \in \mathcal{M}$ with $\mu(E) = \infty$. Then there exists a countable family $\{E_n\}_{n=1}^\infty \subset \mathcal{M}$ such that $\mu(E_n) < \infty$ for any $n \in \mathbb{N}$ and $X = \cup_{n=1}^\infty E_n$. Therefore, we have

$$E = E\bigcap X = \bigcup_{n=1}^\infty \left(E\bigcap E_n\right)$$

and for any $n \in \mathbb{N}$, $\mu(E\cap E_n) \le \mu(E_n) < \infty$ and there there exists a $n \in \mathbb{N}$ such that $\mu(E\cap E_n) > 0$, otherwise $\mu(E) = 0$ which contradicts to $\mu(E) = \infty$. Hence μ is semifinite.

2.15 Let X be an uncountable set and $(X, \mathcal{P}(X), c)$ a measure space with counting measure c. Then c is semifinite, but it is not σ-finite.

2.20 (1) Since $\mu(E \triangle F) = 0$, $\mu(E\setminus F) = \mu(F\setminus E) = 0$ and

$$\begin{aligned}\mu(E) &= \mu((E\cap F)\cup(E\setminus F)) = \mu(E\cap F) + \mu(E\setminus F)\\ &= \mu(E\cap F)\\ &= \mu(F\cap E) + \mu(F\setminus E) = \mu((F\cap E)\cup(F\setminus E))\\ &= \mu(F).\end{aligned}$$

(2) For any $E\in\mathcal{M}$, $\mu(E\triangle E) = \mu(\emptyset) = 0$ and so $E\sim E$, i.e., $\sim$ is reflexive. For any $E, F\in\mathcal{M}$, if $E\sim F$, then $0 = \mu(E\triangle F) = \mu(F\triangle E)$ and so $F\sim E$, i.e., $\sim$ is symmetric. For any $E, F, G\in\mathcal{M}$, if $E\sim F$ and $F\sim G$, then we have

$$\begin{aligned}E\setminus G &= ((E\setminus F)\setminus G)\cup((F\setminus G)\cap E)\\ &\subset (E\setminus F)\cup(F\setminus G),\end{aligned}$$

similarly, we have

$$G\setminus E \subset (F\setminus E)\cup(G\setminus F).$$

Therefore, we see that $\mu(E\triangle G) = 0$ and so $E\sim G$, i.e., $\sim$ is transitive. Hence, $\sim$ is an equivalence relation.

2.22 Let $\nu = \nu^+ - \nu^-$ be the Jordan decomposition of ν. Then for any $E \in \mathcal{M}$, we have

$$\begin{aligned}|\nu(E)| &= |\nu^+(E) - \nu^-(E)| \le \nu^+(E) + \nu^-(E)\\ &= |\nu|(E).\end{aligned}$$

2.24 $\Rightarrow$) It is obvious.

$\Leftarrow$) Suppose that $E \subset X$ and for any $F \in \mathcal{A}$,

$$\mu^*(F) = \mu^*(F \cap E) + \mu^*(F \cap E^c)$$

holds. Let $A \subset X$ with $\mu^*(A) < \infty$ and $\epsilon > 0$ be given. There exists a countable family $\{E_k\}_{k=1}^{\infty} \subset \mathcal{A}$ with $A \subset \cup_{k=1}^{\infty} E_k$ such that

$$\mu^*(A) + \epsilon \ge \sum_{k=1}^{\infty} \mu_*(E_k).$$

Then we obtain that

$$\begin{aligned}\mu^*(A) + \epsilon &\ge \sum_{k=1}^{\infty} \mu_*(E_k) = \sum_{k=1}^{\infty} \mu^*(E_k) = \sum_{k=1}^{\infty} \mu^*(E_k \cap E) + \sum_{k=1}^{\infty} \mu^*(E_k \cap E^c)\\ &\ge \mu^*(A \cap E) + \mu^*(A \cap E^c),\end{aligned}$$

which implies that

$$\mu^*(A) \ge \mu^*(A \cap E) + \mu^*(A \cap E^c).$$

Therefore, E is μ^*-measurable.

2.25 Let $\mathcal{A}$ be an algebra of subsets of X and μ_* be a premeasure defined on $\mathcal{A}$. Let $A \subset X$. Suppose that A is not μ^*-measurable. Then $\mu^*(A) > 0$ and by Exercise **2.24**, there exists a $B \in \mathcal{A} \in \sigma(\mathcal{A}) \subset \mathfrak{M}$ (by Proposition 2.5.24) such that

$$\mu^*(B) < \mu^*(B \cap A) + \mu^*(B \cap A^c).$$

Since $A, A^c, B \in \sigma(\mathfrak{M} \cup \{A\})$, the restriction $\mu^*|_{\sigma(\mathfrak{M} \cup \{A\})}$ of μ^* to the σ-algebra $\sigma(\mathfrak{M} \cup \{A\})$ does not satisfy the countable additivity.

3.1 It is obvious that $\emptyset \in \mathcal{A}$ and $\mathbb{R} = (a, \infty) \in \mathcal{A}$ for $a = -\infty$. Let $E \in \mathcal{A}$. Then by the definition of $\mathcal{A}$, there exist increasing finite sequences $\{a_i\}_{i=1}^n$ and $\{b_i\}_{i=1}^n$ such that $a_{i+1} > b_i$ and

$$E = (a_1, b_1] \cup \cdots (a_n, b_n] \cup A,$$

where $A = \emptyset$ or (a, ∞) for $a \ge b_n$, and so

$$E^c = F \cup (c_1, d_1] \cup \cdots (c_{n-1}, d_{n-1}] \cup G,$$

where $c_i = b_i$, $d_i = a_{i+1}$ for $i = 1, \cdots, n-1$ and

$$F = \begin{cases} \emptyset, & a_1 = -\infty, \\ (-\infty, a_1], & a_1 > -\infty \end{cases} \quad \text{and} \quad G = \begin{cases} (b_n, \infty), & A = \emptyset \\ (b_n, a], & A = (a, \infty) \quad \text{for } a > b_n \\ \emptyset, & A = (a, \infty) \quad \text{for } a = b_n. \end{cases}$$

Therefore, E^c belongs to $\mathcal{A}$ and so $\mathcal{A}$ is closed under complements. Let $E \in \mathcal{A}$ be given by

$$E = I_1 \cup \cdots \cup I_n,$$

which is a finite disjoint union of open-closed intervals, and let I an open-closed interval. Put $\Lambda = \{i \,|\, i \in \{1, \cdots, n\} \text{ and } \overline{I_i} \cap \overline{I} \neq \emptyset\}$. Then

$$E \cup I = \left(\cup_{i \in \{1,\cdots,n\} \setminus \Lambda} I_i\right) \cup \left(\cup_{i \in \Lambda} I_i \cup I\right),$$

which is a finite disjoint union of open-closed intervals and so $E \cup I \in \mathcal{A}$. Therefore, by mathematical induction, we see that union of two elements of $\mathcal{A}$ belongs to $\mathcal{A}$ and so $\mathcal{A}$ is closed under finite unions. Finally, $\mathcal{A}$ is an algebra.

3.2 Let $E \subset \mathbb{R}$ be given. Let $\{I_k\}_{k=1}^{\infty}$ be a countable collection of open-closed intervals I_k such that $E \subset \cup_{k=1}^{\infty} I_k$. Then for any given $\epsilon > 0$, we can construct a countable collection $\{J_k\}_{k=1}^{\infty}$ of open intervals J_k such that $I_k \subset J_k$ and $\ell(J_k) \leq \ell(I_k) + \epsilon/2^k$ for $k = 1, 2, \cdots$, e.g.

$$J_k = (\inf(I_k), \sup(I_k) + \epsilon/2^{k+1}), \qquad k = 1, 2, \cdots.$$

Then $E \subset \cup_{k=1}^{\infty} J_k$ and

$$m^{**}(E) \leq \sum_{k=1}^{\infty} \ell(J_k) \leq \sum_{k=1}^{\infty} \ell(I_k) + \epsilon. \tag{10.13}$$

Therefore, by taking infimum on the right hand side of (10.13), we have

$$m^{**}(E) \leq m^*(E) + \epsilon.$$

Since $\epsilon > 0$ is arbitrary, we have $m^{**}(E) \leq m^*(E)$. Similarly, let $\{J_k\}_{k=1}^{\infty}$ be a countable collection of open intervals J_k such that $E \subset \cup_{k=1}^{\infty} J_k$. Then for any given $\epsilon > 0$, we can construct a countable collection $\{I_k\}_{k=1}^{\infty}$ of open-closed intervals I_k such that $J_k \subset I_k$ and $\ell(I_k) \leq \ell(J_k) + \epsilon/2^k$ for $k = 1, 2, \cdots$, e.g.

$$I_k = (\inf(J_k), \sup(J_k) + \epsilon/2^{k+1}], \qquad k = 1, 2, \cdots.$$

Then $E \subset \cup_{k=1}^{\infty} I_k$ and

$$m^*(E) \leq \sum_{k=1}^{\infty} \ell(I_k) \leq \sum_{k=1}^{\infty} \ell(J_k) + \epsilon. \tag{10.14}$$

Therefore, by taking infimum on the right hand side of (10.14), we have

$$m^*(E) \leq m^{**}(E) + \epsilon.$$

Since $\epsilon > 0$ is arbitrary, we have $m^*(E) \leq m^{**}(E)$. Hence $m^{**}(E) = m^*(E)$.

3.3 (i) If there exists k such that $m^*(E_k) = \infty$, then (3.4) holds as both sides are infinite.

(ii) Suppose that $m^*(E_i) < \infty$ for any i. Then for any $\epsilon > 0$ and any i, there exists a countable collection $\{I_{ik}\}_{k=1}^{\infty}$ of nonempty open, bounded intervals covering E_i such that

$$m^*(E_i) > \sum_{k=1}^{\infty} \ell(I_{ik}) - \frac{\epsilon}{2^i}.$$

Then $\{I_{ik}\}_{1\le i,k<\infty}$ is a countable collection of nonempty open, bounded intervals covering $\cup_{i=1}^{\infty} E_i$ and we have

$$m^*\left(\bigcup_{k=1}^{\infty} E_k\right) \le \sum_{i,k=1}^{\infty} \ell(I_{ik}) = \sum_{i=1}^{\infty}\sum_{k=1}^{\infty} \ell(I_{ik}) < \sum_{i=1}^{\infty}\left(m^*(E_i) + \frac{\epsilon}{2^i}\right)$$
$$= \sum_{i=1}^{\infty} m^*(E_i) + \epsilon,$$

which implies the countable subadditivity (3.4).

3.4 Let $E \subset \mathbb{R}$ with $m^*(E) = 0$. Then for any $A \subset \mathbb{R}$, we have $0 \le m^*(A\cap E) \le m^*(E) = 0$, which implies that $m^*(A \cap E) = 0$ and

$$m^*(A) \ge m^*(A\cap E^{c}) = m^*(A\cap E) + m^*(A\cap E^{c}).$$

Therefore, E is measurable.

3.5 Let E_1 and E_2 be measurable. Then for any $A \subset \mathbb{R}$, we have

$$\begin{aligned} m^*(A) &= m^*(A\cap E_1) + m^*(A\cap E_1^{c}) \\ &= m^*(A\cap E_1) + m^*(A\cap E_1^{c}\cap E_2) + m^*(A\cap E_1^{c}\cap E_2^{c}) \\ &\ge m^*(A\cap(E_1\cup E_2)) + m^*(A\cap(E_1\cup E_2)^{c}), \end{aligned}$$

which implies that $E_1 \cup E_2$ is measurable.

3.6 Since E_k $(k = 1, 2, \cdots, n)$ is measurable, we obtain that

$$\begin{aligned} m^*\left(A\cap\bigcup_{k=1}^{n} E_k\right) &= m^*\left(\left(A\cap\bigcup_{k=1}^{n} E_k\right)\cap E_1\right) + m^*\left(\left(A\cap\bigcup_{k=1}^{n} E_k\right)\cap E_1^{c}\right) \\ &= m^*(A\cap E_1) + m^*\left(A\cap\bigcup_{k=2}^{n} E_k\right) \\ &= m^*(A\cap E_1) + m^*\left(\left(A\cap\bigcup_{k=2}^{n} E_k\right)\cap E_2\right) \\ &\quad + m^*\left(\left(A\cap\bigcup_{k=2}^{n} E_k\right)\cap E_2^{c}\right) \\ &= m^*(A\cap E_1) + m^*(A\cap E_2) + m^*\left(A\cap\bigcup_{k=3}^{n} E_k\right) \\ &\cdots \\ &= \sum_{k=1}^{n} m^*(A\cap E_k), \end{aligned}$$

which gives the proof of (3.7). By taking $A = \mathbb{R}$ in (3.7), we see (3.8).

3.7 For any countable disjoint family $\{E_k\}_{k=1}^{\infty}$ of measurable sets E_k and any subset $A \subset \mathbb{R}$, we put $E = \cup_{k=1}^{\infty} E_k$ and then we see that

$$\begin{aligned} m^*(A) &= m^*\left(A \cap \bigcup_{k=1}^{n} E_k\right) + m^*\left(A \cap \left(\bigcup_{k=1}^{n} E_k\right)^{c}\right) \\ &\geq m^*\left(A \cap \bigcup_{k=1}^{n} E_k\right) + m^*\left(A \cap E^{c}\right). \end{aligned}$$

Therefore, by applying (3.7), we obtain that

$$m^*(A) \geq \sum_{k=1}^{n} m^*\left(A \cap E_k\right) + m^*\left(A \cap E^{c}\right),$$

which, by the countable subadditivity of m^*, implies that

$$\begin{aligned} m^*(A) &\geq \sum_{k=1}^{\infty} m^*\left(A \cap E_k\right) + m^*\left(A \cap E^{c}\right) \\ &\geq m^*\left(A \cap E\right) + m^*\left(A \cap E^{c}\right), \end{aligned}$$

which implies that $E = \cup_{k=1}^{\infty} E_k \in \mathfrak{M}$. For any countable family $\{E_k\}_{k=1}^{\infty}$ of measurable sets E_k, put

$$F_1 = E_1, \qquad F_k = E_k \setminus \bigcup_{i=1}^{k-1} E_i \in \mathfrak{M}, \qquad k = 2, 3, \cdots.$$

Then for $i \neq j$, $F_i \cap F_j = \emptyset$, i.e., $\{F_k\}_{k=1}^{\infty}$ is a countable disjoint family. Therefore, we have

$$\bigcup_{k=1}^{\infty} E_k = \bigcup_{k=1}^{\infty} F_k \in \mathfrak{M}.$$

Hence, $\mathfrak{M}$ is closed under the countable union.

3.8 Since E is measurable, we have

$$m^*(A) = m^*(A \cap E) + m^*(A \cap E^{c}) = m^*(E) + m^*(A \setminus E),$$

which implies that

$$m^*(A \setminus E) = m^*(A) - m^*(E)$$

since $m^*(E) < \infty$.

3.10 From the countable subadditivity (see Proposition 3.1.9) of the Lebesgue outer measure, it is obvious that

$$m\left(\bigcup_{k=1}^{\infty} E_k\right) \leq \sum_{k=1}^{\infty} m(E_k).$$

On the other hand, for any $n \in \mathbb{N}$, by the monotonicity of the Lebesgue outer measure and (3.8), we have

$$m\left(\bigcup_{k=1}^{\infty} E_k\right) = m^*\left(\bigcup_{k=1}^{\infty} E_k\right) \geq m^*\left(\bigcup_{k=1}^{n} E_k\right) = \sum_{k=1}^{n} m^*(E_k) = \sum_{k=1}^{n} m(E_k),$$

which implies tat

$$m\left(\bigcup_{k=1}^{\infty} E_k\right) \geq \sum_{k=1}^{\infty} m(E_k).$$

Therefore, m satisfies the countable additivity (3.10).

3.11 (1) Define

$$C_1 = E_1, \qquad C_k = E_k \setminus E_{k-1}, \qquad k = 2, 3, \cdots.$$

Then $\{C_k\}_{k=1}^{\infty}$ is a countable disjoint family of measurable sets and

$$\bigcup_{k=1}^{\infty} E_k = \bigcup_{k=1}^{\infty} C_k.$$

Therefore, by the countable additivity (see Theorem 3.3.3), we obtain that

$$\begin{aligned} m\left(\bigcup_{k=1}^{\infty} E_k\right) &= m\left(\bigcup_{k=1}^{\infty} C_k\right) = \sum_{k=1}^{\infty} m(C_k) = \lim_{n\to\infty} \sum_{k=1}^{n} m(C_k) = \lim_{n\to\infty} \left(\bigcup_{k=1}^{n} C_k\right) \\ &= \lim_{n\to\infty} m(E_n), \end{aligned}$$

which gives the proof of (3.11).

(2) Define

$$D_1 = \bigcap_{k=1}^{\infty} E_k, \qquad D_k = E_{k-1} \setminus E_k, \qquad k = 2, \cdots.$$

Then $\{D_k\}_{k=0}^{\infty}$ is a countable disjoint family of measurable sets. Therefore, by the countable additivity (see Theorem 3.3.3) and the excision property (see Proposition 3.2.14), we obtain that

$$\begin{aligned} m(E_1) &= m\left(\bigcup_{k=1}^{\infty} D_k\right) = \sum_{k=1}^{\infty} m(D_k) = m(D_1) + \lim_{n\to\infty} \sum_{k=2}^{n} m(D_k) \\ &= m(D_1) + \lim_{n\to\infty} m\left(\bigcup_{k=2}^{n} D_k\right) = m\left(\bigcap_{k=1}^{\infty} E_k\right) + \lim_{n\to\infty} m\left(E_1 \setminus E_n\right) \\ &= m\left(\bigcap_{k=1}^{\infty} E_k\right) + \lim_{n\to\infty} \left(m\left(E_1\right) - m\left(E_n\right)\right) \quad (\text{since } m(E_n) < \infty) \\ &= m\left(\bigcap_{k=1}^{\infty} E_k\right) + m\left(E_1\right) - \lim_{n\to\infty} m\left(E_n\right), \end{aligned}$$

which implies (3.12) since $m(E_1) < \infty$.

Note. Let's consider another proof of (2) in Theorem 3.3.4. Define

$$F_0 = \bigcap_{k=1}^{\infty} E_k, \qquad F_k = E_1 \setminus E_k, \qquad k = 1, 2, \cdots .$$

Then $\{F_k\}_{k=1}^{\infty}$ is an increasing family of measurable sets and so, by (1) and the excision property of m, we have

$$m\left(\bigcup_{k=1}^{\infty} F_k\right) = \lim_{n\to\infty} m(F_n) = m(E_1) - \lim_{n\to\infty} m(E_n).$$

On the other hand, we have

$$\begin{aligned} m\left(\bigcup_{k=1}^{\infty} F_k\right) &= m\left(\bigcup_{k=1}^{\infty} \left(E_1 \bigcap E_k^{\mathrm{c}}\right)\right) = m\left(E_1 \bigcap \left(\bigcup_{k=1}^{\infty} E_k^{\mathrm{c}}\right)\right) \\ &= m\left(E_1 \bigcap \left(\bigcap_{k=1}^{\infty} E_k\right)^{\mathrm{c}}\right) = m\left(E_1 \setminus \bigcap_{k=1}^{\infty} E_k\right) \\ &= m\left(E_1\right) - m\left(\bigcap_{k=1}^{\infty} E_k\right). \end{aligned}$$

Therefore, we have

$$m(E_1) - \lim_{n\to\infty} m(E_n) = m\left(E_1\right) - m\left(\bigcap_{k=1}^{\infty} E_k\right),$$

which, since $m(E_1) < \infty$, gives the proof.

3.12 Put

$$E_0 = \limsup_{n\to\infty} E_n = \bigcap_{n\geq 1} \bigcup_{k\geq n} E_k.$$

Then E_0 is measurable and, by the monotonicity of m and the assumption, for any $n \in \mathbb{N}$, we have

$$m\left(\bigcup_{k\geq n} E_k\right) \leq \sum_{k=n}^{\infty} m(E_k) < \infty.$$

Therefore, by applying Theorem 3.3.4, we have

$$m(E_0) = \lim_{n\to\infty} m\left(\bigcup_{k\geq n} E_k\right) = \lim_{n\to\infty} \sum_{k=n}^{\infty} m(E_k) = 0.$$

Then for any $x \in \mathbb{R} \setminus E_0$, we obtain that

$$\begin{aligned} x \in \mathbb{R} \setminus E_0 \quad &\Longleftrightarrow \quad x \in \left(\bigcap_{n\geq 1} \bigcup_{k\geq n} E_k\right)^{\mathrm{c}} \\ &\Longleftrightarrow \quad x \in \bigcup_{n\geq 1} \bigcap_{k\geq n} E_k^{\mathrm{c}} \\ &\Longleftrightarrow \quad \text{there exists } n_0 \in \mathbb{N} \text{ such that for any } k \geq n_0\,, x \notin E_k, \end{aligned}$$

which gives the proof.

4.1 Since $f^{-1}(\emptyset) = \emptyset \in \mathcal{M}$, $\emptyset \in \mathcal{F}$. Also, since $f^{-1}(Y) = X \in \mathcal{M}$, $X \in \mathcal{F}$. If $F \in \mathcal{F}$, then $f^{-1}(F) \in \mathcal{M}$ and $f^{-1}(F^{c}) = (f^{-1}(F))^{c} \in \mathcal{M}$. Therefore $\mathcal{F}$ is closed under complements. Let $\{F_i\}_{i=1}^{\infty} \subset \mathcal{F}$ be a countable family. Then we have

$$f^{-1}(\cup_{i=1}^{\infty} F_i) = \cup_{i=1}^{\infty} f^{-1}(F_i) \in \mathcal{M},$$

which implies that $\cup_{i=1}^{\infty} F_i \in \mathcal{F}$. Therefore, $\mathcal{F}$ is closed under countable unions. Hence $\mathcal{F}$ is a σ-algebra.

4.5 For the given function f, we have

$$X_{\infty} = \bigcap_{n\in\mathbb{N}} f^{-1}((n,\infty]), \qquad X_{-\infty} = \bigcap_{n\in\mathbb{N}} f^{-1}([-\infty,n))$$

and

$$f^{-1}((c,\infty]) = f^{-1}((c,\infty)) \bigcup X_{\infty} = f_{\mathrm{f}}^{-1}((c,\infty)) \bigcup X_{\infty}.$$

Then the proof is straightforward.

4.11 For each $n \in \mathbb{N}$, put

$$f_n = \chi_{(-\infty,n]}, \qquad f = 1.$$

$\{f_n\}$ converges pointwise on $\mathbb{R}$ to f. But for any Lebesgue measurable set E with finite (Lebesgue) measure,

$$\sup_{x\in E^{c}} \{|f_n(x) - f(x)|\} = 1,$$

which implies that f_n does not converge uniformly on E^{c} to f.

5.1 Consider the function $f : [0,1] \to \mathbb{R}$ defined by

$$f(x) = \begin{cases} 0, & x \in [0,1] \cap \mathbb{Q}, \\ 1, & x \in [0,1] \cap \mathbb{Q}^{c}. \end{cases}$$

Then f is Lebesgue integrable with $\int_0^1 f\, dm = 1$, but not Riemann integrable.

5.2 Consider the function $f = \chi_{\mathbb{N}}$. Then since $\mathbb{N}$ is countable, $m(\mathbb{N}) = 0$ and so $\int_{\mathbb{R}} f(x)dm(x) < \infty$, but $\lim_{|x|\to\infty} f(x) \neq 0$.

5.3 For the sequence $\{f_n\}_{n=1}^{\infty}$ of functions f_n defined by

$$f_n(x) = \frac{1}{\sqrt{2\pi}} e^{-\frac{(x-n)^2}{2}}, \qquad x \in \mathbb{R}.$$

Suppose that $f_n \le g$ a.e on $\mathbb{R}$ for all $n \in \mathbb{N}$ and g is integrable. Then by applying the dominated convergence theorem, we have

$$1 = \lim_{n\to\infty} \int_{\mathbb{R}} f_n\, dm = \int_{\mathbb{R}} \lim_{n\to\infty} f_n\, dm = \int_{\mathbb{R}} 0\, dm = 0,$$

which contradicts to $0 \neq 1$. Therefore, g is not integrable.

5.11 Suppose that f is continuous. Let $\epsilon > 0$ be given. Then f is Riemann integrable and so there exists a step function $\phi = \sum_{i=1}^{l} \alpha_i \chi_{I_i}$, $I_i = [a_i, b_i]$ such that

$$\int_a^b |f - \phi|\, dm < \epsilon.$$

Then we obtain that

$$\begin{aligned}\left|\int_a^b f(x)\sin(nx)dx\right| &\le \int_a^b |(f(x) - \phi(x))\sin(nx)|\, dx + \left|\int_a^b \phi(x)\sin(nx)dx\right| \\ &\le \epsilon + \left|\int_a^b \phi(x)\sin(nx)dx\right|,\end{aligned}$$

and by using the change of variable formula, we obtain that

$$\begin{aligned}\left|\int_a^b \phi(x)\sin(nx)dx\right| &= \left|\sum_{i=1}^{l} \alpha_i \int_{a_i}^{b_i} \sin(nx)dx\right| \\ &\le \frac{1}{n}\sum_{i=1}^{l} |\alpha_i| \left|\int_{na_i}^{nb_i} \sin(y)dy\right| \\ &\le \frac{2lM}{n},\end{aligned}$$

where $M = \max\{|\alpha_i| \,|\, i = 1, 2, \cdots, n\}$. Therefore, we have

$$\lim_{n\to\infty} \left|\int_a^b f(x)\sin(nx)dx\right| \le \epsilon$$

for arbitrary $\epsilon > 0$, and so

$$\lim_{n\to\infty} \int_a^b f(x)\sin(nx)dx = 0.$$

Suppose that $f : [a, b] \to \mathbb{R}$ is measurable and $|f(x)| \le M$ for all $x \in [a, b]$. Then for each $\epsilon > 0$, by the Lusin's theorem (Theorem 4.2.5), there exist a continuous function g on $\mathbb{R}$ and a closed set F contained in $[a, b]$ such that

$$f = g \text{ on } F \text{ and } m([a, b] \setminus F) < \epsilon.$$

Therefore, we have

$$\begin{aligned}\left|\int_a^b f(x)\sin(nx)dx\right| &\le \left|\int_F f(x)\sin(nx)dx\right| + \left|\int_{[a,b]\setminus F} f(x)\sin(nx)dx\right| \\ &\le \left|\int_F g(x)\sin(nx)dx\right| + M\epsilon.\end{aligned}$$

Hence we can see that

$$\lim_{n\to\infty} \int_a^b f(x)\sin(nx)dx = 0.$$

6.1 **(1)** If $\alpha = 0$, then $\alpha\mathcal{F} = \{0\}$ is uniform integrable. Suppose that $\alpha \in \mathbb{R}$ with $\alpha \neq 0$. Let $\epsilon > 0$ be given. Since $\mathcal{F}$ is uniformly integrable over X, there exists a $\delta > 0$ such that for any $f \in \mathcal{F}$ and $A \in \mathcal{M}$ with $\mu(A) < \delta$, $\int_A |f| d\mu < \epsilon/|\alpha|$, and so

$$\int_A |\alpha f| d\mu \leq |\alpha| \int_A |f| d\mu < \epsilon.$$

6.5 **(1)** By the Radon-Nikodym Theorem, for any $E \in \mathcal{M}$, it holds that

$$\mu(E) = \int_E \frac{d\mu}{d\nu} d\nu,$$

and then there exists an increasing sequence $\{\phi_n\}$ of nonnegative simple functions such that $\phi_n = \sum_{i=1}^{k_n} a_{n,i}\chi_{E_{n,i}}$ converges to $\frac{d\mu}{d\nu}$. Therefore, by the monotone convergence theorem, it holds that

$$\begin{aligned}\mu(E) = \int_E \frac{d\mu}{d\nu} d\nu &= \lim_{n\to\infty} \phi_n d\nu = \lim_{n\to\infty} \sum_{i=1}^{k_n} a_{n,i}\nu(E \cap E_{n,i}) \\ &= \lim_{n\to\infty} \sum_{i=1}^{k_n} a_{n,i} \int_{E\cap E_{n,i}} \frac{d\nu}{d\lambda} d\lambda \\ &= \lim_{n\to\infty} \int_E \phi_n \frac{d\nu}{d\lambda} d\lambda \\ &= \int_E \frac{d\mu}{d\nu} \cdot \frac{d\nu}{d\lambda} d\lambda,\end{aligned}$$

which implies the proof of (i) since the Radon-Nikodym derivative is unique.

(2) The proof is a simple application of (i)

(3) By the Radon-Nikodym Theorem, it holds that for any $E \in \mathcal{M}$,

$$\begin{aligned}\int_E \left(\frac{d\mu}{d\lambda} + \frac{d\nu}{d\lambda}\right) d\lambda &= \int_E \frac{d\mu}{d\lambda} d\lambda + \int_E \frac{d\nu}{d\lambda} d\lambda \\ &= \mu(E) + \nu(E) \\ &= \int_E \frac{d(\mu+\nu)}{d\lambda} d\lambda,\end{aligned}$$

which implies the proof of (iii).

6.7 **(1)** By Jordan's theorem, $f = f_1 - f_2$ for two increasing functions f_1 and f_2. Therefore, we have that $|f| \leq f_1 + f_2$ and $|f(x)| \leq f_1(b) + f_2(b)$ for all $x \in I$, and so f is bounded. Also, since every monotone function is measurable and all linear combinations of measurable functions are measurable, f is measurable.

(2) By Jordan's theorem, $f = f_1 - f_2$ for two increasing functions f_1 and f_2. Since every monotone function is continuous except at the points of a countable subset of $[a, b]$, there exist two countable subsets C_1 and C_2 of $[a, b]$ such that for any $x \in [a, b] \setminus C_1$ and $y \in [a, b] \setminus C_2$, f_1 and f_2 are continuous at x and y, respectively. Then $C = C_1 \cup C_2$ is countable and for any $z \in [a, b] \setminus C$, f is continuous at z.

6.13 If $g_1 + h_1 = f = g_2 + h_2$, where h_1, h_2 are singular, then we have

$$(g_2 - g_1)' = (h_1 - h_2)' = 0,$$

which, by Proposition 6.6.8, implies that $g_2 - g_1$ is a constant function.

6.14 Since f is the uniform limit of continuous functions on $[a, b]$, f is continuous on $[a, b]$ and then $f \in \mathcal{RS}(\alpha, I)$. By the linearity of Riemann-Stieltjes integral (Proposition 6.7.3) and Corollary 6.7.7, we obtain that

$$\begin{aligned}\left|\int_I f_n d\alpha - \int_I f d\alpha\right| &= \left|\int_I (f_n - f) d\alpha\right| \\ &\leq \sup_{x\in I} |f_n(x) - f(x)| TV(\alpha_I) \\ &\to 0\end{aligned}$$

as $n \to \infty$ since f_n converges uniformly to f on I, which implies that

$$\lim_{n\to\infty} \int_I f_n d\alpha = \int_I d\alpha.$$

6.15 Suppose that α is increasing and absolutely continuous, then by Lebesgue theorem, α is almost everywhere differentiable on I and α' is Lebesgue integrable. On the other hand, since $f : I = [a, b] \to \mathbb{R}$ is continuous, f is bounded. Therefore, $f\alpha'$ is Lebesgue integrable. Define a signed measure ν_α on $\mathcal{B}(I)$, the Borel σ-algebra of Borel subsets of I, by

$$\nu_\alpha(B) = \int_B \alpha' dm, \qquad B \in \mathcal{B}(I).$$

Then ν_α is absolutely continuous with respect to m and $\alpha' = d\nu_\alpha/dm$. On the other hand, for any $\epsilon > 0$, there exists a partition $\mathcal{P}_\epsilon$ of I such that for any partition $\mathcal{P} : a = x_0 < x_1 < \cdots < x_n = b$ finer than $\mathcal{P}_\epsilon$ and any $x_{k-1} \leq \tau_k \leq x_k$,

$$\left|\sum_{k=1}^{n} f(\tau_k)\Delta\alpha_k - \int_I f\, d\alpha\right| < \frac{\epsilon}{2}.$$

By Corollary 6.6.10, we see that

$$\Delta\alpha_k = \alpha(x_k) - \alpha(x_{k-1}) = \int_{x_{k-1}}^{x_k} \alpha'\, dm = \nu_\alpha([x_{k-1}, x_k]), \qquad k = 1, 2, \cdots, n.$$

Since there exists a partition $\mathcal{P} : a = y_0 < y_1 < \cdots < y_l = b$ finer than $\mathcal{P}_\epsilon$ such that

$$\left|\sum_{k=1}^{l} m_k \nu_\alpha([x_{k-1}, x_k]) - \int_I f\, d\nu_\alpha\right| < \frac{\epsilon}{2},$$

where $m_k = \min_{y\in[y_{k-1}, y_k]}\{f(y)\}$ and $\int_I f\, d\nu_\alpha$ is called the **Lebesgue-Stieltjes integral**. Therefore, for $y_k' \in [y_{k-1}, y_k]$ with $f(y_k') = m_k$, we obtain that

$$\left|\int_I f\, d\nu_\alpha - \int_I f\, d\alpha\right| \leq \left|\sum_{k=1}^{l} m_k \nu_\alpha([x_{k-1}, x_k]) - \int_I f\, d\nu_\alpha\right|$$

$$+\left|\sum_{k=1}^{l} f(y_k')\,(\alpha(x_k)-\alpha(x_{k-1})) - \int_I f\,d\nu_\alpha\right|$$
$$< \frac{\epsilon}{2}+\frac{\epsilon}{2}=\epsilon,$$

which implies that

$$\int_I f\,d\alpha = \int_I f\,d\nu_\alpha,$$

and hence, by Radon-Nykodym theorem, we have

$$\int_I f\,d\alpha = \int_I f\,d\nu_\alpha = \int_I f\left(\frac{d\nu_\alpha}{dm}\right)dm = \int_I f\,\alpha'\,dm.$$

Let α is absolutely continuous. Then $\alpha=\alpha_1-\alpha_2$ for two increasing absolutely continuous functions α_1 and α_2, and so we have

$$\int_I f\,d\alpha = \int_I f\,d\alpha_1 - \int_I f\,d\alpha_2 = \int_I f\,\alpha_1'\,dm - \int_I f\,\alpha_2'\,dm = \int_I f\,\alpha'\,dm.$$

7.2 (i) [Reflexive] The reflexivity is obvious.

(ii) [Symmetric] Suppose that $\|\cdot\|$ and $\|\cdot\|'$ are equivalent. Then there exist constants C, C' with $0 < C \le C'$ such that

$$C\|v\| \le \|v\|' \le C'\|v\|, \qquad v \in V,$$

which implies that

$$\frac{1}{C'}\|v\|' \le \|v\| \le \frac{1}{C}\|v\|', \qquad v\in V.$$

for $0 < 1/C' \le 1/C$.

(ii) [Transitive] Suppose that $\|\cdot\|$ and $\|\cdot\|'$ are equivalent, and $\|\cdot\|'$ and $\|\cdot\|''$ are equivalent. Then there exist constants C_i $(i=1,2,3,4)$ with $0<C_1\le C_2$ and $0<C_3\le C_4$such that

$$C_1\|v\| \le \|v\|' \le C_2\|v\|,\quad C_3\|v\|' \le \|v\|'' \le C_4\|v\|', \qquad v\in V,$$

which implies that

$$C_1C_3\|v\| \le C_3\|v\|' \le \|v\|'' \le C_4\|v\|' \le C_2C_4\|v\|,$$

and so $\|\cdot\|$ and $\|\cdot\|''$ are equivalent

7.4 **(1)** The proof is straightforward.

(2) Consider the identity operator $I : V \to V$. Then for any $u=\alpha_1e_1+\cdots+\alpha_ne_n \in V$, we obtain that

$$\|u\| = \left\|\sum_{k=1}^{n}\alpha_ke_k\right\| \le \left(\max_{1\le k\le n}\{\|e_k\|\}\right)\sum_{k=1}^{n}|\alpha_k|$$

$$= \left(\max_{1 \le k \le n} \{ \|e_k\| \} \right) \|u\|_1,$$

which implies that the identity map $I : (V, \|\cdot\|_1) \to (V, \|\cdot\|)$ is continuous, and the norm $\|\cdot\| : (V, \|\cdot\|) \to \mathbb{R}$ is continuous. Therefore, the norm $\|\cdot\| = \|\cdot\| \circ I : (V, \|\cdot\|_1) \to \mathbb{R}$ is continuous. Put

$$S = \{v \in V \mid \|v\|_1 = 1\} \cong \bigcup_{(\epsilon_1, \cdots, \epsilon_n) \in \{1,-1\}^n} \{(x_1, \cdots, x_n) \mid \epsilon_1 x_1 + \cdots + \epsilon_n x_n = 1\}$$

and then S is compact in V. Therefore, $C_1 := \inf_{v \in S}\{\|v\|\}$ and $C_2 : \sup_{v \in S}\{\|v\|\}$ are finite, and so for any $\|v\|_1 = 1$,

$$C_1 \le \|v\| \le C_2,$$

which implies that for any $v \in V$,

$$C_1\|v\|_1 \le \|v\| \le C_2\|v\|_1.$$

Hence $\|\cdot\|$ and $\|\cdot\|_1$ are equivalent.

(3) By **(2)**, $\|\cdot\|$ and $\|\cdot\|_1$ are equivalent and $\|\cdot\|_1$ and $\|\cdot\|'$ are equivalent. Therefore, by the transitivity of the equivalent relation $\sim$ given as in Exercise **7.2**, $\|\cdot\|$ and $\|\cdot\|'$ are equivalent.

7.7 Since S is total, $S^\perp = \{0\}$. Therefore, we have

$$H = S^\perp \oplus S^{\perp\perp} = S^{\perp\perp}.$$

7.8 Let $h \in H$ be given. Then for any $n \in H$, we obtain that

$$0 \le \left\| h - \sum_{k=1}^{n} \langle h, e_k \rangle e_k \right\|^2 = \|h\|^2 - \sum_{k=1}^{n} |\langle h, e_k \rangle|^2,$$

which implies that

$$\sum_{k=1}^{n} |\langle h, e_k \rangle|^2 \le \|h\|^2,$$

and by taking supremun in n on the left hand side, we have

$$\sum_{k=1}^{\infty} |\langle h, e_k \rangle|^2 \le \|h\|^2.$$

On the other hand, for any square summable sequence $\{\alpha_n\}_{n=1}^\infty$, i.e., $\sum_{n=1}^{\infty} |\alpha_n|^2 < \infty$, we can easily see that $\left\{ \sum_{k=1}^{n} \alpha_k e_k \right\}_{n=1}^{\infty}$ is a Cauchy sequence in H and the series $\sum_{k=1}^{\infty} \alpha_k e_k$ converges in H since H is complete. Therefore, the series:

$$\sum_{k=1}^{\infty} \langle h, e_k \rangle e_k$$

converges in H. For each $n \in \mathbb{N}$, we have

$$\left\langle h - \sum_{k=1}^{\infty} \langle h, e_k \rangle e_k, e_n \right\rangle = \langle h, e_n \rangle - \langle h, e_n \rangle = 0.$$

Therefore, $h - \sum_{k=1}^{\infty} \langle h, e_k \rangle e_k \in \{e_n \,|\, n \in \mathbb{N}\}^{\perp}$ and $\{e_n \,|\, n \in \mathbb{N}\}^{\perp} = \{0\}$ since $\{e_n \,|\, n \in \mathbb{N}\}$ is total, and so we have

$$h = \sum_{k=1}^{\infty} \langle h, e_k \rangle e_k.$$

Moreover, we obtain that

$$\|h\|^2 = \langle h, h \rangle = \sum_{l=1}^{\infty} \sum_{m=1}^{\infty} \langle h, e_l \rangle \overline{\langle h, e_m \rangle} \langle e_l, e_m \rangle = \sum_{n=1}^{\infty} |\langle h, e_n \rangle|^2.$$

7.9 Suppose that $\{x_n\}_{n=1}^{\infty} \subset X$ be a sequence converging to $x \in X$. Let U be an open subset of Y containing $f(x)$. Since f is continuous, $f^{-1}(U)$ is open in X containing x. Since $\{x_n\}_{n=1}^{\infty}$ converges to x, there exists $N \in \mathbb{N}$ such that for any $n \geq N$, $x_n \in f^{-1}(U)$ and so $f(x_n) \in U$ for any $n \geq N$. Therefore, $\{f(x_n)\}_{n=1}^{\infty}$ converges to $f(x)$ and so f is sequentially continuous.

7.10 Let U be an open subset of Y. Suppose that $f^{-1}(U)$ is nonempty and not open. Then there exists $x \in f^{-1}(U)$ such that x is not an interior point of $f^{-1}(U)$. Since X is first countable, there exists a countable open basis $\mathcal{B}(x)$ of x. Put

$$\mathcal{B}(x) = \{V_1, V_1, \cdots, V_n, \cdots\}.$$

Then by taking

$$O_n = \bigcap_{k=1}^{n} V_n, \qquad n \in \mathbb{N}.$$

We have a descending family $\{O_n\}_{n=1}^{\infty}$ of open subsets containing x which becomes an open (neighborhood) basis of x. Since x is not an interior point of $f^{-1}(U)$, for each $n \in \mathbb{N}$, there exists $x_n \in O_n$ such that $x_n \notin f^{-1}(U)$. Then $\{x_n\}_{n=1}^{\infty}$ converges to $x \in X$. In fact, for any open subset W of X containing x, there exists $N \in \mathbb{N}$ such that $O_N \subset W$ (since $\{O_n\}_{n=1}^{\infty}$ is an open (neighborhood) basis of x). Therefore, for any $n \geq N$, $x_n \in O_n \subset O_N \subset W$. But, since $x_n \notin f^{-1}(U)$, $f(x_n) \notin U$ for all $n \in \mathbb{N}$, and so $\{f(x_n)\}_{n=1}^{\infty}$ does not converge to x, which contradicts the sequentially continuity of f. Hence $f^{-1}(U)$ is open and f is continuous.

7.11 Let (X, τ) be a topological space and $\mathcal{B}$ be a countable basis for the topology of X. Let $x \in X$. Then the collection

$$\mathcal{B}(x) = \{B \in \mathcal{B} \,|\, x \in B\}$$

is countable and an open (neighborhood) basis of x. In fact, every open subset of X is a countable union of elements in $\mathcal{B}$, Therefore, for any open subset V of X containing $x \in X$, there exists $B \in \mathcal{B}$ such that $x \in B$ and $B \subset V$. Hence $\mathcal{B}(x)$ is a countable open (neighborhood) basis of x.

7.12 For each $k \in K$, we define a continuous linear functional $f_k : H \to \mathbb{C}$ by

$$f_k(h) = \langle Th, k \rangle_K, \qquad h \in H. \tag{10.15}$$

In fact, by applying the Cauchy-Schwarz inequality, we obtain that

$$|f_k(h)| = |\langle Th, k \rangle_K| \le \|Th\|_K \, \|k\|_K \le \|T\|_{\mathrm{OP}} \, \|h\|_H \, \|k\|_K . \tag{10.16}$$

Then by the Riesz representation theorem, there exists a unique $h_k \in H$ such that

$$f_k(h) = \langle h, h_k \rangle, \qquad h \in H.$$

Now, we define an operator $T^* : K \to H$ by $T^*k = h_k$ for $k \in K$. Then by the definition, we have

$$\langle h, T^*k \rangle = \langle h, h_k \rangle = f_k(h) = \langle Th, k \rangle, \qquad h \in H, \quad k \in K.$$

and we obtain that

$$\begin{aligned}\langle h, T^*(\alpha k_1 + \beta k_2) \rangle &= \langle Th, \alpha k_1 + \beta k_2 \rangle = \overline{\alpha} \langle Th, k_1 \rangle + \overline{\beta} \langle Th, k_2 \rangle \\ &= \overline{\alpha} \langle h, T^*k_1 \rangle + \overline{\beta} \langle h, T^*k_2 \rangle \\ &= \langle h, \alpha T^*k_1 \rangle + \langle h, \beta T^*k_2 \rangle,\end{aligned}$$

which implies that T^* is linear. Also, from (10.16) we obtain that

$$\|T^*k\|_H = \|T\|_{\mathrm{OP}} \|k\|_K, \qquad k \in K,$$

which implies that $\|T^*\|_{\mathrm{OP}} \le \|T\|_{\mathrm{OP}}$, and so T^* is bounded.

7.13 **(1)** We obtain that

$$\begin{aligned}\|T^*\|_{\mathrm{OP}} &= \sup_{\|k\|_K = 1} \{\|T^*k\|\} = \sup_{\|k\|_K = 1} \sup_{\|h\|_H = 1} \{|\langle h, T^*k \rangle_H|\} \\ &= \sup_{\|k\|_K = 1} \sup_{\|h\|_H = 1} \{|\langle Th, k \rangle_K|\} \\ &= \|T\|_{\mathrm{OP}},\end{aligned}$$

which implies that $\|T^*\|_{\mathrm{OP}} = \|T\|_{\mathrm{OP}}$.

(2) For any $h \in H$ and $k \in K$, we obtain that

$$\langle T^{**}h, k \rangle_K = \langle h, T^*k \rangle_H = \langle Th, k \rangle_K,$$

which implies that $T^{**} = T$.

(3) For any $h \in H$ and $k \in K$, we obtain that

$$\langle h, (\alpha T)^*k \rangle_H = \langle \alpha Th, k \rangle_K = \alpha \langle Th, k \rangle_K = \langle h, \overline{\alpha} T^*k \rangle_H,$$

which implies that $(\alpha T)^* = \overline{\alpha} T^*$.

(4) For any $h \in H$ and $l \in L$, we obtain that

$$\langle h, (ST)^*l \rangle_H = \langle STh, l \rangle_L = \langle Th, S^*l \rangle_K = \langle h, T^*S^*l \rangle_H,$$

which implies that $(ST)^* = S^*T^*$.

7.16 **(1)** If $\int_X |g|^q\,d\mu = \infty$, then since $q < 0$, $\left(\int_X |g|^q\,d\mu\right)^{1/q} = 0$ and so the inequality is obvious. Suppose that $\int_X |g|^q\,d\mu < \infty$ and so $|g| > 0$ almost everywhere on X. Put

$$u = |fg|^p, \quad v = |g|^{-p}.$$

Then by the Hölder's inequality, we obtain that

$$\begin{aligned}\int_X |f|^p\,d\mu = \int_X uv\,d\mu &\le \left(\int_X u^{1/p}\,d\mu\right)^p \left(\int_X v^{1/(1-p)}\,d\mu\right)^{1-p} \\ &= \left(\int_X |fg|\,d\mu\right)^p \left(\int_X |g|^r\,d\mu\right)^{-p/r},\end{aligned}$$

where the equality holds only for $\alpha|fg| = \beta|g|^r$ with $(\alpha,\beta) \ne (0,0)$, which implies that

$$\int_X |fg|\,d\mu \ge \left(\int_X |f|^p\,d\mu\right)^{1/p} \left(\int_X |g|^r\,d\mu\right)^{1/r}.$$

(2) For any measurable functions $f, g \ge 0$ and for $r = -p/(1-p)$, by applying (2), we obtain that

$$\begin{aligned}\|f+g\|_p^p = \int_X (f+g)^p\,d\mu &= \int_X f(f+g)^{p-1}\,d\mu + \int_X g(f+g)^{p-1}\,d\mu \\ &\ge \left(\int_X f^p\,d\mu\right)^{1/p} \left(\int_X (f+g)^{(p-1)r}\,d\mu\right)^{1/r} \\ &\quad + \left(\int_X g^p\,d\mu\right)^{1/p} \left(\int_X (f+g)^{(p-1)r}\,d\mu\right)^{1/r} \\ &= (\|f\|_p + \|g\|_p)\,\|f+g\|_p^{p-1},\end{aligned}$$

where the equality holds only for $\alpha f(f+g)^{p-1} = \beta(f+g)^{(p-1)r} = \beta(f+g)^p$, equivalently, $\alpha f = \beta g$ with $(\alpha,\beta) \ne (0,0)$, which implies that

$$\|f+g\|_p \ge \|f\|_p + \|g\|_p.$$

Hence $\|\cdot\|_p$ does not satisfy the triangle inequality for seminorm and so $\|\cdot\|_p$ is not a seminorm.

7.17 $\Rightarrow$) Suppose that $x \in \overline{E}$. Since X is first countable, there exists a countable open basis $\mathcal{B}(x)$ of x. Put

$$\mathcal{B}(x) = \{V_1, V_1, \cdots, V_n, \cdots\}.$$

Then by taking

$$O_n = \bigcap_{k=1}^{n} V_n, \qquad n \in \mathbb{N}.$$

We have a descending family $\{O_n\}_{n=1}^\infty$ of open subsets containing x which becomes an open (neighborhood) basis of x. Since $x \in \overline{E}$, for each $n \in \mathbb{N}$, there exists $x_n \in O_n \cap E$. Then $\{x_n\}_{n=1}^\infty \subset E$ converges to $x \in X$. In fact, for any open subset W of X containing x, there exists $N \in \mathbb{N}$ such that $O_N \subset W$ (since $\{O_n\}_{n=1}^\infty$ is an open (neighborhood) basis of x). Therefore, for any $n \ge N$, $x_n \in O_n \subset O_N \subset W$.

$\Leftarrow$) Suppose that $\{x_n\}_{n=1}^{\infty} \subset E$ converges to x. Let W be an open subset of X containing x. Then there exists $N \in \mathbb{N}$ such that for any $n \geq N$, $x_n \in W$, which implies that $W \cap E \neq \emptyset$. Therefore, $x \in \overline{E}$.

8.1 **(1)** Let $B, C \in \mathcal{A}$ be nonempty disjoint elements and $B, D \in \mathcal{B}$ be nonempty disjoint elements. Then $A \times B$, $C \times D \in \mathcal{A} \times \mathcal{B}$, but

$$(A \times B) \cup (C \times D) \notin \mathcal{A} \times \mathcal{B}.$$

(2) Note that $\mathcal{M}_0$ is the family of all finite unions of measurable rectangles. For a finite family $\{E_i\}_{i=1}^{l} \subset \mathcal{M}_0$, we have

$$E_i = \bigcup_{k=1}^{m_i} A_{ik} \times B_{ik}, \qquad i = 1, 2, \cdots, l, \quad A_{ik} \in \mathcal{A}, \quad B_{ik} \in \mathcal{B}$$

and we obtain that

$$\bigcup_{i=1}^{l} E_i = \bigcup_{i=1}^{l} \left(\bigcup_{k=1}^{m_i} A_{ik} \times B_{ik} \right) = \bigcup_{i=1}^{l} \bigcup_{k=1}^{m_i} A_{ik} \times B_{ik} \in \mathcal{M}_0,$$

and so $\mathcal{M}_0$ is closed under finite unions. Let $E \in \mathcal{M}_0$. Then

$$E = \bigcup_{k=1}^{m} A_k \times B_k, \qquad A_k \in \mathcal{A}, \quad B_k \in \mathcal{B},$$

and we obtain that

$$\begin{aligned} E^{\mathrm{c}} &= \bigcap_{k=1}^{m} (A_k \times B_k)^{\mathrm{c}} \\ &= \bigcap_{k=1}^{m} ((A_k^{\mathrm{c}} \times B_k) \cup (A_k \times B_k^{\mathrm{c}}) \cup (A_k^{\mathrm{c}} \times B_k^{\mathrm{c}})) \in \mathcal{M}_0. \end{aligned}$$

Also, it is obvious that $X \times Y, \emptyset \in \mathcal{M}_0$. Therefore, $\mathcal{M}_0$ is an algebra. Let $\mathcal{F}$ be an algebra containing all rectangles. Then for any $E \in \mathcal{M}_0$ with

$$E = \bigcup_{k=1}^{m} A_k \times B_k, \qquad A_k \in \mathcal{A}, \quad B_k \in \mathcal{B},$$

and $E \in \mathcal{F}$ since $A_k \times B_k$ is a measurable rectangle and $\mathcal{F}$ is closed under finite unions. Therefore, $\mathcal{M}_0 \subset \mathcal{F}$ and $\mathcal{M}_0$ is the smallest algebra containing all measurable rectangles of $X \times Y$.

(3) We prove it by induction. We first have that

$$(A \times B) \cup (C \times D) = (A \times B) \cup ((A \cap C) \times (D \setminus B)) \cup ((C \setminus A) \times D)$$

for any $A, C \in \mathcal{A}$ and $B, D \in \mathcal{B}$. Let $E \in \mathcal{M}_0$ with

$$E = \bigcup_{k=1}^{m} A_k \times B_k, \qquad A_k \in \mathcal{A}, \quad B_k \in \mathcal{B}.$$

Suppose that $\bigcup_{k=1}^{m-1} A_k \times B_k$ is represented as a finite disjoint union of measurable rectangles:

$$\bigcup_{k=1}^{m-1} A_k \times B_k = \bigcup_{i=1}^{l} C_i \times D_i.$$

Then we obtain that

$$\begin{aligned}\bigcup_{k=1}^{m} A_k \times B_k &= \left(\bigcup_{i=1}^{l} C_i \times D_i\right) \cup (A_m \times B_m) = \bigcup_{i=1}^{l} ((C_i \times D_i) \cup (A_m \times B_m)) \\ &= \bigcup_{i=1}^{l} ((C_i \times D_i) \cup ((C_i \cap A_m) \times (B_m \setminus D_i)) \cup ((A_m \setminus C) \times B_m)),\end{aligned}$$

where $C = \cup_{i=1}^{l} C_i$.

8.2 Let $\mathcal{F}$ be a σ-algebra. Let $\{A_n\}_{n=1}^{\infty}$ be an ascending family of $\mathcal{F}$. It is obvious that

$$\bigcup_{n=1} A_n \in \mathcal{F}$$

since $\mathcal{F}$ is a σ-algebra and so $\mathcal{F}$ is closed under countable unions. Let $\{D_n\}_{n=1}^{\infty}$ be a descending family of $\mathcal{F}$. Then we have

$$\bigcap_{n=1}^{\infty} D_n = \left(\bigcup_{n=1}^{\infty} D_n^{\mathrm{c}}\right)^{\mathrm{c}} \in \mathcal{F}$$

since $\mathcal{F}$ is a σ-algebra and so $\mathcal{F}$ is closed under complements and countable unions.

8.3 Let $\{\mathcal{M}_\lambda\}_{\lambda\in\Lambda}$ be a family of monotone classes. Put

$$\mathcal{M} = \bigcap_{\lambda\in\Lambda} \mathcal{M}_\lambda.$$

Let $\{A_n\}_{n=1}^{\infty}$ be an ascending family of $\mathcal{M}$. Then for any $\lambda \in \Lambda$, $\{A_n\}_{n=1}^{\infty}$ be an ascending family of the monotone class $\mathcal{M}_\lambda$ and so $\cup_{n=1}^{\infty} A_n \in \mathcal{M}_\lambda$ for any $\lambda \in \Lambda$. Therefore, $\cup_{n=1}^{\infty} A_n \in \mathcal{M}$. Let $\{D_n\}_{n=1}^{\infty}$ be a descending family of $\mathcal{M}$. Then for any $\lambda \in \Lambda$, $\{A_n\}_{n=1}^{\infty}$ be a descending family of the monotone class $\mathcal{M}_\lambda$ and so $\cap_{n=1}^{\infty} A_n \in \mathcal{M}_\lambda$ for any $\lambda \in \Lambda$. Therefore, $\cap_{n=1}^{\infty} A_n \in \mathcal{M}$.

8.4 Let $\mathfrak{M}$ be the class of all monotone classes containing $\mathcal{C}$. Then by Exercise **8.2**, the smallest σ-algebra $\sigma(\mathcal{C})$ containing $\mathcal{C}$ is a monotone class and so $\mathfrak{M} \neq \emptyset$. Also, by Exercise **8.3**,

$$\bigcap_{\mathcal{M}\in\mathfrak{M}} \mathcal{M}$$

is a monotone class containing $\mathcal{C}$ and by the construction, it is the smallest monotone class containing $\mathcal{C}$.

8.5 Let $\mathcal{M}(\mathcal{S})$ be the monotone class generated by $\mathcal{S}$. Since $\sigma(\mathcal{S})$ is the σ-algebra containing $\mathcal{S}$, by Exercise **8.2**, $\sigma(\mathcal{S})$ is a monotone class. On the other hand, by the definition, $\mathcal{M}(\mathcal{S})$ is the smallest monotone class containing $\mathcal{S}$. Hence $\mathcal{M}(\mathcal{S}) \subset \sigma(\mathcal{S})$.

8.6 By applying the Tonelli's theorem (see Theorem 8.2.6), we obtain that

$$\begin{aligned}\int_A f\, d(m\times m) &= \int_0^1 \int_{\mathbb{R}} f(x,y)\, dm(y)dm(x) = \int_0^1 \int_{\mathbb{R}} e^{-\frac{x^2}{2y^2}}\, dm(x)dm(y)\\ &= \int_0^1 \sqrt{2\pi y^2} dm(y)\\ &= \frac{\sqrt{2\pi}}{2}.\end{aligned}$$

9.1 $\Rightarrow$) Suppose that $\{f_n\}_{n=1}^{\infty}$ converges to f in measure. Let $\epsilon, \delta > 0$ be given. Then from the convergence in measure, we have

$$\lim_{n\to\infty} \mu\left(\{x\in X \,|\, |f_n(x) - f(x)| > \epsilon\}\right) = 0,$$

and so there exists $N \in \mathbb{N}$ such that for any $n \geq N$,

$$\mu\left(\{x\in X \,|\, |f_n(x) - f(x)| > \epsilon\}\right) < \delta.$$

$\Leftarrow$) Let $\epsilon > 0$ be given. Then for any $\delta > 0$, there exists $N \in \mathbb{N}$ such that for any $n \geq N$,

$$\mu\left(\{x\in X \,|\, |f_n(x) - f(x)| > \epsilon\}\right) < \delta,$$

which implies that

$$\limsup_{n\to\infty} \mu\left(\{x\in X \,|\, |f_n(x) - f(x)| > \epsilon\}\right) < \delta$$

for arbitrary $\delta > 0$ and so

$$\lim_{n\to\infty} \mu\left(\{x\in X \,|\, |f_n(x) - f(x)| > \epsilon\}\right) = 0,$$

and $\{f_n\}_{n=1}^{\infty}$ converges to f in measure.

9.2 $\Leftarrow$) By (iii) in Theorem 9.1.3.

$\Rightarrow$) Suppose that $\{f_n\}_{n=1}^{\infty}$ converges to a measurable function f in measure. Then for given $\epsilon > 0$,

$$\lim_{n\to\infty} \mu\left(\{x\in X \,|\, |f_n(x) - f(x)| > \epsilon/2\}\right) = 0.$$

Therefore, we obtain that

$$\begin{aligned}&\lim_{n,m\to\infty} \mu\left(\{x\in X \,|\, |f_n(x) - f_m(x)| > \epsilon\}\right)\\ &\quad\leq \lim_{n,m\to\infty} \mu\left(\{x\in X \,|\, |f_n(x) - f_m(x)| > \epsilon/2\}\right)\\ &\qquad + \lim_{n,m\to\infty} \mu\left(\{x\in X \,|\, |f_n(x) - f_m(x)| > \epsilon/2\}\right)\\ &\quad= 0,\end{aligned}$$

and so $\{f_n\}_{n=1}^{\infty}$ is a Cauchy sequence in measure.

9.3 Let $\epsilon > 0$ be given. We first note that

$$\{x \in X \mid |f(x)| = \infty\} = \bigcap_{k=1}^{\infty} \{x \in X \mid |f(x)| > k\}.$$

Since f is finite a.e., by the continuity of measure ($\mu(X) < \infty$), we obtain that

$$\begin{aligned} 0 = \mu(\{x \in X \mid |f(x)| = \infty\}) &= \mu\left(\bigcap_{k=1}^{\infty} \{x \in X \mid |f(x)| > k\}\right) \\ &= \lim_{k\to\infty} \mu\left(\{x \in X \mid |f(x)| > k\}\right), \end{aligned}$$

which that there exists $K \in \mathbb{N}$ such that for any $k \geq K$,

$$\mu\left(\{x \in X \mid |f(x)| > k\}\right) < \epsilon.$$

Therefore, by taking $M \in \mathbb{N}$ with $M \geq K$ and $E = \{x \in X \mid |f(x)| > M\}$, we have

$$\mu(E) < \epsilon, \qquad |f| \leq M \quad \text{on } E^{\mathrm{c}}.$$

9.4 **(1)** Let $\epsilon > 0$ be given. Note that

$$||f_n| - |f|| \leq |f_n - f|\,.$$

Therefore, we obtain that

$$\begin{aligned} \lim_{n\to\infty} \mu\left(\{x \in X \mid ||f_n(x)| - |f(x)|| > \epsilon\}\right) &\leq \lim_{n\to\infty} \mu\left(\{x \in X \mid |f_n(x) - f(x)| > \epsilon\}\right) \\ &= 0, \end{aligned}$$

and so $\{|f_n|\}_{n=1}^{\infty}$ converges to $|f|$ in measure.

(2) Let $\epsilon > 0$ be given. Then it holds that

$$\begin{aligned} &\mu\left(\{x \in X \mid |(f_n(x) + g_n(x)) - (f(x) + g(x))| > \epsilon\}\right) \\ &\quad \leq \mu\left(\{x \in X \mid |f_n(x) - f(x)| > \epsilon/2\}\right) + \mu\left(\{x \in X \mid |g_n(x) - g(x)| > \epsilon/2\}\right) \\ &\quad \to 0 \end{aligned}$$

as $n \to \infty$. Therefore, $\{f_n + g_n\}_{n=1}^{\infty}$ converges to $f + g$ in measure.

(3) If $\alpha = 0$, then the statement is obvious. Suppose that $\alpha \neq 0$. Let $\epsilon > 0$ be given. Then it holds that

$$\begin{aligned} \lim_{n\to\infty} \mu\left(\{x \in X \mid |\alpha(f_n(x) - f(x))| > \epsilon\}\right) &= \lim_{n\to\infty} \mu\left(\{x \in X \mid |f_n(x) - f(x)| > \epsilon/|\alpha|\}\right) \\ &= 0. \end{aligned}$$

Therefore, $\{\alpha f_n\}_{n=1}^{\infty}$ converges to αf in measure.

(4) The proof is immediate from Theorem 9.1.7 by taking $h(x) = x^2$. However, we can prove directly as following. Let $\epsilon > 0$ be given. Since $\{f_n\}_{n=1}^{\infty}$ converges to f in measure, f is finite a.e. Therefore, by the previous exercise (Exercise **3**), for any

$\lambda > 0$ there exists $M > 0$ and a measurable set E such that $\mu(E) < \lambda$ and $|f| < M$ on E^{c}. Note that

$$\begin{aligned} \left|f_n(x)^2 - f(x)^2\right| &= |f_n(x) + f(x)|\,|f_n(x) - f(x)| \\ &\leq (|f_n(x) - f(x)| + 2\,|f(x)|)\,|f_n(x) - f(x)| \\ &= |f_n(x) - f(x)|^2 + 2|f(x)|\,|f_n(x) - f(x)|\,. \end{aligned}$$

Hence, we obtain that

$$\begin{aligned} \mu\left(\left\{x \in X \,\middle|\, \left|f_n(x)^2 - f(x)^2\right| > \epsilon\right\}\right) \leq \mu\left(\left\{x \in X \,\middle|\, |f_n(x) - f(x)| > \sqrt{\epsilon/2}\right\}\right) \\ + \mu\left(\{x \in X \mid 2|f(x)|\,|f_n(x) - f(x)| > \epsilon/2\}\right). \end{aligned}$$

On the other hand, we obtain that

$$\begin{aligned} &\mu\left(\left\{x \in X \,\middle|\, 2|f(x)|\,|f_n(x) - f(x)| > \frac{\epsilon}{2}\right\}\right) \\ &\quad = \mu\left(\left\{x \in E^{\mathrm{c}} \,\middle|\, 2|f(x)|\,|f_n(x) - f(x)| > \frac{\epsilon}{2}\right\}\right) \\ &\quad + \mu\left(\left\{x \in E \,\middle|\, 2|f(x)|\,|f_n(x) - f(x)| > \frac{\epsilon}{2}\right\}\right) \\ &\leq \mu\left(\left\{x \in E^{\mathrm{c}} \,\middle|\, |f_n(x) - f(x)| > \frac{\epsilon}{4M}\right\}\right) + \lambda \end{aligned}$$

for arbitrary $\lambda > 0$. Hence we have

$$\begin{aligned} \mu\left(\left\{x \in X \,\middle|\, \left|f_n(x)^2 - f(x)^2\right| > \epsilon\right\}\right) \leq \mu\left(\left\{x \in X \,\middle|\, |f_n(x) - f(x)| > \sqrt{\epsilon/2}\right\}\right) \\ + \mu\left(\left\{x \in E^{\mathrm{c}} \,\middle|\, |f_n(x) - f(x)| > \frac{\epsilon}{4M}\right\}\right) + \lambda \end{aligned}$$

for arbitrary $\lambda > 0$, which implies that

$$\lim_{n\to\infty} \mu\left(\left\{x \in X \,\middle|\, \left|f_n(x)^2 - f(x)^2\right| > \epsilon\right\}\right) \leq \lambda$$

for arbitrary $\lambda > 0$, and so

$$\lim_{n\to\infty} \mu\left(\left\{x \in X \,\middle|\, \left|f_n(x)^2 - f(x)^2\right| > \epsilon\right\}\right) = 0.$$

Hence $\{f_n^2\}_{n=1}^{\infty}$ converges to f^2 in measure.

(5) Note that

$$f_n g_n = \frac{1}{4}\left((f_n + g_n)^2 - (f_n - g_n)^2\right).$$

Therefore, by applying **(2)**, **(3)** and **(4)**, we see that $\{f_n g_n\}_{n=1}^{\infty}$ converges to fg in measure.

9.5 **(1)** By an elementary computation, we can prove that for any nonnegative real numbers a, b and c with $a + b \geq c$, it holds that

$$\frac{a}{1+a} + \frac{b}{1+b} \geq \frac{c}{1+c}.$$

Therefore, for any $f, g, h \in L^0(X)$, we obtain that

$$\gamma(f,h) = \int_X \frac{|f-h|}{1+|f-h|} d\mu \leq \int_X \frac{|f-g|}{1+|f-g|} d\mu + \int_X \frac{|g-h|}{1+|g-h|} d\mu$$
$$= \gamma(f,g) + \gamma(g,h),$$

which is the triangle inequality. It is obvious that $\gamma(f,g) \geq 0$ and $\gamma(f,g) = 0$ if and only if $f = g$ as in the quotient space. Also, $\gamma(f,g) = \gamma(g,f)$. Therefore, γ is a metric.

(2) $\Rightarrow$) Let $\epsilon > 0$ be given. For each $n \in \mathbb{N}$, we put

$$E_n = \{x \in X \,|\, |f_n(x) - f(x)| > \epsilon\}.$$

Then we obtain that

$$\gamma(f_n, f) = \int_{E_n} \frac{|f_n - f|}{1+|f_n - f|} d\mu + \int_{E_n^c} \frac{|f_n - f|}{1+|f_n - f|} d\mu$$
$$\leq \mu(E_n) + \int_{E_n^c} \epsilon d\mu \leq \mu(E_n) + \mu(X)\epsilon,$$

which implies that

$$\lim_{n\to\infty} \gamma(f_n, f) \leq \mu(X)\epsilon$$

for arbitrary $\epsilon > 0$. Hence $\lim_{n\to\infty} \gamma(f_n, f) = 0$.

$\Leftarrow$) By the assumption and the dominated convergence theorem, we obtain that

$$0 = \lim_{n\to\infty} \gamma(f_n, f) = \int_X \left(\lim_{n\to\infty} \frac{|f_n - f|}{1+|f_n - f|} \right) d\mu,$$

which implies that $\{f_n\}_{n=1}^{\infty}$ converges pointwise a.e. on X to f. Hence, by Theorem 9.1.5, $\{f_n\}_{n=1}^{\infty}$ converges to f in measure.

9.6 Let $\{f_n\}_{n=1}^{\infty}$ be a sequence of measurable functions and f, g be measurable functions. Suppose that $\{f_n\}_{n=1}^{\infty}$ converges to f and g almost uniformly. Then for any $k \in \mathbb{N}$, there exists $E_k, F_k \in \mathcal{M}$ such that $\mu(E_k) < 1/k$, $\mu(F_k) < 1/k$ and $\{f_n\}_{n=1}^{\infty}$ converges uniformly on E_k^c and F_k^c to f and g, respectively, which implies that $f = g$ on $(E_k \cup F_k)^c$ and so $f = g$ on $\bigcup_{k=1}^{\infty} (E_k \cup F_k)^c = \left(\bigcap_{k=1}^{\infty} (E_k \cup F_k)\right)^c$. On the other hand, we have

$$\mu\left(\bigcap_{k=1}^{\infty} (E_k \cup F_k)\right) = \lim_{k\to\infty} \mu(E_k \cup F_k) \leq= \lim_{k\to\infty} \frac{2}{k} = 0.$$

Hence $f = g$ a.e.

9.7 By applying the Cauchy-Schwarz inequality, we obtain that

$$\lim_{n\to\infty} \left| \int_X f_n^2 \, d\mu - \int_X f^2 \, d\mu \right| \leq \lim_{n\to\infty} \int_X \left| f_n^2 - f^2 \right| d\mu$$

$$
\begin{aligned}
&= \lim_{n\to\infty} \int_X |f_n + f|\,|f_n - f|\,d\mu \\
&\le \lim_{n\to\infty} \int_X \left(|f_n - f|^2 + 2|f|\,|f_n - f|\right) d\mu \\
&\le \lim_{n\to\infty} \int_X |f_n - f|^2\,d\mu \\
&\qquad + 2\lim_{n\to\infty} \left(\int_X |f|^2\,d\mu\right)^{1/2} \left(\int_X |f_n - f|^2\,d\mu\right)^{1/2} \\
&= 0,
\end{aligned}
$$

which implies that

$$\lim_{n\to\infty} \int_X f_n^2\,d\mu = \int_X f^2\,d\mu.$$

10.1 The proof is a simple modification of the proof of Theorem 10.1.5. Let $V_0 = U + \mathbb{C}w$ and $\delta > 0$ be given. Consider the bounded linear functional $L_0 : V_0 \to \mathbb{C}$ defined by

$$L_0(u + \alpha w) = \alpha\delta, \qquad u \in U, \quad \alpha \in \mathbb{C}.$$

Then by the construction, we have $L_0(w) = \delta$, and for $\alpha \in \mathbb{C}$ with $\alpha \neq 0$, we have

$$|L_0(u + \alpha w)| = |\alpha|\delta \le \frac{\delta}{d(w,U)}|\alpha|\,\|w + \alpha^{-1}u\| = \frac{\delta}{d(w,U)}\|u + \alpha w\|,$$

which implies that $\|L_0\| \le \delta/d(w,U)$. On the other hand, there exists a sequence $\{u_n\}_{n=1}^{\infty} \subset U$ such that

$$d(w,U) = \lim_{n\to\infty} \|w - u_n\|,$$

and then we obtain that

$$\delta = \lim_{n\to\infty} L_0(w - u_n) \le \lim_{n\to\infty} \|L_0\|\|w - u_n\| = \|L_0\| d(w,U),$$

which implies that $\|L_0\| \ge \delta/d(w,U)$ and so $\|L_0\| = \delta/d(w,U)$. Therefore, by Hahn-Banach Theorem (Theorem 10.1.4), there exists a bounded linear functional L on V such that $Lv = L_0 v$ for all $v \in V_0$ and so $L = 0$ on U, $Lw = L_0 w = \delta$ and $\|L\| = \|L_0\| = \delta/d(w,U)$.

10.2 Let U be a linear subspace of a normed space $(V, \|\cdot\|)$ and let $w \in V$ be fixed. Prove that $w \in \overline{U}$ if and only if for bounded linear functional $L \in V^*$ vanishing on U, $L(w) = 0$ since $w \in \overline{U}$.

$\Rightarrow$) Let L be a bounded linear functional on V such that $L = 0$ on U. Then by the continuity of L, $L = 0$ on $\overline{U}$ and so $L(w) = 0$.

$\Leftarrow$) Suppose that $w \notin \overline{U}$. Then we have

$$d(w,U) = \inf\{\|w - u\| \mid u \in U\} > 0.$$

Therefore, by Exercise **10.1**, there exists a bounded linear functional L on V such that $L = 0$ on U and $L(w) = \delta$ for $\delta > 0$, which contradicts the assumption.

10.3 $\Rightarrow$) Let L be a bounded linear functional on V such that $L = 0$ on U. Then since U is dense in V, $L = 0$ on V.

$\Leftarrow$) Suppose that U is not dense in V, i.e., there exists $w \in V \setminus \overline{U}$. By the contrapositive statement of Exercise **10.2**, there exists a bounded linear functional $L \in V^*$ vanishing on U such that $L \neq 0$ on $V \setminus \overline{U}$, which contradicts the assumption. Therefore, U is dense in V.

10.4 The proof is immediate from Exercise **10.1** by taking $U = \{0\}$.

10.5 The proof is immediate from Exercise **10.4**. In fact, since $z - w \neq 0$, by Exercise **10.4**, there exists a bounded linear functional $L \in V^*$ such that $L(z - w) = \delta$ for any given $\delta > 0$ and so $L(z) \neq L(w)$.

10.6 Let $x \in X$. Then $x \notin \mathring{A}$ if and only if for any open set U containing x, $A^c \cap U \neq \emptyset$ if and only if $x \in \overline{A^c}$. Therefore, $\mathring{A} = \emptyset$ if and only if $\overline{A^c} = X$ if and only if A^c is dense in X.

10.7 Let V be a nonempty open subset in U. Then V is a nonempty open subset in X and since E is dense in X, $E \cap V \neq \emptyset$, and so we have

$$(E \cap U) \cap V = E \cap V \neq \emptyset.$$

Hence $E \cap U$ is dense in U.

10.8 Let O be an open dense subset of X. Suppose that $x \in \mathring{\overline{O^c}} = \mathring{O^c}$. Then there exists an open set V containing x such that

$$V \subset O^c \quad \Leftrightarrow \quad V \cap O = \emptyset,$$

and so $x \notin \overline{O}$, which contradicts that $\overline{O} = X$. Hence $\mathring{\overline{O^c}} = \emptyset$ and so O^c is nowhere dense.

10.9 Let X be a locally compact Hausdorff space and $\{U_n\}_{n=1}^{\infty}$ be a countable family of open dense subsets of X. Suppose U is a nonempty open set in X. Since U_1 is an open dense subset of X, $U \cap U_1 \neq \emptyset$ and so there exists $x_1 \in U \cap U_1$ and there exists a compact neighborhood W_1 of x_1 such that $x_1 \in \mathring{W_1} \subset W_1$. Put

$$V_1 = \mathring{W_1} \cap U \cap U_1$$

and then V_1 is an open subset and $x_1 \in V_1$. Since U_2 is dense in X, $V_1 \cap U_2 \neq \emptyset$ and so there exists $x \in V_1 \cap U_2$ and there exists a compact neighborhood W_2 of x_2 such that $x_2 \in \mathring{W_2} \subset W_2$. Put

$$V_2 = \mathring{W_2} \cap V_1 \cap U_2$$

and then V_2 is an open subset and $x_2 \in V_2$. By continuing this process, we construct a sequence $\{x_n\}_{n=1}^{\infty}$ and an decreasing family $\{V_n\}_{n=1}^{\infty}$ such that $x_n \in V_n \subset W_1$ for all $n \in \mathbb{N}$. Therefore, $\{V_n\}_{n=1}^{\infty}$ satisfies the finite intersection property, i.e., for any finite subset I of $\mathbb{N}$,

$$\bigcap_{i \in I} V_i \neq \emptyset,$$

and W_1 is compact. Hence,

$$\emptyset \neq \bigcap_{n \in \mathbb{N}} V_n \subset U \cap \left(\bigcap_{n \in \mathbb{N}} U_n \right).$$

Hence X is a Baire space.

10.10 We obtain that

$$\|J_v(F)\| = \|F(v)\| \leq \|F\| \|v\|, \qquad F \in V^*,$$

which implies that $\|J_v\| \leq \|v\|$. On the other hand, by Exercise **10.4**, there exists $G \in V^*$ such that $\|G\| = 1$ and $G(v) = \|v\|$. Therefore, we have

$$\|J_v\| = \sup_{\|F\| \leq 1} \{|J_v(F)|\} = \sup_{\|F\| \leq 1} \{|F(v)|\} \geq |G(v)| = \|v\|.$$

Hence we have $\|J_v\| = \|v\|$.

10.11 Let $B = \{v \in V \mid \|v\| \leq 1\}$. For each $v \in B$, we consider the linear functional $J_{Tv} : W^* \to \mathbb{C}$ defined by $J_{Tv}(F) = F(Tv)$. Then for each $F \in W^*$, we obtain that

$$\sup_{v \in B} \{|J_{Tv}(F)|\} \leq \sup_{v \in B} \{|(F \circ T)(v)|\} \leq \|F \circ T\|.$$

Since W^* is a Banach space, by applying the uniform boundedness principle (Theorem 10.3.1) we prove that there exists a constant $C \geq 0$ such that

$$\sup_{v \in B} \{\|J_{Tv}\|\} \leq C.$$

On the other hand, by Exercise **10.10**, we know that $\|J_{Tv}\| = \|Tv\|$ and so we have

$$\|T\| = \sup_{v \in B} \{\|Tv\|\} = \sup_{v \in B} \{\|J_{Tv}\|\} \leq C,$$

which implies that T is bounded.

10.12 For each $v \in V$ and $F \in W^*$, by assumption, we have

$$\sup_{\lambda \in \Lambda} \{|J_{T_\lambda v}(F)|\} = \sup_{\lambda \in \Lambda} \{|F(T_\lambda v)|\} < \infty.$$

Since W^* is a Banach space, by applying the uniform boundedness principle (Theorem 10.3.1) and Exercise **10.10**, we prove that there exists a constant $C_v \geq 0$ such that

$$\sup_{\lambda \in \Lambda} \{\|T_\lambda v\|\} = \sup_{\lambda \in \Lambda} \{\|J_{T_\lambda v}\|\} \leq C_v.$$

Therefore, by applying the uniform boundedness principle once again,

$$\sup_{\lambda \in \Lambda} \{\|T_\lambda\|\} < \infty.$$

10.13 By the assumption, the identity operator

$$I : (V, \|\cdot\|_1) \to (V, \|\cdot\|_2)$$

is bounded and bijection. Therefore, by the bounded inverse theorem (Theorem 10.4.3), the identity operator

$$I : (V, \|\cdot\|_2) \to (V, \|\cdot\|_1)$$

is bounded and so there exists a constant $K > 0$ such that

$$\|v\|_1 \leq K\|v\|_2, \qquad v \in V.$$

Hence, we have

$$\frac{1}{K}\|v\|_1 \leq \|v\|_2 \leq C\|v\|_1, \qquad v \in V,$$

and so $\|\cdot\|_1$ and $\|\cdot\|_2$ are equivalent.

Bibliography

[1] R. B. Ash: "Real Analysis and Probability," Academic Press, New York, 1972.

[2] S. K. Berberian: "Introduction to Hilbert Space," Oxford University Press, New York, 1961.

[3] G. de Barra: "Measure Theory and Integration," John Wiley & Sons, New York, 1981.

[4] R. G. Bartle: "The Elements of Integration," John Wiley & Sons, New York, 1966.

[5] M. Capiński and E. Kopp: "Measure, Integral and probability," Springer, London, 1999.

[6] G. B. Folland: "Real Analysis: Modern Techniques and Their Applications, 2nd Edition," John Wiley & Sonc, INC, 1999.

[7] H. Royden and P. Fitzpatrick: "Real Analysis, 4th Edition," Prentice Hall, 2010.

[8] W. Rudin: "Principle of Mathematical Analysis, 3rd Edition," McGraw-Hill, 1976.

[9] W. Rudin: "Real and Complex Analysis," McGraw-Hill, 1987.

[10] A. Torchinsky: "Real Variables," Addison-Wesley Publishing Company, Redwood City, 1988.

Index

REAL ANALYSIS

ISBN 979-11-6073-103-3

KM KYUNG MOON SA
174, Wausan-ro Mapo-gu Seoul, Korea
Tel: (82) 2 332 2004
Fax: (82) 2 336 5193

Printed in Korea

Price ₩ 20,000